T0282424

Basic Science Methods
for Clinical Researchers

Basic Science Methods for Clinical Researchers

Edited by

Morteza Jalali MB ChB PhD (Cantab) MRes BSc MRCS (Eng) FCPS
University of Oxford, Oxford, United Kingdom

Francesca Y.L. Saldanha MB BChir MA (Cantab)
Whitehead Institute, Massachusetts Institute of Technology, Cambridge, MA, United States

Mehdi Jalali MB ChB (Hons) MPhil
University of Liverpool, Liverpool, United Kingdom

ACADEMIC PRESS

An imprint of Elsevier

Academic Press is an imprint of Elsevier
125 London Wall, London EC2Y 5AS, United Kingdom
525 B Street, Suite 1800, San Diego, CA 92101-4495, United States
50 Hampshire Street, 5th Floor, Cambridge, MA 02139, United States
The Boulevard, Langford Lane, Kidlington, Oxford OX5 1GB, United Kingdom

British Library Cataloguing-in-Publication Data
A catalogue record for this book is available from the British Library

Library of Congress Cataloging-in-Publication Data
A catalog record for this book is available from the Library of Congress

ISBN: 978-0-12-803077-6

For Information on all Academic Press publications
visit our website at https://www.elsevier.com/books-and-journals

Working together
to grow libraries in
developing countries

www.elsevier.com • www.bookaid.org

Publisher: Mica Haley
Editorial Project Manager: Lisa Eppich
Production Project Manager: Edward Taylor
Designer: Mark Rogers

Typeset by MPS Limited, Chennai, India

Dedication

To all current clinician scientists in training, without whom this book would never have been inspired.

To our unwaveringly supportive families, without whom this book would not have been possible.

To you, the reader, may this book guide, excite, and encourage your pursuit of tomorrow's advancements in healthcare that stem from basic scientific research.

Contents

2.   Methods of Cloning

Alessandro Bertero, Stephanie Brown and Ludovic Vallier

3.   Whole-Mount In Situ Hybridization and a Genotyping Method on Single *Xenopus* Embryos

*Martyna Lukoseviciute, Robert Lea, Shoko Ishibashi
and Enrique Amaya*

4. Microarrays: An Introduction and Guide to Their Use

Frederick D. Park, Roman Sasik and Tannishtha Reya

5. Analysis of Human Genetic Variations Using DNA Sequencing

Gregory A. Hawkins

6.   Western Blot

Tomasz Gwozdz and Karel Dorey

7.   The Enzyme-linked Immunosorbent Assay: The Application of ELISA in Clinical Research

Jefte M. Drijvers, Imad M. Awan, Cory A. Perugino,
Ian M. Rosenberg and Shiv Pillai

8. Immunofluorescence

Sonali Joshi and Dihua Yu

9. Cell Culture: Growing Cells as Model Systems in vitro

Charis-P. Segeritz and Ludovic Vallier

10.   Flow Cytometry

Eleni Chantzoura and Keisuke Kaji

11. Transfection

Semira Sheikh, Amanda S. Coutts and Nicholas B. La Thangue

12. In Vivo Animal Modeling: *Drosophila*

Michael F. Wangler and Hugo J. Bellen

13. Zebrafish as a Research Organism: *Danio rerio* in Biomedical Research

John Collin and Paul Martin

14. Xenopus as a Model Organism for Biomedical Research

Shoko Ishibashi, Francesca Y.L. Saldanha and Enrique Amaya

15. Basic Mouse Methods for Clinician Researchers: Harnessing the Mouse for Biomedical Research

Laurens J. Lambert, Mandar D. Muzumdar,
William M. Rideout III and Tyler Jacks

Appendix A. Legal Framework on the Scientific Use of Animals in Research

David Pettitt, Adam Pettitt, MacKenna Roberts, James Smith and David Brindley

Appendix B. Regulatory Frameworks for Stem Cell Research

Eleanor Jane Budge and Morteza Jalali

Appendix C. Using Multiple Experimental Methods to Address Basic Science Research Questions
Justyna Zaborowska and Morteza Jalali

List of Contributors

Enrique Amaya University of Manchester, Manchester, United Kingdom

Imad M. Awan Ragon Institute of MGH, MIT and Harvard, Cambridge, MA, United States

Hugo J. Bellen Department of Molecular and Human Genetics, Baylor College of Medicine (BCM), Houston, TX, United States; Jan and Dan Duncan Neurological Research Institute, Texas Children's Hospital, Houston, TX, United States; Howard Hughes Medical Institute, Houston, TX, United States

Alessandro Bertero University of Cambridge, Cambridge, United Kingdom

David Brindley University of Oxford, Oxford, United Kingdom; University College London, London, United Kingdom; Harvard Stem Cell Institute, Cambridge, MA, United States; USCF-Stanford Center of Excellence in Regulatory Science and Innovation (CERSI), San Fransisco, CA, United States

Stephanie Brown University of Cambridge, Cambridge, United Kingdom

Eleanor Jane Budge University of Oxford, Oxford, United Kingdom

Eleni Chantzoura University of Edinburgh, Edinburgh, United Kingdom

John Collin University of Bristol Medical School, Bristol, United Kingdom

Amanda S. Coutts University of Oxford, Oxford, United Kingdom

Karel Dorey University of Manchester, Manchester, United Kingdom

Jefte M. Drijvers Ragon Institute of MGH, MIT and Harvard, Cambridge, MA, United States

Tomasz Gwozdz University of Manchester, Manchester, United Kingdom

Gregory A. Hawkins Wake Forest School of Medicine, Winston-Salem, NC, United States

Shoko Ishibashi University of Manchester, Manchester, United Kingdom

Tyler Jacks David H. Koch Institute for Integrative Cancer Research, Massachusetts Institute of Technology, Cambridge, MA, United States; Howard Hughes Medical Institute, Massachusetts Institute of Technology, Cambridge, MA, United States

Mehdi Jalali University of Liverpool, Liverpool, United Kingdom

Morteza Jalali University of Oxford, Oxford, United Kingdom

Sonali Joshi University of Texas MD Anderson Cancer Center, Houston, TX, United States

Keisuke Kaji University of Edinburgh, Edinburgh, United Kingdom

Nicholas B. La Thangue University of Oxford, Oxford, United Kingdom

Laurens J. Lambert David H. Koch Institute for Integrative Cancer Research, Massachusetts Institute of Technology, Cambridge, MA, United States

Robert Lea University of Manchester, Manchester, United Kingdom

Martyna Lukoseviciute University of Manchester, Manchester, United Kingdom

Paul Martin University of Bristol Medical School, Bristol, United Kingdom

Mandar D. Muzumdar David H. Koch Institute for Integrative Cancer Research, Massachusetts Institute of Technology, Cambridge, MA, United States; Dana-Farber Cancer Institute, Boston, MA, United States; Harvard Medical School, Boston, MA, United States

Frederick D. Park University of California San Diego School of Medicine, La Jolla, CA, United States; Sanford Consortium for Regenerative Medicine, La Jolla, CA, United States

Cory A. Perugino Ragon Institute of MGH, MIT and Harvard, Cambridge, MA, United States

Adam Pettitt Royal Veterinary College, London, United Kingdom

David Pettitt University of Oxford, Oxford, United Kingdom

Shiv Pillai Ragon Institute of MGH, MIT and Harvard, Cambridge, MA, United States

Tannishtha Reya University of California San Diego School of Medicine, La Jolla, CA, United States; Sanford Consortium for Regenerative Medicine, La Jolla, CA, United States

William M. Rideout III David H. Koch Institute for Integrative Cancer Research, Massachusetts Institute of Technology, Cambridge, MA, United States

MacKenna Roberts University of Oxford, Oxford, United Kingdom

Ian M. Rosenberg Ragon Institute of MGH, MIT and Harvard, Cambridge, MA, United States

Francesca Y.L. Saldanha Massachusetts Institute of Technology, MA, United States

Roman Sasik University of California San Diego School of Medicine, La Jolla, CA, United States

Charis-P. Segeritz University of Cambridge, Cambridge, United Kingdom

Semira Sheikh University of Oxford, Oxford, United Kingdom

James Smith University of Oxford, Oxford, United Kingdom

Ludovic Vallier University of Cambridge, Cambridge, United Kingdom; Wellcome Trust Sanger Institute, Hinxton, United Kingdom

Michael F. Wangler Department of Molecular and Human Genetics, Baylor College of Medicine (BCM), Houston, TX, United States; Jan and Dan Duncan Neurological Research Institute, Texas Children's Hospital, Houston, TX, United States

Dihua Yu University of Texas MD Anderson Cancer Center, Houston, TX, United States

Justyna Zaborowska University of Oxford, Oxford, United Kingdom

Foreword

Professor Dame Sally C. Davies FRS FMedSci
Chief Medical Officer, UK Department of Health, London, England

This first edition of *Basic Science Methods for Clinical Researchers* represents a unique approach for the training of clinicians who endeavor to translate basic scientific research into real healthcare benefits for patients.

There is now international recognition of the need for clinicians who are dually trained in medicine and science, and thus exceptionally equipped to harness the translational potential of laboratory discoveries and permit their safe delivery to the bedside. This welcomed and timely volume comes as we celebrate a decade since our National Institute for Health Research (NIHR) was established here in the United Kingdom, an organization which sets out to drive research from bench to bedside by supporting the most talented investigators. *Basic Science Methods for Clinical Researchers* will support the endeavors of such investigators and those aspiring to be, enjoying expert contributions from research laboratories at the world's foremost universities.

Accessible and concise in its format, this book will be embraced as a trusted bench and home study companion by clinicians and medical students globally, helping to facilitate the development of tomorrow's leaders in discovery science. I applaud the Editors on achieving their goal of putting together the best possible resource for those embarking on laboratory-based research and wish all those who are pursuing a career in translational science great success.

Preface

Biomedical research drives healthcare innovation in the prevention, diagnosis, and treatment of a wide array of illnesses. There is growing recognition of the need for clinicians who are scientifically trained (clinician scientists) to take laboratory discoveries from bench to bedside, illuminating new therapies for patients. The global move toward recruiting such a cohort of individuals, dually trained in medicine and science, presents an interesting paradox. There is an expectation that by the end of medical school, clinicians are well-versed in cellular and molecular biology and related experimental techniques. However, increasingly pressurized undergraduate medical school curricula now lean toward early patient exposure and clinical case-based learning, resulting in an ever-diminishing emphasis on basic science. This is later compounded by a lack of exposure to basic science due to the heavy time commitment demanded by clinical work. Thus the aspiring contemporary clinician scientist faces difficult challenges in learning to simultaneously navigate both the medical and scientific worlds. The onus is now, more than ever, on clinicians to independently pursue their interest in basic science and proactively seek research training opportunities outside of formalized clinical training. With the foundations of basic science and good laboratory practice no longer being provided in medical school, there is an inevitably steep learning curve when transitioning from the clinical to the laboratory environment. It can leave many feeling overwhelmed and without the necessary skills required to address scientific research questions experimentally.

Basic Science Methods for Clinical Researchers addresses the specific challenges faced by clinicians without a conventional science background. It is the first book of its kind to support the research endeavors of aspiring clinician scientists, in an engaging and accessible format. It seeks to ease the "culture shock" of the medicine to science transition and provide a platform for understanding how core experimental methods can be applied in biomedical research. We hope that *Basic Science Methods for Clinical Researchers* will be a vital companion for healthcare professionals globally, who are undertaking laboratory-based basic scientific research. It aims to support clinicians in the pursuit of their academic interests and in making an original contribution to their chosen fields. In this way, *Basic Science Methods for Clinical Researchers* will facilitate the necessary development of tomorrow's clinician scientists and future leaders in translational medicine.

About This Book

Basic Science Methods for Clinical Researchers was carefully conceptualized and designed by Morteza Jalali and his co-editors, who together assembled a host of world-class contributors from leading higher education institutions to provide unique and expert perspectives in their scientific fields of interest. All those involved in this project were inspired by the common frustrations faced by today's clinician scientists in training and were united in their pursuit of creating an invaluable resource for those undertaking laboratory-based science.

Basic Science Methods for Clinical Researchers provides a foundation in the principles and practice of core experimental techniques commonly used to answer questions in basic scientific research. The book is organized around four research themes pertaining to key biological molecules, from genes to proteins to cells, and model organisms. All the chapters follow a standardized format and take a uniform approach to describing, explaining, and critiquing the featured research method. Each includes protocols, techniques for troubleshooting common problems, and importantly an explanation of the advantages and limitations of a technique in generating conclusive data. In this way the reader is left with a powerful understanding of each of the techniques covered. At the end of this book there are three appendices which provide resources for practical research methodology including legal frameworks for using stem cells and animals in the laboratory, ethical considerations, and good laboratory practice (GLP).

Basic Science Methods for Clinical Researchers is unique to the market, providing enough detail to permit the new and inexperienced researcher to approach a given research technique with confidence, as well as serving as a useful refresher and quick reference point for those who are more seasoned in their research career. Each chapter underwent multiple rounds of editing and revisions until all contributors were happy with the end result—an engaging and accessible text that inspires and gives confidence to anyone wishing to undertake basic science training and produce high-quality research. Those who have been involved in this project from around the world, sincerely hope that you enjoy reading *Basic Science Methods for Clinical Researchers* and welcome your feedback for future editions, in order to help maintain this book as a premier text for aspiring clinician scientists. May *Basic Science Methods for Clinical Researchers* be an indispensable companion for this most rewarding and exciting journey between medicine and science.

Chapter 1

The Polymerase Chain Reaction: PCR, qPCR, and RT-PCR

Mehdi Jalali[1], Justyna Zaborowska[2] and Morteza Jalali[2]
[1]University of Liverpool, Liverpool, United Kingdom; [2]University of Oxford, Oxford, United Kingdom

Chapter Outline

Objectives

- Describe the scientific principles behind PCR
- Provide PCR laboratory protocols
- List PCR applications
- Describe a typical scenario in which qRT-PCR might be used
- Discuss key limitations and troubleshooting

Basic Science Methods for Clinical Researchers. DOI: http://dx.doi.org/10.1016/B978-0-12-803077-6.00001-1
1

INTRODUCTION

The polymerase chain reaction (PCR) was developed in the 1980s by Dr. Kary Mullis. The technique has been compared to a "molecular photocopier" owing to its ability to recognize a specific sequence of DNA, and rapidly and accurately synthesize a high number of copies [1]. It has revolutionized molecular biology, and in particular genetic manipulations, the diagnosis of genetic and infectious diseases, genotyping and DNA forensics. It is considered one of the greatest scientific discoveries of the 20th century. To date, a variety of spin-off techniques based on the original PCR method have been developed. Real-time PCR, also known as quantitative PCR (qPCR), combines PCR amplification and detection in a single step. Another technique known as reverse transcription polymerase chain reaction (RT-PCR) uses RNA as the nucleic acid starting template.

IN PRINCIPLE

PCR resembles an in vitro and elementary form of DNA replication, a physiological process used by all living cells to duplicate their genetic material prior to cell division [2]. It involves repeated cycles of heating and cooling of a reaction mixture containing DNA template, DNA polymerase, primers, and nucleotides (Table 1.1). DNA template is the DNA containing the target sequence. Primers are short chains of nucleotides which locate the specific target DNA of interest and bind to it upon cooling, through complementary base pairing. They act as a starting point for DNA polymerase to create the new complementary strand. DNA polymerase is an enzyme that synthesizes new strands of DNA complementary to the target sequence.

Each cycle of PCR consists of three steps (Fig. 1.1):

- Denaturation—reaction mixture heated to over 90°C to unwind double helix of DNA by breaking apart hydrogen bonds.
- Primer annealing—reaction mixture cooled to 45–65°C to allow for primer annealing. Forward and reverse primers hybridize through complementary base pairing to opposite strands of the DNA. They must be complementary to the 3′ ends of the antiparallel strand of template DNA.
- Extension—the reaction mixture is heated to 72°C toward the optimal temperature for DNA polymerase enzyme activity. Polymerase binds to the primer-template hybrid complex and then assembles a new complementary strand using the free nucleotides in the reaction mixture.

Following extension, the reaction is returned to the denaturation step and PCR continues. Each cycle approximately doubles the amount of DNA, as a new strand of DNA subsequently acts as a template for replication in the following cycle. This results in an exponential increase in quantity of DNA. For example, after 6 cycles there are 2^6 copies. A total of 25–40 PCR cycles is carried out depending on the expected yield of the PCR product [2].

TABLE 1.1 Components of PCR and reaction guidelines for reaction optimization

Components of PCR	Guidelines
DNA TEMPLATE	• Use high quality, purified DNA templates. • Use 1 ng–1 µg of genomic templates per 50 µL reaction. • Contaminating DNA reduces efficiency.
DNA POLYMERASE	• The most commonly used enzyme for PCR experiments is Taq DNA Polymerase; use 0.5–2 U per 50 µL reaction. • The optimal activity for Taq DNA Polymerase is 75°C. As such high temperature can cause the primer and template to dissociate, a slightly lower temperature of 72°C is used instead. • Magnesium is required as a co-factor for Taq DNA Polymerase activity; use 1.5–2.0 mM for Taq Polymerase
PRIMERS	• Length of 20–30 nucleotides • GC content 40–60% • Primer pairs should have similar melting temperatures (T_m 42–65°C) which differ by no more than 5°C. • The primer cannot be self-complementary or complementary to another primer in the reaction because hairpins or primer dimers can be formed
dNTPs	• The concentration of each dNTP in the PCR reaction mixture should be equal

Template DNA contains the DNA sequence that will be amplified by PCR. DNA polymerases copy DNA molecules during the PCR reaction. Primers are short single-stranded DNA molecules that bind by complementary base pairing to opposite DNA strands. The melting temperature (Tm) of the primer is the temperature at which one-half of the double-stranded DNA dissociates to become single stranded [3].

ANALYSIS OF THE PCR PRODUCT

The PCR product or "amplicon" can be visualized and analyzed with the use of agarose gel electrophoresis, which separates DNA products on the basis of size and charge (Fig. 1.2). qPCR, which will be discussed later in this chapter, does not require such postamplification analyses and instead the product is analyzed throughout the reaction in "real-time." Gel electrophoresis involves the separation of charged molecules in an electrical field on an agarosegel, followed by staining with ethidium bromide. PCR by-products such as primer dimers appear as diffuse, smudgy bands near the bottom of the gel. The other way to validate a PCR reaction is to directly sequence the amplicon.

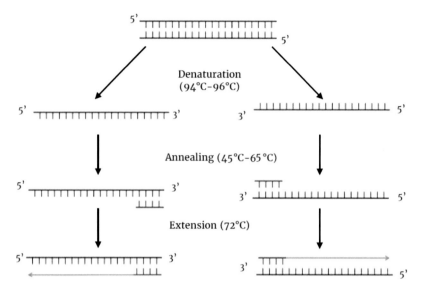

FIGURE 1.1 First cycle of PCR. During the denaturation step, DNA is heated to above 90°C and the two strands of the DNA target sequence separate. The temperature of the reaction is then cooled to 45–65°C and primers anneal to their complementary sequence in the template DNA. In the extension step, the reaction is heated to 72°C to allow the DNA polymerase to synthesize a new DNA strand complementary to the template strand.

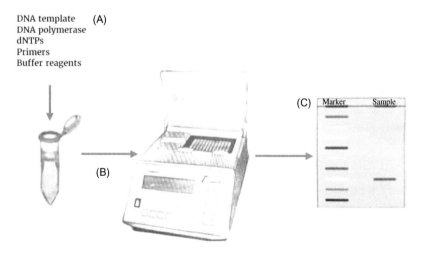

FIGURE 1.2 PCR experimental design. (A) The polymerase chain reaction (PCR) requires several components including DNA template, DNA polymerase, deoxynucleotide triphosphates (dNTPs) and oligonucleotide primers. (B) The prepared PCR reactions are placed in the thermal cycler. (C) The size of a DNA fragment can be estimated by gel electrophoresis. In an electric field, negatively charged DNA molecules will migrate toward the positive electrode.

PHASES OF PCR

The PCR reaction can be split into three phases—exponential, linear, and plateau (Fig. 1.3). During the exponential phase, reaction components are in excess and there is exact doubling of product at each cycle. During the linear stage, the reaction components start to run out and consequently the reaction slows down. In the plateau phase, the reaction stops and no more product is generated.

The depletion in reagents will occur at varying rates due to the different reaction kinetics in each PCR tube. Rates of depletion will start to vary in the linear phase, and each sample will plateau at a different point. Therefore, even replicate DNA template samples can end up with different copy numbers in the plateau phase, despite starting with the same quantity [4]. The exponential phase is optimal for data analysis, as it yields high-quality quantitative data. Conventional PCR quantifies data from the linear and plateau phases. Therefore, data from conventional PCR can only be considered semi-quantitative at best, and allows detection of only a tenfold change in gene expression. The PCR variant named real-time PCR measures at the exponential phase of the PCR reaction, allowing detection of twofold change in gene expression [5].

VARIATIONS OF PCR METHOD: QUANTITATIVE PCR

In recent years, a technological innovation of PCR, qPCR, has become increasingly important in clinical diagnostics. qPCR allows for detection and

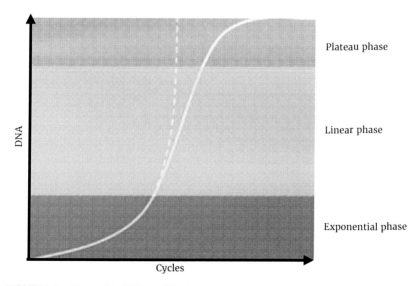

FIGURE 1.3   Phases of the PCR amplification curve. The exact doubling of product accumulates at every cycle during the exponential phase. In the linear phase the reaction is slowing as components are being consumed. In the plateau phase the reaction stops and no more products are being made.

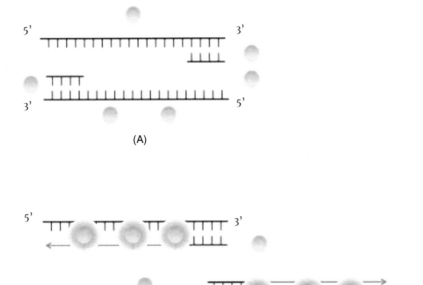

(A)

(B)

FIGURE 1.4 Action of SYBR Green I Dye. (A) When DNA is denatured, SYBR Green I Dye floats free and emits low fluorescence. (B) SYBR Green Dye binds to the double-stranded product and fluoresces.

quantification of the target DNA, as the reaction progresses. A variety of fluorescent chemistries correlate PCR product concentration to fluorescence intensity [6]. The most commonly used fluorescent DNA binding dye is SYBR Green I. It emits fluorescence when bound to double-stranded DNA and the intensity of fluorescence increases proportionally to the concentration of PCR product (Fig. 1.4).

SYBR Green I can bind to any double-stranded DNA, including nonspecific products such as primer dimers [7]. Therefore, it is important to carefully design primers that only bind to the selected target sequence. In addition, a "melt curve" is required when using dye-based methods to validate the results and ensure specific product. At the end of the PCR run, the reaction mixture is exposed to a temperature gradient from around 60°C to 95°C and fluorescence readings are continually collected. At a certain temperature, the amplified product will fully dissociate. This results in a rapid fall in fluorescence emission, as SYBR Green I dissociates.

Double-stranded DNA melts at different temperatures according to its length, GC content, and the presence of base pair mismatches and secondary structures (Fig. 1.5). Therefore, it is possible to validate how many products of amplification are present in a sample. If binding is specific, a single tight peak should appear on the graph representing the specific amplicon of interest. Primer dimers instead appear as shorter broader waves at lower temperatures [8] (Fig. 1.6).

Obtained qPCR fluorescence data can be plotted on an amplification graph with cycle number on the X-axis and fluorescence on the Y-axis (Fig. 1.7). Reactions are characterized by the PCR cycle at which amplification of a PCR product is first detected [9]. The baseline in the amplification plot is the

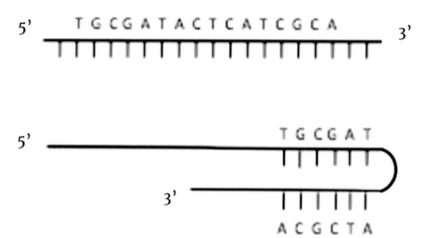

FIGURE 1.5 Hairpin structure. Owing to the presence of complementary sequences within the length of single-stranded nucleic acid sequences, a secondary hairpin structure might form.

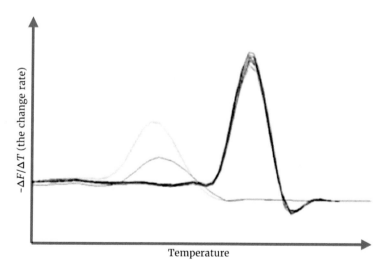

FIGURE 1.6 Melt curve analysis from a qPCR assay. A uniform melt curve with a single tight peak means that only the target DNA of interest has been generated. A smaller peak to the left may correspond to the dissociation curve for primer dimers.

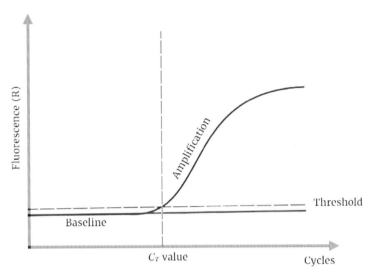

FIGURE 1.7 qPCR amplification plot. Baseline-subtracted fluorescence versus number of PCR cycles. The threshold cycle (C_T) is the cycle number at which the fluorescent signal of the reaction crosses the established threshold line.

average background. The threshold is the level of fluorescence above the baseline. It is usually calculated automatically and corresponds to the three standard deviations above the mean baseline values, within the exponential phase. The threshold cycle (C_T value) is defined as the cycle in which the fluorescent signal crosses the threshold, exceeding background level. Accordingly, the greater the quantity of target DNA in the sample, the sooner the fluorescence crosses the threshold, yielding a lower C_T [9].

There are two major methods for normalization of qPCR assays: absolute quantification and relative quantification [10]. Whether absolute or relative quantification is required depends on the experimental question that needs answering. For instance, absolute quantification is used to measure the exact number of target molecules in a sample, for example the number of viral particles in a given amount of blood. Alternatively, relative quantification might be used to compare gene expression between samples, and to calculate the fold difference in expression.

VARIATIONS OF PCR METHOD: REVERSE TRANSCRIPTION PCR

Reverse transcription (RT)-PCR is used to amplify RNA targets. The RNA template is converted into complementary (c)DNA by the enzyme reverse transcriptase. The cDNA serves later as a template for exponential amplification using PCR. RT-PCR can be undertaken in one or two steps. One-step RT-PCR combines the RT reaction and PCR reaction in the same tube. Only sequence-specific primers may be used. During two-step RT-PCR, the synthesized cDNA is transferred into a second tube for PCR. Oligo (dT), random hexamer or gene-specific primers can be used. Oligo (dT) primers are generally preferred as they hybridize to the 3′ poly (A) tails in mRNAs (transcribed gene sequences), whereas random primers prime anything including ribosomal RNA [11]. There are advantages and disadvantages to both methods. One-step reactions are easier to set up and ideal for high throughput screening. Two-step reactions are ideal for detection of several messages from a single RNA sample.

IN PRACTICE

This section provides detailed experimental protocols for qRT-PCR using SYBR Green I. We begin with total RNA extraction, followed by RT-PCR. cDNA is then used as a template for qPCR with gene-specific primers. When working with RNA, the researcher must maintain an RNase-free environment, which means wearing gloves, keeping tube lids covered, using RNase-free pipette tips and microcentrifuge tubes, and RNase-free water. Control samples should be used for both RT and PCR. The no- template control helps to detect contamination with template or amplicons from previous reaction steps. In addition, a no-reverse-transcriptase control detects contaminating DNA within the RNA sample. The positive control consists of a source of quantified DNA/RNA with the target sequence and provides assurance that the reaction is working if there is absence of gene expression in the target cells.

EXTRACTION OF TOTAL RNA

RNA can be extracted in 30–60 minutes using the method presented below, which has been adapted from Chomczyński and Sacchi 1987 [12]. This protocol utilizes TRIzol Reagent, which is designed to isolate high-integrity total RNA.

Homogenization
1. For tissues, add 1 mL of TRIzol reagent per 50–100 mg of tissue sample. Homogenize sample using, for example, a power homogenizer.

   For cells grown in suspension, centrifuge at $300\times g$ for 5 minutes to pellet the cells and pour off supernatant. Add 0.75 mL of TRIzol Reagent per 0.25 mL of sample.

   For cells grown in a monolayer, remove growth media from culture dish and rinse cells with ice-cold phosphate-buffered saline. Add 1 mL of TRIzol Reagent per 10 cm^2 of culture dish surface area.

   TIP: Thoroughly homogenize your samples. Fragments left undisrupted during homogenization step represent RNA lost. Also, rinse your homogenizer in-between samples to prevent cross-contamination.
2. Re-suspend the lysate using a 1 mL pipette tip.
3. Incubate mixture at room temperature for 5 minutes.

Phase separation

4. While working in the fume hood, add 200 μL of chloroform per 2 mL of TRIzol Reagent.
5. Vortex samples for 15 seconds and incubate them at room temperature for 3 minutes.
6. Centrifuge the samples at 12,000 g for 15 minutes at 4°C. The mixture separates into a lower red phenol-chloroform phase, an interphase, and a colorless upper aqueous phase where RNA remains.
7. Transfer the upper aqueous phase to a fresh tube. Make sure you do not draw any of the interphase or organic layer into the pipette.

RNA precipitation

8. Add 500 μL of 100% isopropanol and incubate samples at room temperature for 10 minutes.
9. Centrifuge at 12,000× g for 10 minutes at 4°C. The RNA will form a gel-like precipitate at the bottom and side of the tube.

RNA wash and resuspension

10. Remove supernatant and wash precipitate by adding 1 mL of 75% ethanol. Mix the samples by vortexing.

    TIP: Don't forget to vortex! This allows ethanol to penetrate the nucleic acid pellet to dissolve any residual salt.
11. Centrifuge at 7500× g for 5 minutes at 4°C.
12. Remove the supernatant. Allow a few minutes for any remaining ethanol to dry.
13. Re-suspend pellet in RNase-free water (20–50 μL).

At this point, the extracted RNA may be treated with DNase enzyme to prevent genomic DNA contamination. In particular, for downstream RT-PCR, co-amplification of genomic DNA can lead to nonspecific results. For DNase I digestion, use 1 U of DNase I per 1 μg–5 μg of total RNA. Keep total volume of 50 μL. Incubate for 10 minutes at +37°C. Finally, the RNA sample must be purified of DNases, to avoid any destruction of cDNA generated subsequently in the RT-PCR step.

Assessing RNA quantity, purity and integrity

Nucleic acids can be quantified by UV absorption using a spectrophotometer such as the "Nanodrop." Ultraviolet absorbance at 260 nm is used to measure the amount of nucleic acid in the sample. An A260 reading of 1.0 is equivalent to around 40 μg/mL of RNA. The RNA purity is determined from the relative absorbance at 260 and 280 nm. A ratio $A_{260/280}$ greater than 1.8 is satisfactory.

To determine RNA integrity or quality one can look at the intensity of rRNA bands on denaturing agarose gels. Alternatively, the use of a bioanalyzer automatically measures the sizes of the rRNA bands using laser and fluorescent technology, and attributes an RNA integrity number (RIN) to the sample. The RIN is a standardized measure that may be compared between different RNA samples [13].

For short-term storage RNA can be kept at −20°C; for long-term storage at −80°C.

SYNTHESIS OF cDNA BY REVERSE TRANSCRIPTION

The protocol below has been adapted from Invitrogen [11,14]. RT can be undertaken in 90–120 minutes depending on the number of samples. For cDNA synthesis use high-integrity RNA.

> *Denaturing step*
> 1. Combine the following in a 0.5 mL Eppendorf tube:
>   Total RNA (1 pg—5 µg)
>   50 µM oligo (dT) primer—1 µL
>   10 mM dNTP mix—1 µL
>   RNase-free water—make up to 10 µL final volume
>   TIP: Oligo (dT) is a specific priming method. However, if the message is long or does not have a poly (A) tail use a random primer instead. For one-step RT-PCR use gene-specific primers only.
> 2. Incubate at 65°C for 5 minutes.
> 3. Briefly centrifuge to bring the solution to the bottom of the tube and place on ice.
>   *Annealing step*
> 4. Prepare the cDNA synthesis master mix. For each reaction:
>   10× RT reaction buffer—2 µL
>   25 mM MgCl$_2$—4 µL
>   0.1 M DTT—2 µL
>   RNase inhibitor (e.g., RNasin 20–40 U/µL)—1 µL
>   Reverse transcriptase (e.g., M-MLV)—1 µL
> 5. Add 10 µL of the cDNA synthesis mix to the RNA/primer mixture. Include a no-reverse transcription control.
> 6. Mix gently by pipetting and briefly centrifuge to collect the sample.
>   *cDNA synthesis*
> 7. Incubate at 50°C for 50 minutes.
>   *Terminate reaction*
> 8. Heat to 85°C for 5 minutes.
> 9. Place the sample on ice for 5 minutes.
> 10. Briefly centrifuge to collect the sample.
>   *RNA removal step*
> 11. Add 1 µL of RNase H to the tube and incubate at 37°C for 20 minutes.

RNase H degrades RNA template from the RNA/cDNA hybrid complex. cDNA can be used immediately for PCR amplification or stored at −20°C or −80°C.

REAL-TIME PCR AMPLIFICATION

The protocol presented below [11] requires 30–60 minutes for reaction set up, around 3 hours to run the PCR program, and 1–2 hours for data analysis. It is based on the Quantitect SYBR Green PCR kit sold by Qiagen. The reaction mix

TABLE 1.2   Typical program for real-time cycler [13]

Step	Time	Temperature
Initial denaturation	15 min	95°C
Denaturation	30 sec	95°C
Annealing	30 sec	45°C to 65°C (5° below primer Tm)
Extension	30 sec	72°C
Repeat cycles	35–45 cycles (depends on the amount of template)	
Final extension	5 min	72°C
Melt curve		45°C to 95°C

contains HotStar*Taq* DNA polymerase, QuantiTect SYBR Green PCR Buffer, dNTP mix, SYBR Green I, ROX passive dye, and 5 mM $MgCl_2$.

1. Place these components on ice: Reaction mix (2×), 50 μM forward primer, 50 μM reverse primer, RNase-free water, cDNA.
   For each sample, the master mix consists of the following components:
   Reaction mix (2×)—12.5 μL (final concentration 1×)
   Forward primer—0.5 μL (final concentration 1 μM)
   Reverse primer—0.5 μL (final concentration 1 μM)
   RNase-free water—bring up to 25 μL
   cDNA—1–5 μL
2. Calculate the total volume of master mix required according to the number of samples that need to be analyzed. Samples should be run in triplicate. Mix the master mix thoroughly and dispense equal aliquots into each PCR tube.
   TIP: Add 10% extra volume to your master mix to allow for pipetting error. Remember to include a positive control and no-template negative control.
3. Add cDNA to each reaction tube and cap tubes.
4. Place the tubes in the thermal cycler and set the PCR machine according to the manufacturer's instructions. Example of typical program is shown in Table 1.2.
5. Run the PCR reaction in order to determine C_T values for each sample.
6. The PCR specificity can be examined on 3% agarose gel using 5 μL from each reaction.

DATA ANALYSIS: GENERATING A STANDARD CURVE

A standard curve must be generated before analyzing results by absolute quantification. The curve is used as a reference standard for extrapolating quantitative information for DNA/RNA targets of unknown concentration [7]. A standard curve is also necessary when analyzing results using the Pfaffl relative quantification method, in that amplification efficiency must be recorded for each primer set in order to make valid comparisons between different samples. The comparative C_T method does not require use of a standard curve.

In order to determine amplification efficiencies for the various primer sets follow the steps below. This protocol was adapted from Fraga et al. [11].

1. Place the following reaction components on ice: Reaction mix (2×), 50 μM forward primer, 50 μM reverse primer, RNase-free water.
2. Prepare a serial dilution of the cDNA preparation.
   There should be at least five data points representing at least five dilutions.
3. Prepare a PCR 50× master mix in 1.5 mL microcentrifuge tubes as described above.
4. Mix the master mix thoroughly and dispense equal aliquots into each PCR tube.
5. Add the serially diluted template to each reaction tube as needed.
   Remember to include a negative control.
6. Place the tubes in the thermal cycler and set the PCR machine according to the manufacturer's instructions.
7. Run the PCR program to determine C_T values for each dilution.
8. Plot the C_T values versus the logarithm of the concentration or the copy number of the template.
   The line drawn through these points should have a high R^2 value (>0.99).
9. Once the standard curve is generated, the following equation is used to determine amplification efficiency (E): $E = 10^{(-1/\text{slope})} - 1$.

   This is repeated for all primer pairs.

RELATIVE QUANTIFICATION (PFAFFL METHOD)

The "Pfaffl" calculation provides a fold change for gene expression, and uses one or more reference genes to normalize for any variability in reaction kinetics between the real-time PCR reactions [11,15].

1. Calculate difference in C_T values between experimental and control samples, for both target gene and housekeeping gene:

$$\Delta C_T = C_{T\,\text{control}} - C_{T\,\text{experimental}}$$

2. Calculate amplification efficiency for each primer set (see section above). Convert each amplification efficiency into Pfaffl efficiency (E_P) by adding 1.0.
   An efficiency of 100% (or 1.0) is equal to a 2.0-fold increase per cycle.
3. Calculate the fold change in gene expression between experimental and control samples, for both target gene and housekeeping gene:

$$\text{Fold change between samples} = (E_P)^{\Delta C_T}$$

If difference in C_T values between experimental and control samples is $\Delta C_T = 2.2$ and the amplification efficiency of the primer set is 93% ($E_P=1.93$), then the fold change for these two samples is $(1.93)^{2.2} = 4.25$.

4. Divide the fold change for the target gene by the fold change for the housekeeping gene. Fold change for housekeeping gene should be 1 or very close to 1.

If experimental target $C_T = 26.4$; experimental housekeeping $C_T = 27.1$; control target $C_T = 28.7$; control housekeeping $C_T = 27.2$.

Efficiency of both primer sets for target gene is 94% ($E_P = 1.94$) and for housekeeping gene is 98% ($E_P = 1.98$).

$$\Delta C_T(\text{target}) = C_T \text{ control} - C_T \text{ experimental} = 28.7 - 26.4 = 2.3$$

$$\Delta C_T (\text{housekeeping}) = C_T \text{ control} - C_T \text{ experimental} = 27.2 - 27.1 = 0.1$$

$$\text{Fold change normalized} = (1.94)^{2.3}/(1.98)^{0.1} = 4.59/1.07 = 4.29$$

This corresponds to a 4.29-fold increase in target gene expression in the experimental samples compared with the control samples.

APPLICATIONS OF PCR

PCR has numerous applications in research, biology, medicine, and archeology. Some of the most common applications are listed below:

- PCR can be used to investigate how gene expression changes with cell differentiation, environmental changes, or exposure to various drugs.
- It plays a major role in cloning and sequencing.
- It is used for characterization and detection of infectious disease organisms.
- With genetic testing, prospective parents might be tested to see whether they are carriers of a particular genetic disease.
- DNA profiling can be utilized to identify individuals from samples of their DNA. Crucial forensic evidence in a crime scene may be present in very small quantities, for example one human hair. These can be rapidly amplified with PCR, and then compared with the suspects DNA, or with historical DNA in the police database.
- PCR plays a major role in the production of recombinant proteins such as insulin.
- Environmental applications of PCR include testing water purity, and in archaeology to amplify DNA often badly degraded by time and the elements [16].

SCENARIO

A researcher wants to establish whether there is a difference in the expression of p53 tumor suppressor gene between benign and malignant tissue. The evidence suggests that GAPDH, a housekeeping gene, is expressed equally between the tissues. RNA is extracted from a similar amount of benign and malignant tissue. RNA samples of high purity and quality are taken for further analysis using the two-step RT method. First, RT of the RNA into cDNA is performed using oligo dT or a mixture of random primers. Second, a proportion of the cDNA sample is amplified by real-time PCR using a SYBR Green PCR kit. The cDNA pool generated in the RT step can be stored for long periods and used for future experiments.

The "melt curve" obtained from the real-time PCR analysis shows a single tight peak, which suggests that primer annealing is specific. The amplification plot reveals that the C_T values are lower for the benign cells compared to malignant cells. This means that the fluorescence is detected earlier, due to a higher amount of cDNA. The Pfaffl equation is then used to analyze the C_T values of the samples, and results are normalized to the housekeeping gene (GAPDH) values. In conclusion, there is 4.2-fold decrease in p53 expression in the malignant cells compared to the benign cells.

KEY LIMITATIONS

Contamination of the sample with DNA can be problematic and produce misleading results. Also, in order to design primers for PCR, the researcher must possess some knowledge of the DNA sequence prior to the procedure. Moreover, the primers can anneal nonspecifically to sequences similar to target DNA. Lastly, imperfect purification of nucleic acids can leave traces of various substances that can inhibit PCR reactions.

TROUBLESHOOTING

Little or No PCR Product

- The gene of interest might be expressed transiently or only in certain tissues.
- It may also be due to poor primer design or low-quality template.

Include positive PCR control. Check primer sequences are correct. Also, several reactions can be set up with variable concentrations of template.

Efficiency of Reaction Is Too Low

- May be due to poor primer design, or insufficient concentrations of primer, magnesium, or polymerase enzyme.
- Improper storage of fluorescent probes (for example temperature, length of storage, exposure to light), may cause photobleaching, meaning that fluorescence is reduced and not easily detected by the real-time cycler.

Ensure annealing temperature is appropriate to the Tm of the primers, and that primers have similar annealing temperatures. Additionally, review amplification plot—outliers and whole dilution sets may be removed from the standard curve to improve efficiency. Re-designing the primers might also help.

Efficiency of Reaction Is Too High

- Efficiency of greater than 110% is usually due to inhibitors in the DNA sample.

Try to remove the highest template concentration from the analysis and recreate the standard curve. Re-purify the template, try additional drying time to remove ethanol or add more washes to remove chaotropic salts [16].

Multiple Bands On Gel, Or Multiple Peaks On Melting Curve

- Can result from contaminating DNA. In conventional PCR, an extra step of postamplification processing may be a source of contamination.
- Can result from nonspecific primer hybridization.
- Improper storage of fluorescent probes causes degradation and leakage of dye, and may lead to background noise with rough amplification curves.
- Template concentration may be too low.

Optimize PCR to remove primer dimers. For instance, raise the annealing temperature, lower primer concentration, raise template concentration, ensure magnesium concentration is not too high, use "hot-start" polymerase, set up reaction on ice, use software to evaluate tendency for dimerization. It may be necessary to redesign the primers. Additionally, try nested PCR, which involves two runs of PCR with a different primer in each run. It is unlikely that any of the unwanted amplicon will be complementary to two different primers in two successive runs of PCR.

No RT Control Produces a Band

- The RNA may be contaminated with amplicon in the reagent, or genomic DNA from the RNA preparation step.

Repeat reaction with new reagents or use RNase-free DNase to digest the RNA preparation.

Negative PCR Control Produces a Band

- Possibly due to amplicon contamination of reagents.

Repeat the PCR reaction with new reagents.

CONCLUSION

PCR is a highly accurate and rapid method for duplicating genetic material. The discovery of thermostable polymerase enzymes has permitted the automation of PCR, thus reducing the manpower required to conduct these experiments. With the advent of qPCR, amplified products may also be quantified accurately. PCR has numerous important and diverse applications spanning research, medicine, law, ecology, and archaeology. There are some limitations to the technique including unwanted amplification of contaminated material, but on the whole PCR has become indispensible to the researcher and has been truly revolutionary to the life sciences. Many variants of PCR are continually being developed, including digital PCR, which permits faster and more precise results. The appliances currently required to undertake PCR are bulky and expensive, but newer more miniature devices are being developed to allow the benefits of PCR to be taken out of the laboratory.

REFERENCES

[1] Mullis KB. The unusual origin of the polymerase chain reaction. Sci Am 1990 Apr;262(4): 56–61. 4–5. PubMed PMID: 2315679.

[2] Baynes J, Dominiczak MH. Medical biochemistry. Elsevier Health Sciences; 2009.

[3] Rychlik W, Spencer W, Rhoads R. Optimization of the annealing temperature for DNA amplification in vitro. Nucleic Acids Res 1990;18(21):6409–12.

[4] Heid CA, Stevens J, Livak KJ, Williams PM. Real time quantitative PCR. Genome Res 1996 Oct;6(10):986–94. PubMed PMID: 8908518.

[5] Reischl U, Kochanowski B. Quantitative PCR. Quantitative PCR Protocols. Springer; 1999.3.30

[6] Higuchi R, Fockler C, Dollinger G, Watson R. Kinetic PCR analysis: real-time monitoring of DNA amplification reactions. Biotechnology 1993;11:1026–30.

[7] Ponchel F, Toomes C, Bransfield K, Leong FT, Douglas SH, Field SL, et al. Real-time PCR based on SYBR-Green I fluorescence: an alternative to the TaqMan assay for a relative quantification of gene rearrangements, gene amplifications and micro gene deletions. BMC Biotechnol 2003 Oct 13;3:18. PubMed PMID: 14552656. Pubmed Central PMCID: 270040.

[8] Pals G, Pindolia K, Worsham MJ. A rapid and sensitive approach to mutation detection using real-time polymerase chain reaction and melting curve analyses, using BRCA1 as an example. Molecular Diagnosis 1999;4(3):241–6.

[9] Wong ML, Medrano JF. Real-time PCR for mRNA quantitation. BioTechniques 2005;39(1):75.

[10] Sellars MJ, Vuocolo T, Leeton LA, Coman GJ, Degnan BM, Preston NP. Real-time RT-PCR quantification of Kuruma shrimp transcripts: a comparison of relative and absolute quantification procedures. J Biotechnol 2007;129(3):391–9.

[11] Fraga D, Meulia T, Fenster S. Real-Time PCR. Current Protocols Essential Laboratory Techniques. 2008:10.3. 1-.3. 40.

[12] Chomczynski P, Sacchi N. Single-step method of RNA isolation by acid guanidinium thiocyanate-phenol-chloroform extraction. Anal Biochem 1987;162(1):156–9.

[13] Manchester K. Use of UV methods for measurement of protein and nucleic acid concentrations. BioTechniques 1996;20(6):968.

[14] https://tools.thermofisher.com/content/sfs/manuals/superscriptIIIfirststrand_pps.pdf.

[15] Pfaffl MW. A new mathematical model for relative quantification in real-time RT–PCR. Nucleic Acids Res 2001;29(9) e45-e.

[16] Hongbao M. Development application of polymerase chain reaction (PCR). J Am Sci 2005;1(3):1–15.

SUGGESTED FURTHER READING

[1] Nolan T, Hands RE, Bustin SA. Quantification of mRNA using real-time RT-PCR. Nat Protoc 2006;1.3:1559–82.

[2] Innis MA, Gelfand DH, Sninsky JJ, White TJ. PCR protocols: a guide to methods and applications. Academic Press; 2012.

[3] Logan JM, Edwards KJ, Saunders NA. Real-time PCR: current technology and applications. : Horizon Scientific Press; 2009.

[4] Dheda K, et al. The implications of using an inappropriate reference gene for real-time reverse transcription PCR data normalization. Anal Biochem 2005;344.1:141–3.

GLOSSARY

Amplicon   Piece of DNA or RNA that is the result of replication or amplification techniques

Annealing temperature   Temperature at which primer anneals to template DNA

Antiparallel   Two opposite strands run parallel but in opposite directions

Complementary   Specific hydrogen-bonding patterns according to Watson–Crick hypothesis of DNA structure. AT and GC pairs

Deoxyribonuclease   Degrades DNA

Hot-start   Variant of PCR that uses a Taq polymerase which is inactivated at lower temperatures, thus reducing nonspecific binding

Housekeeping gene   Also, known as an internal control

Melting temperature   The temperature at which half of the DNA is single-stranded

Multiplex   More than one nucleotide sequence is amplified in the same PCR reaction using multiple primers

Nucleotide   The "building blocks" of nucleic acids such as DNA and RNA

Primer dimer   Primers that have inappropriately hybridized to one another instead of to the DNA template

Ribonuclease   Degrades RNA

LIST OF ACRONYMS AND ABBREVIATIONS

AT	adenine-Thymine
Bp	base pair
BSA	bovine serum albumin
cDNA	complementary DNA
C_T	threshold value
dNTP	nucleoside triphosphate
DNase	deoxyribonuclease
FRET	fluorescence resonance energy transfer
GAPDH	glyceraldehyde 3-phosphate dehydrogenase
GC	guanine-Cytosine
mRNA	messenger RNA
NTC	no template control
R^2	correlation coefficient
RNase	ribonuclease
RT	reverse transcriptase
RT-PCR	reverse transcription PCR
Tm	melting temperature

Chapter 2

Methods of Cloning

Alessandro Bertero[1], Stephanie Brown[1] and Ludovic Vallier[1,2]

[1]University of Cambridge, Cambridge, United Kingdom; [2]Wellcome Trust Sanger Institute, Hinxton, United Kingdom

Chapter Outline

Basic Science Methods for Clinical Researchers. DOI: http://dx.doi.org/10.1016/B978-0-12-803077-6.00002-3

Objectives

- Introducing the concept of molecular cloning.
- Presenting the various molecular cloning techniques and their relative strengths and weaknesses.
- Describing step-by-step the methods involved in an experiment of "traditional" molecular cloning.
- Providing a broad outline of the applications of molecular cloning to address biomedical research questions.
- Exemplifying the use of molecular cloning to address a specific research question.

INTRODUCTION

Molecular cloning is the set of experimental techniques used to generate a population of organisms carrying the same molecule of recombinant DNA (see Glossary). This is first assembled *in vitro* and then transferred to a host organism that can direct its replication in coordination with its growth. This is usually achieved in an easy-to-grow, nonpathogenic laboratory bacterial strain of *Escherichia coli*. A single modified *E. coli* cell carrying the desired recombinant DNA can easily be grown in an exponential fashion to generate virtually unlimited identical copies of this DNA. As such, molecular cloning can be seen as an "*in vivo* polymerase chain reaction (PCR)," in which a desired piece of DNA can be isolated and expanded. However, molecular cloning allows more flexibility, better fidelity, higher yields, and lower costs than a PCR.

The development of molecular cloning techniques started with the discovery of bacterial enzymes known as "restriction endonucleases," which cleave DNA molecules at specific positions that are defined by their sequence. These restriction endonucleases allow researchers to break up large DNA fragments into smaller pieces that are then joined with other DNA molecules (vectors) using an enzyme called DNA ligase. The most commonly used vectors are known as plasmids, which are small circular DNA molecules physically distinct from the chromosomal DNA and capable of independent replication.

Restriction endonucleases generate either "sticky ends," in which the DNA fragment has a single-stranded overhang (either on the 3' or 5' ends, see Glossary), or "blunt ends," in which no overhang is present (Fig. 2.1). Both these types of ends can be joined together (ligated), and each has its own associated advantages and disadvantages. For a sticky-end fragment ligation to be successful, the two overhangs to be joined must have complementary Watson–Crick base pairing. However, this is not a requirement of a blunt-end ligation, making it much more flexible. On the other hand, blunt-end ligation is much less efficient than sticky-end ligation due to a lack of binding stability of the two fragments. Importantly, sticky ends can be enzymatically converted into blunt ends (either by "filling in" missing nucleotides, or by removing the overhangs), and vice versa (by using 5'-3' or 3'-5' exonucleases to create new overhangs). These cut-and-paste approaches are still widely used today and are commonly referred to as "traditional" (or conventional) cloning.

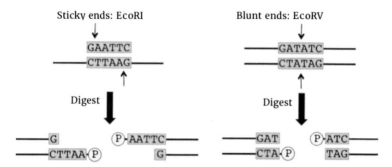

FIGURE 2.1 Type of DNA ends generated by restriction enzymes. Representative examples of restriction enzymes generating sticky or blunt ends. The arrows indicate the cut sites. Phosphate groups attached to the 5' ends after restriction digestion are indicated in yellow.

Apart from this strategy, many other molecular cloning techniques have been developed, and nowadays there are many alternatives to perform a molecular cloning experiment (Fig. 2.2), each presenting distinct advantages and disadvantages (Table 2.1):

- PCR cloning involves the direct ligation of a PCR-generated DNA fragment without using restriction enzymes to cut the insert. One of the most commonly used PCR cloning method takes advantage of an adenine (A) residue that is added by the *Taq* polymerase at the 3' ends of the DNA fragments during the amplification process. These "A-tailed" products can be directly ligated with "T-tailed" vectors. This method is therefore known as "TA" cloning.
- Ligation independent cloning (LIC) is usually carried out by adding short sequences of DNA to the fragment to be cloned that are homologous to the destination vector (this is easily accomplished by using modified primers during the PCR amplification). Complementary cohesive ends between the vector and insert are then formed by using enzymes with 3' to 5' exonuclease activity (which chew back 3' ends to create 5' overhangs), and the resulting two molecules are then mixed together and annealed. The resulting plasmid has four single-stranded DNA nicks that are efficiently repaired by the host organism. Importantly, the resulting product does not contain any new restriction enzyme sites, nor other unwanted sequences, and is therefore "scar-free."
- Seamless cloning is a group of techniques that allow sequence-independent and scar-free insertion of one or more DNA fragments into a vector. The most well known of these methods is the Gibson Assembly Method, in which up to 10 fragments can be easily combined. Similar to LIC, this relies on the addition of regions of homology at each end of the fragments to be cloned. Then, the combined action of an exonuclease (which chews back 5' ends to create 3' compatible overhangs), a DNA polymerase (which fills in gaps in the annealed fragments), and a DNA ligase (which seals the nicks in the assembled DNA) allows the generation of the recombinant DNA.

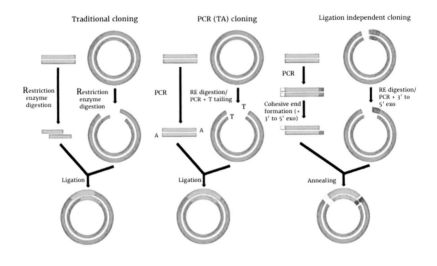

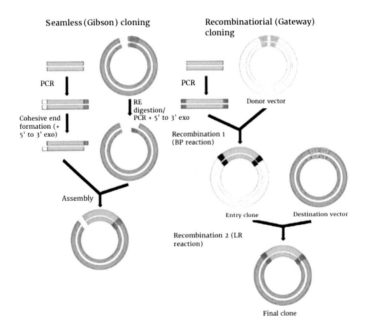

FIGURE 2.2 Overview of the main molecular cloning methodologies. Schematic examples of cloning approaches. Refer to the text for detailed explanations.

TABLE 2.1 Comparison of the Main Molecular Cloning Methodologies

Cloning Method	Cost	Sequence Dependency	Throughput	Assembly of Multiple Fragments	Directional Cloning	Need for Dedicated Vectors	Examples of Commercially Available Products
Traditional cloning (restriction enzyme-based)	Low	Yes (restriction enzyme sites)	Low to mid (can be increased by using ligation adapters)	Difficult for more than two fragments	Possible	No	–
PCR cloning	Medium (vectors)	No	High	Challenging (requires special modifications)	Difficult	Yes (for certain applications)	TOPO TA
Ligation independent cloning	Medium (reagents)	Limited (vector)	Low	Yes	Yes	No	In-Fusion
Seamless cloning	High (reagents)	No	Low	Yes	Yes	No	Gibson assembly GeneArt
Recombinatorial cloning	High (reagents and vectors)	No	High	Challenging (requires special modifications)	Yes	Yes	Gateway Echo Cloning Creator

- Recombinatorial cloning uses site-specific DNA recombinases, enzymes capable of "swapping" pieces of DNA between two molecules containing the appropriate sequences (known as recombination sites). The most widely used system in this category is the Gateway Cloning System (Life Technologies/Invitrogen), which relies on two proprietary enzyme mixes ("BP Clonase" and "LR Clonase") to swap a DNA fragment across various recombination sites. First the appropriate recombination sites are inserted by PCR on either side of the insert to be cloned, and then this is recombined with a Donor vector to create an Entry clone. This Entry clone is recombined again with a Destination vector (the required final vector) to make the final construct. Importantly, a large collection of Entry clones is already available on the market to facilitate Gateway cloning.

Regardless of the technique used to generate the recombinant DNA, molecular cloning is arguably the cornerstone of most biomedical sciences research labs. Indeed, the ability to isolate and expand a specific fragment of DNA that can be then introduced into a secondary host is often the first crucial step in both basic and translational scientific studies. The rest of the chapter will focus only on the methodologies associated with traditional cut-and-paste molecular cloning. This method has still many advantages over the more recently derived techniques (like the very low costs and simplicity of execution), and the concepts and technical skills that the reader will acquire by learning traditional cloning will be instrumental to then expand his/her study into other molecular cloning techniques.

IN PRINCIPLE

A traditional molecular cloning experiment can be divided into nine steps (Fig. 2.3):

1. Selection of the host organism
2. Selection of cloning vector
3. Preparation of the vector
4. Preparation of the insert
5. Generation of the recombinant DNA
6. Introduction of the recombinant DNA into the host organism
7. Selection of the clones of organisms containing the vectors
8. Screening for clones with the desired recombinant DNA molecules
9. Expansion and isolation of the recombinant DNA

Importantly, before performing a cloning experiment, it is always recommended to perform an *in silico* simulation of the procedure using dedicated software for DNA sequence manipulation (several free and commercial options are available). This same software is also useful to align DNA sequences and create publication-quality plasmid maps.

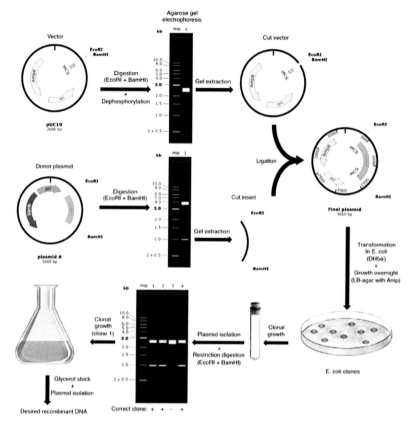

FIGURE 2.3 Overview of a traditional cloning experiment. Schematic outline of the steps involved in a traditional cloning experiment. Refer to the text for detailed explanations.

Selection of the Host Organism

As mentioned above, the most commonly used host organisms are nonpathogenic laboratory strains of *E. coli*. Several *E. coli* strains are available on the market, and these have been genetically engineered for optimal performance in certain applications and should be chosen accordingly (Table 2.2).

Selection of the Cloning Vector

Bacterial plasmids are by far the most commonly used cloning vectors, given their simplicity of use and the fact that they are appropriate for most common cloning experiments as they can hold up to 20 kilo base pairs (kb) of foreign

TABLE 2.2 Examples of the Most Common *E. coli* Strains Used for Molecular Cloning

Strain	Key Attribute	Optimized for:
BL21(DE3)	Expresses the T7 RNA Polymerase under the lacZ promoter (inducible by the lactose analog IPTG)	General expression of recombinant proteins
BL21(DE3) pLysE*	Lower basal expression levels of T7 RNA polymerase compared to BL21(DE3)	Expression of toxic recombinant proteins
DB3.1	Mutation in gyrA gene makes it resistant to toxin from ccdB gene	Propagation of plasmids expressing the ccdB gene (Gateway system)
DH5α	–	General cloning procedures
JM110	Lacks DNA methyltransferases	Growth of plasmids that must not be methylated
Origami2 (DE3)	Enhanced activity of enzymes that facilitate protein folding (reductases)	Expression of poorly soluble proteins
Rosetta2 (DE3)	Contains additional tRNAs for rare codons that are poorly expressed in *E. Coli*	Optimized expression of eukaryotic proteins by bypassing codon usage bias problems
Stbl2	Lacks an enzyme involved in DNA recombination (recA)	Growth of plasmids with high potential to recombine (like lenti- and retroviral plasmids)
XL10 Gold	Exhibit the Hte (high transformation efficiency) phenotype	Transformation of large plasmids and preparation of DNA libraries

DNA. In its simplest form a plasmid must contain the two following DNA elements (Fig. 2.4):

- Origin of replication (ORI). This recruits the DNA replication machinery and allows the propagation of the plasmid. Different types of ORI exist and they can affect the number of plasmid copies per bacterial cell.
- Selectable marker. This allows the selection of plasmid-containing bacteria. The most commonly used are drug resistance genes such as those that confer ampicillin, kanamycin, or chloramphenicol resistance.

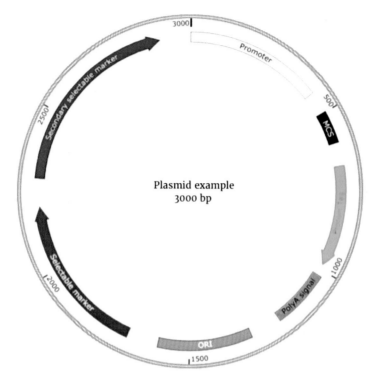

FIGURE 2.4   Schematic structure of a representative basic bacterial plasmid which contains the most common features. Refer to the text for detailed explanations of each feature.

In addition to these two features there are several accessory elements that are found in most common plasmids (Fig. 2.4):

- Multiple cloning site (MCS). A region engineered to contain multiple restriction enzyme sites to facilitate the cloning procedure.
- Promoter. This drives the expression of the cloned DNA. RNA Polymerase II or III promoters are respectively used for protein coding complementary DNA (cDNA) or short noncoding RNA.
- Protein tag. This is fused to the cloned DNA to generate chimeric proteins with specific properties, for example a green fluorescent protein (GFP) to allow easy monitoring of the protein localization.
- Poly-adenylation signal. This is located after the cloned DNA and induces termination of the messenger RNA (mRNA) transcription by poly-adenylation.
- Secondary selectable marker. This allows for further selection of organisms that contain the plasmid. Drug resistance genes (such as for Neomycin or Puromycin resistance) are commonly used to select mammalian cells.

Finally, specialized vectors exist that allow the generation of recombinant viral particles that can be used to transduce mammalian cell types with the cloned DNA at high efficiency. Table 2.3 reports some of the most commonly used plasmids.

TABLE 2.3 Examples of Commonly Used Bacterial Plasmids

Aim	Key Features	Plasmid
Mammalian protein expression	CMV promoter; Flag and HA tags.	pCDNA3 Flag HA
Transgenic mice generation	Neomycin resistance gene	pBigT
Retroviral-mediated mammalian protein expression	Retroviral packaging sequences	pBABE
Lentiviral-mediated mammalian shRNA expression	Lentiviral packaging sequences; U6 promoter	pLKO.1
Recombinant protein production and purification from bacteria	Glutathione S-transferase (GST) tag	pGEX
Recombinant protein production and purification from bacteria	Maltose binding protein (MBP) tag	pMAL
Simple cloning experiment	—	pUC19

Preparation of the Vector

The cloning vector can be prepared by two main methods:

- Restriction digestion. Blunt or sticky-ends can be generated depending on the restriction enzyme used (Fig. 2.1, also see Glossary). Once digested, the vector is purified by size-selection using agarose gel electrophoresis.
- PCR. This generates blunt ends that can be further cut by restriction enzymes as described above. Compared to the previous method, amplifying the vector allows for more flexibility (as new restriction sites or other desired sequences can be introduced at the end of the vector by using modified primers containing such sequences), but the resulting vector has the potential to contain mutations introduced during the PCR and hence must be sequenced to exclude this. Before ligation, the vector must be purified.

If the prepared vector has compatible ends that have the potential to re-ligate without the presence of the desired insert, the phosphate groups need to be removed from the 5' ends to prevent this from happening. Indeed, DNA ligase requires a phosphate group to be present on the 5' end of a DNA fragment in order to covalently link two fragments together. DNA ends generated by restriction enzymes retain phosphate groups on the 5' ends (Fig. 2.1); however, they can be removed *in vitro* using enzymes such as alkaline phosphatase. PCR products do not have phosphate groups on the 5' ends.

Preparation of the Insert

The DNA to be cloned (insert) can come from multiple sources:

- Genomic DNA. This can be purified from any organism of interest using appropriate protocols. Before cloning, it is either cut with a restriction enzyme or mechanically sheared to produce fragments with the desired size. The resulting DNA is used to prepare whole genomic DNA libraries.
- Complementary DNA (cDNA). This is double-stranded DNA obtained from cellular mRNA by means of reverse transcription. This can be used to generate cDNA libraries containing all expressed genes in a certain cell type.
- Plasmid DNA. This already contains the sequence of interest that needs to be transferred to a new vector. The sequence is first removed from this donor plasmid using appropriate restriction enzymes, then size-selected before cloning.
- PCR product. This can be obtained using appropriate primers from genomic DNA, cDNA, or a plasmid. Restriction sites or other desired sequences can be added to the ends of the PCR product by using primers that include such sequences. Blunt-ended PCR products can either be used directly or cut with restriction enzymes. Size selection of the fragment of interest is usually performed.
- Synthetic DNA. This can be created *in vitro* without using any preexisting DNA template and hence allows the highest level of flexibility in generating a desired sequence. Before cloning, these fragments are treated as for PCR products. Sequencing of the cloned product is always required as mutations are often introduced during the DNA synthesis procedure.

Regardless of its source, the insert DNA must contain ends that are compatible with the ones of the vector prepared as described above. Moreover, if a dephosphorylated vector is to be used, it is essential for the insert to be phosphorylated on each 5' end. When the insert has been digested by restriction enzymes, the 5' ends retain these phosphate groups. However, PCR-derived products or synthetic DNA fragments must be phosphorylated *in vitro* using enzymes such as T4 polynucleotide kinase before ligation.

Generation of the Recombinant DNA

Once prepared, the vector and inserts are mixed and the adenosine triphosphate (ATP)-dependent DNA ligation reaction is performed using enzymes such as the T4 DNA ligase.

Introduction of the Recombinant DNA into the Host Organism

E. coli can be made competent for DNA uptake in two main ways:

- Chemical competence. This is achieved by pretreating the cells with chemicals (often calcium chloride) under cold conditions, followed by a short pulse of heat shock. This procedure increases the permeability of the cell membrane to DNA. Plasmids up to 10 kb can be efficiently introduced using this method, and given its simplicity this is the most commonly used technique to introduce recombinant DNA into bacteria.

- Electroporation. The cells are subjected to a brief electric shock that generates small pores into the cell surface, thus allowing plasmid DNA to enter. This technology has a higher efficiency compared to chemical transformation and is used mostly to introduce very large plasmids.

Selection of the Clones of Organisms Containing the Vectors

Bacteria are cultured on semi-solid agar and media petri dishes containing the appropriate selection agent (usually an antibiotic). This allows selection of the organisms that carry the drug resistance DNA marker and therefore the desired plasmid. After overnight incubation, individual bacteria grow up into visible colonies of several million cells, which can be isolated, expanded, and analyzed.

Screening for Clones with the Desired Recombinant DNA Molecules

This can be achieved by three main strategies:

- Colony PCR screening. The presence of the desired insert is identified by performing a PCR directly on a bacterial colony.
- Restriction digestion screening. The plasmid is first isolated from the bacterial clone and then subjected to a restriction digestion that generates fragments of DNA of a certain size only if the desired recombinant DNA is present.
- DNA sequencing. This is performed on isolated plasmids and used to confirm the presence of the insert and its correct sequence.

Expansion and Isolation of the Recombinant DNA

After screening, the bacterial clone carrying the desired recombinant DNA is expanded in liquid culture to amplify the plasmid. The amplified plasmid can be extracted and purified using appropriate protocols or commercial kits.

IN PRACTICE

By now, it will be clear to the reader that it is not possible to have a protocol that can be universally applied to any traditional cloning experiment. In this section, we will provide a "basic protocol" that can be used as a practical guide and tailored to different experimental requirements. Application of this protocol will require the use of basic molecular biology techniques that will be only briefly summarized here. More detailed protocols for these techniques can be found in other chapters of this text, in other laboratory manuals [1], or from the suppliers of commercial molecular cloning products/kits [2–5].

The aim of this general protocol is to transfer a 1 kb DNA fragment from a donor plasmid (plasmid A, size of 5 kb) into a destination plasmid for a different downstream application. The DNA of interest can be removed by digestion with two commonly used restriction enzymes (EcoRI and BamHI, see Fig. 2.3).

Selection of the Host Organism

For this example a DH5α *E. coli* strain will be used.

Selection of Cloning Vector

For this example a pUC19 plasmid vector will be used.

Preparation of the Vector

- Cut 10 µg of pUC19 by mixing with EcoRI and BamHI in the appropriate digestion buffer and incubating at 37°C for 2 hours.
  NOTE: Refer to the enzyme supplier for recommended digestion conditions.
- Run the vector on a 1% agarose gel for DNA electrophoresis and extract the size-fractionated vector using a commercially available gel-extraction kit (or an equivalent in-house protocol).

Preparation of the Insert

- Cut 10 µg of plasmid A with EcoRI and BamHI as described above.
- Size-fractionate the insert from the vector backbone using a 1% agarose gel.

Generation of the Recombinant DNA

- Quantify the concentration of vector and insert previously prepared by measuring the absorbance of the DNA solution at 260 nm using a spectrophotometer.
- Prepare the ligation mix in 10 µL by adding 50 ng of vector, 50 ng of insert, the T4 DNA ligase, and the T4 DNA ligase buffer. Incubate the mix for 2 hours at room temperature.

NOTE: Refer to the enzyme supplier for recommended ligation conditions. In this example the cut pUC has a size of 2665 bp while the fragment is 1000 bp long. To calculate an equimolar quantity of the two fragments use the formula shown below:

$$2665 \, bp/50 \, ng = 1000 \, bp/X ng.$$

$$(1000 \, bp * 50 \, ng)/2665 = 18.76 \, ng.$$

It is usually recommended to use a 1:3 vector to insert molar ratio, as such use 18.76 ng*3 = 56.28 ng (rounded down to 50 ng).

- OPTIONAL: perform a negative control ligation by adding 50 ng of vector but no insert.
  NOTE: this optional reaction can be useful to estimate the level of background due to contaminating uncut plasmids or ligation by-products and therefore the efficiency of the cloning experiment.

Introduction of the Recombinant DNA into the Host Organism

- Transform 50 μL of chemically competent DH5α *E. Coli* with 5 μL of the ligation.
  NOTE: refer to the supplier for the recommended transformation protocol.

Selection of the Clones of Organisms Containing the Vectors

- Plate the transformed *E. coli* onto Luria Bertani (LB) broth-agar plates containing 100 μg/mL ampicillin and incubate overnight at 37°C in a humidified bacterial incubator.
  NOTE: in this case ampicillin is used as pUC19 carries an ampicillin resistance gene.

Screening for Clones with the Desired Recombinant DNA Molecules

- Individually pick 8 bacterial colonies using a sterile tip and inoculate them in 5 mL of LB supplemented with 100 μg/mL ampicillin in a microbiology tube. Grow them for 16 hours at 37°C with agitation (225 revolutions per minute, rpm) in a bacterial orbital shaker.
  NOTE: if a negative control ligation was performed, the number of clones screened can be adjusted to take into account the level of background colonies observed.
- Transfer 1 mL of the liquid culture to a fresh tube and store at 4°C as stock and backup.
- Extract the plasmid from the remaining liquid culture using a commercial kit (or an equivalent in-house protocol).
- Digest the extracted plasmids using EcoRI and BamHI as described above.
- Run the digested plasmids on a 1% agarose gel to identify clones containing the insert.
  NOTE: in this case correct clones will have two bands (the vector at 2.6 kb, and the insert at 1 kb), while incorrect clones from self-ligated vectors will have only one band (at 2.6 kb). See Fig. 2.3.

Expansion and Isolation of the Recombinant DNA

- Inoculate 500 μL of the bacterial liquid culture previously stored at 4°C into 250 mL of LB supplemented with 100 μg/mL of ampicillin in a microbiology vessel. Grow for 16 hours at 37°C in agitation (225 rpm) in a bacterial orbital shaker.

- Save 1 mL of liquid culture, add 1 mL of 50% glycerol, mix, and store at −80°C as long-term bacterial stock.
- Extract the plasmid from the rest of the liquid culture using a commercial kit (or an equivalent in-house protocol).
- The recombinant DNA of interest is now ready for any downstream application.

APPLICATIONS

Molecular cloning is arguably one of the cornerstones of most modern biomedical basic studies and translational applications. Initially developed to study a single DNA sequence, molecular cloning techniques now allow an unprecedented ability to generate complex combinations of DNA fragments.

Study of Gene Function

With regard to basic research studies, on top of being an important tool to determine the sequence of a particular DNA fragment, cloning is essential to characterize the function of both genes and noncoding elements of the genome. Gene function can be investigated by cloning a cDNA into an expression vector to induce overexpression in a target organism (gain of function studies), or by cloning a specific short-hairpin RNA (shRNA), a sequence capable of suppressing the expression of the gene of interest using the micro RNA (miRNA) pathway (loss of function studies) [6]. Other methods to suppress gene function include the use of programmable genome editing tools to generate knock-out cells or organisms by disrupting a gene sequence. These tools include Zinc-Finger Nucleases (ZFNs), transcription activator-like (TAL) effector nucleases (TALENs), or CRISPR/Cas9 nucleases [7], all of which must first be cloned into specific vectors. Moreover, gene function can also be assessed by introducing specific mutations through site-directed mutagenesis techniques, or by generating protein truncation mutants, both of which rely on molecular cloning procedures [8]. Among other examples, all these technologies are key to the production of transgenic animals models of human diseases that can be used to find novel therapeutic targets and screen for potential drugs [9].

Study of Genomic Regulatory Regions

The function of noncoding elements can also be characterized by cloning putative gene promoters, enhancers, or silencers into specific vectors that allow measuring their ability to regulate gene transcription [10]. This is can be done both *in vitro* and *in vivo* by measuring the activity of a reporter gene (such as luciferase, β-galactosidase, or GFP) cloned downstream of the genomic element of interest. In the case of established gene regulatory elements, similar reporter constructs can be instead used to visualize a certain cell type or biological process. Furthermore, by using tissue-specific promoters in the context of programmable gene-editing technologies (such as Cre/lox-mediated genetic recombination), it is possible to create lineage tracing tools that are invaluable to modern developmental biology as they allow the study of the specification process of a cell type of interest from its early progenitors [11].

Translational Applications

Molecular cloning is also a key technique in the context of biomedical translational applications. For example, it is key for the production of recombinant proteins for therapeutic or diagnostic uses. Notable examples of these are growth factors (such as insulin, erythropoietin, or growth hormone), enzymes, antibodies, and vaccines [12]. These are commonly produced and then purified from *E. coli*, yeasts, or insect cells that are transformed with the appropriate expression vectors. Molecular cloning is also used to generate transgenic plants and animals with the aim of improving their nutritional value (e.g., by introducing vitamins or essential nutrients), or to introduce genes that code for useful pharmaceuticals (a process known as molecular farming [13]). Finally, gene therapy applications rely on complex molecular cloning experiments. In these cases, a gene of interest is inserted into a modified virus with the aim of delivering it into a patient to induce a therapeutic response [14].

SCENARIO

In this section, we will provide a case study of how a series of simple traditional cloning experiments can lead to the generation of an extremely powerful tool to address a complex biological question.

Let's imagine that we recently performed a screening experiment aimed at identifying novel regulators of liver differentiation. Among these potential targets, one transcription factor (TF X) appears very interesting as our screening suggests that high levels of TF X might be capable of inhibiting hepatocyte specification. As such, our aims are to: (1) study the effect of overexpressing TF X during liver development with regards to its specification; (2) determine the transcriptional changes induced by TF X overexpression.

To address these two biological questions, we decide to generate a plasmid that allows: (1) hepatocyte-specific overexpression of TF X, in order to exclude potential confounding effects due to TF X overexpression in other cell types and to avoid premature expression of TF X at early stages of hepatocyte development; (2) co-expression of an enhanced GFP (EGFP) reporter gene, in order to be able to isolate TF X-overexpressing cells by flow-cytometry to perform detailed transcriptional analysis in a pure population; (3) selection of mammalian cells carrying the plasmid, to allow the generation of a stable transgenic organism for our studies.

Therefore, we perform the following cloning procedures:

- Cloning of a hepatocyte-specific promoter (apolipoprotein A-II, APOA-II [15]) into a vector containing a mammalian secondary selectable marker (pBigT, containing Neomycin resistance gene). We amplify the APOA-II promoter from genomic DNA, then clone this into the pBigT vector using the traditional cloning method described earlier. This generates a pBigT-APOAII plasmid.
- Cloning of TF X into the pBigT-APOAII plasmid. TF X is amplified by PCR from cDNA obtained from liver cells in which TF X is expressed. This cDNA is cloned downstream of the APOA-II promoter to generate pBigT-APOAII-TFx.
- Cloning of the EGFP reporter gene. In order for the EGFP to be expressed from the APOA-II promoter together with TF X we take advantage of an internal ribosome entry site (IRES) sequence. This allows for translation initiation in the middle of an mRNA and generates an open reading frame containing TF X-IRES-EGFP. For the cloning, we cut an IRES-EGFP sequence from a preexisting plasmid (pIRES-EGFP-puro) and ligate it downstream to TF X in order to generate pBigT-APOAII-TFx-IRES-EGFP. Note that the pBigT already contains a human beta globin poly-adenylation sequence (bGH pA) that terminates the transcription of the cloned DNA (Fig. 2.5).

Once generated, this plasmid can then be stably introduced into a host mammalian organism to finally address our biological questions. For example, human pluripotent stem cells can be genetically modified to integrate such plasmid, and then *in vitro* hepatocyte differentiation can be performed to study the role of TF X in human liver development [16]. Moreover, transgenic mice carrying our plasmid can also be generated to validate the function of TF X during *in vivo* liver development [9].

KEY LIMITATIONS

While being widely used because of its simplicity and low costs, the traditional cloning procedure presented in this chapter has some limitations (Table 2.1). First, it relies on the use of specific restriction enzymes, which create a sequence dependence that can complicate a cloning experiment. Secondly, throughput is limited unless specific modifications (such as the use of ligation adapters) are put in place. However, this also often results in lower cloning efficiency. Finally, the assembly of multiple fragments by traditional cloning either involves multiple steps (which is time-consuming), or can be very inefficient (especially when more than two inserts have to be ligated to a vector at the same time). As discussed earlier, these various limitations can be easily overcome by more expensive or more complicated molecular cloning techniques (Table 2.1).

Overall, molecular cloning does not suffer from major limitations. Indeed, careful planning of a cloning experiment will generally result in the predicted positive outcome. However, problems can arise with regard to the biological applications of cloning experiments. For example, overexpression of a transgene

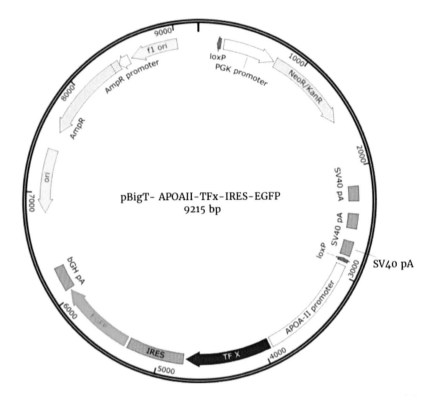

FIGURE 2.5 Example of an application of molecular cloning experiments. Structure of the pBigT-APOAII-TFx-IRES-EGFP plasmid described in the text. The two loxP sequences can be used to eliminate the Neomycin resistance cassette using Cre-mediated recombination. The three SV40 polyadenylation sequences are used both as terminators for the Neomycin gene and as insulators between the two expression cassettes on the vector.

can be challenging due to the transcriptional silencing of the promoter used. This is particularly common for viral promoters such as CMV when used in certain cell types such as stem cells. Moreover, the generation of complex vectors harboring multiple promoters and transgenes can result in suboptimal expression of such elements due to so-called "promoter interference" mechanisms [17]. Finally, complex vectors are also subjected to size limits.

TROUBLESHOOTING

The most common problems that can be encountered during a traditional molecular cloning experiments are summarized in Table 2.4.

TABLE 2.4 Troubleshooting of Cloning Experiments

Stage	Common Problem	Possible Cause and Potential Solutions
Preparation of the vector/insert	Incomplete or no digestion	• Digestion is inhibited by DNA methylation. Check if the enzyme used is methylation-sensitive, and if so use a DNA methyltransferase-free *E. coli* strain to grow the plasmids. • DNA is contaminated with salt, solvents, or other inhibitors of the restriction enzyme. Clean up the plasmids using phenol-chloroform extraction or commercial kits. • Presence of slow-sites or supercoiled DNA result in less-efficient digestion. Increase the enzyme units used and/or the incubation time.
Preparation of the vector/insert	Unexpected bands	• Enzymes show nonspecific cleavage (star activity) under conditions of low ionic strength, high pH, high (>5%) glycerol concentration, high enzyme concentration, and prolonged incubation times. If possible use enzymes modified to lack star activity (commercially available); otherwise, decrease glycerol content, enzyme units used in the digestion, and perform shorter incubation times. • Partial digestion results in bands of unexpected size. Follow recommendations for incomplete digestion described above.
Introduction of the recombinant DNA into the host organism	No colonies or very few colonies	• Inefficient transformation. Test transformation using an uncut plasmid and calculate the transformation efficiency. If this is too low re-prepare competent cells or by new commercial ones. If the construct to be transformed is > 10 kb consider performing electrophoration. • Ligase is not functional. Test the ligase on a plasmid digested with a single enzyme that generates sticky ends. If no colonies are obtained, buy a new ligase and ligase buffer. Note that ligase buffer must not be frozen-thawed multiple times as it contains ATP. • Inefficient ligation conditions. Vary the vector to insert molar ratio from 1:1 to 1:20, test different temperatures (4°C, 16°C, and 25°C) and increase the ligation time to overnight (16 h). If this does not give better results, the vector or insert might be contaminated with ligation inhibitors: consider clean up. • The DNA is toxic to the cells (only for bacterial expression vectors). Incubate the plates at low temperature (25°C), or use *E. coli* with more tight transcriptional control over the toxic DNA.
Screening for clones with the desired recombinant DNA molecules	High background of colonies without the insert	• Dephosphorylation of the vector was insufficient. Increase the phosphatase units used and/or the incubation time. • Contamination of uncut plasmid. Follow recommendations for incomplete digestion described above.
Expansion and isolation of the recombinant DNA	Low plasmid yield	• Incomplete antibiotic selection. Use fresh antibiotic and incubate cells for not more than 16 h to avoid degradation of the antibiotic. • Inappropriate origin of replication. For large plasmids (>15 kb) using a low-copy ORI might be required. In this case, larger cultures will also be needed.

CONCLUSIONS

In this chapter, we provided an overview of the various molecular cloning techniques and described their application to modern biomedical research. Moreover, we provided a general cloning procedure to introduce the basic concepts involved in molecular cloning experiments. The reader should now have the sufficient background to directly apply these procedures in his research questions, as well as to expand his understanding of other molecular cloning techniques using the resources recommended in the chapter.

ACKNOWLEDGMENTS

We thank Dr. Fotis Sampaziotis and Simon Rodler for their critical comments on this chapter.

REFERENCES

[1] Green MR, Sambrook J. Molecular Cloning: A Laboratory Manual, 4th ed. Cold Spring Harbor Laboratory Press; 2012.

[2] New England Biolabs Cloning and Synthetic Biology [Internet]. Available from: https://www.neb.com/applications/cloning-and-synthetic-biology#tabselect0.

[3] Promega Cloning Applications Guide [Internet]. Available from: https://www.promega.co.uk/resources/product-guides-and-selectors/protocols-and-applications-guide/cloning/.

[4] Sigma Aldrich Nucleic Acid Purification Resources [Internet]. Available from: http://www.sigmaaldrich.com/life-science/molecular-biology/dna-and-rna-purification.html.

[5] QIAGEN DNA Sample Technologies [Internet]. Available from: https://www.qiagen.com/gb/products/catalog/sample-technologies/dna-sample-technologies/.

[6] Kappel S, Matthess Y, Kaufmann M, Strebhardt K. Silencing of mammalian genes by tetracycline-inducible shRNA expression. Nat Protoc 2007 Jan;2(12):3257–69.

[7] Kim H, Kim J-S. A guide to genome engineering with programmable nucleases. Nat Rev Genet 2014 May;15(5):321–34.

[8] Zawaira A, Pooran A, Barichievy S, Chopera D. A discussion of molecular biology methods for protein engineering. Mol Biotechnol 2012 May;51(1):67–102.

[9] Mouldy S. Transgenic Animal Models in Biomedical Research. Methods Mol Biol 2006 Oct 15;360:163–202. New Jersey: Humana Press.

[10] Jiang T, Xing B, Rao J. Recent developments of biological reporter technology for detecting gene expression. Biotechnol Genet Eng Rev 2008 Jan;25:41–75.

[11] Kretzschmar K, Watt FM. Lineage tracing. Cell 2012 Jan 20;148(1–2):33–45. Elsevier Inc.

[12] Soler E, Houdebine L-M. Preparation of recombinant vaccines. Biotechnol Annu Rev 2007 Jan;13:65–94.

[13] Twyman RM, Stoger E, Schillberg S, Christou P, Fischer R. Molecular farming in plants: host systems and expression technology. Trends Biotechnol 2003 Dec 12;21(12):570–8. Elsevier.

[14] Kay MA. State-of-the-art gene-based therapies: the road ahead. Nat Rev Genet 2011 May;12(5):316–28. Nature Publishing Group, a division of Macmillan Publishers Limited. All Rights Reserved.

[15] Yang G, Si-Tayeb K, Corbineau S, Vernet R, Gayon R, Dianat N, et al. Integration-deficient lentivectors: an effective strategy to purify and differentiate human embryonic stem cell-derived hepatic progenitors. BMC Biol 2013 Jan;11:86.

[16] Sampaziotis F, Segeritz C-P, Vallier L. Potential of human induced pluripotent stem cells in studies of liver disease. Hepatology 2014 Dec 11;62(1):303–11.

[17] Shearwin KE, Callen BP, Egan JB. Transcriptional interference--a crash course. Trends Genet 2005 Jun;21(6):339–45.

GLOSSARY

Molecular cloning   Procedure by which a DNA fragment of interest is isolated and expanded in a host organism.

Recombinant DNA   DNA molecule that contains sequence originating from different organisms. Usually refers to the product of a molecular cloning experiment.

Vector   DNA molecule that is able to self-propagate in a host organism and is consequently used as a carrier of a desired DNA fragment.

Insert   DNA molecule of interest to be cloned into a vector.

Plasmid   Common cloning vectors able to self-propagate in bacteria.

Restriction enzyme   Endonuclease capable of cutting a DNA fragment at a specific sequence (restrction site).

3'-end   Three prime end: single-stranded DNA terminus consisting of the third carbon of deoxyribose, which is normally linked to an hydroxyl group.

5'-end   Five prime end: single-stranded DNA terminus consisting of the fifth carbon of deoxyribose, which can be linked to a phosphate group.

3'-5' and 5'-3'   Conventional terminology that defines directionality in a single strand of DNA fragment with respect to its ends.

Sticky end   An overhanging piece of single-stranded DNA at the end of a double-stranded DNA fragment. The term *sticky* is used because a complementary piece of single-stranded DNA from another fragment can lead to semi-stable binding of the two molecules.

Blunt end   End of a double-stranded DNA fragment that does not contain overhangs.

Compatible ends   Ends of two DNA fragments that can be efficiently ligated (either blunt ends or sticky ends with complementary overhangs).

Ligase   Enzyme capable of forming covalent bonds between two single-stranded DNA fragments located in close proximity.

Transformation   Bacterial uptake of DNA from the extracellular environment.

Chapter 3

Whole-Mount In Situ Hybridization and a Genotyping Method on Single *Xenopus* Embryos

Martyna Lukoseviciute, Robert Lea, Shoko Ishibashi
and Enrique Amaya
University of Manchester, Manchester, United Kingdom

Chapter Outline

Objectives

- Introduce the theory behind whole-mount in situ hybridization.
- Provide a "ready-to-use" laboratory protocol for WMISH.
- Provide key modifications to the standard WMISH protocol, which enables genotyping of single embryos after WMISH.
- Present a case study using our modified WMISH method, including the genotyping method, on single heterozygous *spib* mutant embryos.

INTRODUCTION

Whole-mount in situ hybridization (WMISH) is a powerful tool that aids the temporal and spatial dissection of gene expression in whole organisms. This method is based on the annealing of a labeled antisense nucleic acid probe to

Basic Science Methods for Clinical Researchers. DOI: http://dx.doi.org/10.1016/B978-0-12-803077-6.00003-5

41

complementary mRNA sequences in fixed tissues that can be visualized by various methods. Nonradioactive WMISH was first developed for *Drosophila* embryos and was soon adapted for other model organisms, including *Xenopus* [1,2]. It is a tremendously useful technique for transcript analysis for several reasons. First of all, even though some high-throughput techniques such as microarrays, enable a transcriptome-wide analysis, the results are usually presented as an average of a cellular population with the outcome providing limited spatial resolution of the expression patterns. In contrast, WMISH offers a single-cell resolution, high sensitivity, and a definite visualization of gene expression domains in whole-embryos that all together allow anatomically significant interpretations [3,4]. Additionally, WMISH can also be more advantageous than well-established immunohistochemistry methods. First, WMISH probes are usually more specific than antibodies, reducing both background noise and false positive results. Furthermore, probes can be derived more straight-forwardly than antibodies. Nevertheless, combined results of WMISH and immunohistochemistry methods facilitate a direct spatiotemporal comparison between transcription and translation of genes of interest [5].

IN PRINCIPLE

The principle of WMISH is based on labeling a complementary RNA probe with a hapten, such as digoxigenin (DIG), fluorescein, or biotin, which can be visualized after hybridization to a targeted mRNA [6] (Fig. 3.1). DIG is the most common hapten used for probe labeling, because it is naturally produced in *Digitalis* plants and hence is absent in animal cells. On the other hand, biotin is endogenous in animal liver and kidney cells, and fluorescein is suitable only for highly abundant mRNA detection due to its lower sensitivity [7]. Probe synthesis is carried out by cloning a cDNA of the desired targeted gene transcript into a transcription vector so that the insert is flanked by promoters for different RNA polymerases, typically T3, T7, or SP6. In this way, using one of these RNA polymerases either sense or antisense strand of an insert can be synthesized [8]. An antisense transcript serves as an RNA probe and a sense transcript can be used as a negative control. Probes are synthesized in the presence of uridine-5′-triphosphate nucleotides conjugated to DIG, which is efficiently incorporated by T3, T7, and SP6 RNA polymerases [2]. Upon hybridization of a labeled probe, the embryos are treated with single-strand specific RNAases in order to eliminate nonhybridized single-stranded antisense probe, and then the remaining double-stranded hybrids can be visualized with an anti-DIG antibody conjugated to a fluorochrome, such as alkaline phosphatase (AP) or horserad-ish peroxidase (HPO), together with appropriate chromogenic substrates. This generates a visible signal that corresponds to the location of the specific mRNA being assessed [2].

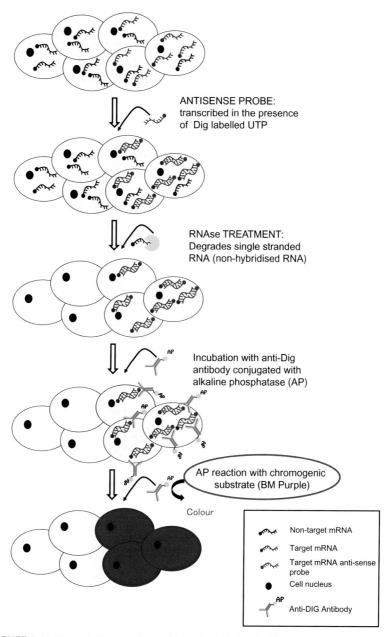

FIGURE 3.1 Pictorial diagram of the principle behind the WMISH protocol. Diagram depicting six cells, where three express a gene of interest, depicted in red. All cells contain thousands of mRNA transcripts to genes, that are not of interest, depicted in black. In order to identify which cells express the gene of interest (in red), an antisense Dig labeled probe complimentary to the gene of interest is synthesized in vitro, depicted in blue. The antisense probe is hybridized to the sense probe in situ, and all remaining single-stranded RNAs are digested by incubating the embryos with single-strand specific RNAses. This removes all mRNAs that are not of interest (black) and any nonhybridized antisense Dig labeled probe. Then the embryos are incubated with AP conjugated antibodies against the Dig hapten, and finally, the cells that expressed the gene of interest are visualized by adding a chromogenic substrate, which changes color in the presence of AP. Thus, the three cells that expressed the gene of interest turns purple.

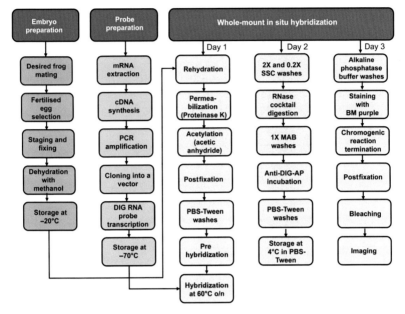

FIGURE 3.2 Flowchart of the multiday WMISH protocol.

WMISH is a fairly laborious procedure that involves multiple washing and incubation steps (Fig. 3.2). As a result, the process takes 3–5 days to complete. However, some designed improvements, such as the use of portable baskets and racks, significantly reduce the amount of manual efforts required for the procedure [9]. The key steps of WMISH are embryo fixation, embryo permeabilization, probe hybridization, antibody incubation followed by staining, and also bleaching if pigmented embryos are used. The fixation steps are needed to preserve embryos from decay and keep RNAs in their initial expression locations throughout the WMISH treatment. This is usually done with an aldehyde, such as formaldehyde, which cross-links lysine residues in the proteins [10]. The passage of the probe and antibody into the embryo cells is ensured by the cellular permeabilization with carefully optimized proteinase K treatments and multiple washes of phosphate-buffered saline containing detergent Tween-20 (PBS-Tween) [2]. The sensitivity and specificity of the RNA probe is increased by acetylation with acetic anhydride that blocks positively charged groups within the tissues and thus reduces charged probe binding to basic proteins in the embryo [7,11]. Ultimately, the pigment of embryos has to be bleached in order to visualize staining thoroughly for an accurate gene expression analysis. This is usually achieved with a bleaching solution containing hydrogen peroxide, formamide, and methanol [12]. Alternatively, albino mutant strains can be used to omit the bleaching stage.

IN PRACTICE—XENOPUS WHOLE-MOUNT IN SITU PROTOCOL—BASKETS

1. **Materials**
   1.1 **Fixation materials**
      1. Wheaton glass scintillation vials (20 mL VWR)
      2. 0.1X Marc's Modified Ringer (MMR): 10 mM NaCl, 0.2 mM KCl, 0.1 mM $MgCl_2$, 0.2 mM $CaCl_2$, 0.5 mM HEPES, pH 7.5. Prepare a 10X stock solution, adjust to pH 7.5 with NaOH. Sterilize the 10X solution by autoclaving.
      3. MEMFA fixative: 0.1 M MOPS pH 7.4, 2 mM EGTA, 1 mM $MgSO_4$, 3.7% formaldehyde. MEMFA should be made fresh from stock solutions of 1 M MOPS pH 7.4, 0.5 M EGTA pH 8, and 1 M $MgSO_4$. Fresh formaldehyde should always be used. Stock solutions of MOPS should be kept away from light.
      4. 100% methanol.
      5. Bouin's post-hybridization fixative: 10% formaldehyde, 5% glacial acetic acid in H_2O.
   1.2 **Probe transcription materials**
      1. 3 M sodium acetate (Sigma)
      2. Ethanol (Fisher)
      3. Diethylpyrocarbonate (DEPC) water (Invitrogen)
      4. RNA polymerase T3, T7, or SP6 (Promega)
      5. 5X transcription buffer supplied with polymerase (Promega)
      6. RNase inhibitor (Roche)
      7. Ribonucleoside triphosphate set lithium salts (Roche)
      8. Turbo DNase (Ambion)
      9. Micro Bio-spin columns (Biorad Cat. N: 732-6221)
   1.3 **WMISH reagents and materials**
      1. Hybridization baskets 9 mm, 55 μm mesh (Intavis Cat. N: 16.33025)
      2. Rack 40 wells (Intavis Cat. N: 33008)
      3. Gilson tip box lid (Gilson Cat. N: F167100)
      4. 7 mL plastic vials (Greiner)
      5. 100% methanol
      6. Proteinase K 20 mg/mL (Ambion)
      7. 0.1 M triethanolamine (pH 7.5)
      8. Acetic anhydride
      9. 4% formaldehyde (Baker) in 1X PBS 0.1% Tween-20
      10. Hybridization Buffer: 50% formamide (Ambion), 5X SSC, 1 mg/ml Torula RNA (Sigma), 100 μg/mL heparin (Sigma), 1X Denhardt's solution (Sigma), 0.1% Tween-20, 0.1% CHAPS (Sigma).
      11. 2X SSC: 300 mM NaCl, 30 mM sodium citrate. Prepare a 20X stock solution, adjust to pH 7.0 with NaOH. Sterilize the 20X solution by autoclaving.

12. 0.2X SSC.
13. RNase cocktail (Ambion). The solution contains 500 U/mL RNase A and 20,000 U/mL RNase T1.
14. 1X MAB: 100 mM maleic acid, 150 mM NaCl. Prepare a 2X stock solution, adjust to pH 7.5 with NaOH. Sterilize the 2X solution by autoclaving.
15. 10% blocking reagent (BR) (Roche).
16. 10% heat-treated lamb serum (HTLS) (Sigma).
17. 1X PBS 0.1% Tween-20 (Sigma).
18. 1X PBS
19. AP buffer: 100 mM Tris, pH 9.5, 50 mM $MgCl_2$, 100 mM NaCl, 0.1% Tween-20.
20. Probes: DIG-11-UTP (Roche Cat. N: 11209256910)
21. Antibodies: Anti-DIG-alkaline phosphatase (Anti-DIG-AP) (Roche Cat. N: 11093274910).
22. Chromagenic reagent:
    i. NBT: Nitro blue tetrazolium (Sigma). Dissolve at 75 mg/mL in 70% Dimethylformamide. Wrap the tube in foil and store at −20°C.
    ii. BCIP: 5-Bromo-4-chloro-3-indolyl phosphate (Roche).
23. Bouin: 9% formaldehyde and 5% glacial acetic acid.
24. Bleaching solution: 10% hydrogen peroxide, 5% formamide, and 85% methanol OR bleaching solution if genotyping will be performed: 5% hydrogen peroxide in 1X PBS.

1.4 Genotyping reagents and materials

1. Water bath
2. Polymerase chain reaction (PCR) thermal-cycler
3. Gel loading tank
4. UV gel imager
5. 1.5 mL Eppendorf tubes
6. 0.2 mL PCR tubes
7. Microcentrifuge
8. Lysis buffer: 10 mM Tris–HCl pH 8.0, 1 mM EDTA, 0.3% Tween-20, 0.3% NP40
9. Proteinase K 20 mg/mL (Ambion)
10. Taq DNA polymerase
11. 10X Taq DNA polymerase buffer
12. A mix of dNTPs
13. Forward and reverse primers, 10 µM each
14. Nuclease-free water (Ambion)
15. DNA ladder
16. 10X loading dye
17. Ethidium bromide
18. Agarose powder

19. 0.5X tris-acetate-EDTA (TAE) buffer
20. Exonuclease I (ExoI) (NEB)
21. Thermosensitive alkaline phosphatase (TSAP) (Promega)

2. Method for WMISH

2.1 Fixation

1. Embryos are collected at the desired developmental stage and fixed with fresh MEMFA for 1 hour at room temperature in 20 mL glass vials.
2. Samples are then washed 3 × 10 minutes in 100% methanol and can be stored for several months at −20°C.

2.2 Probe Transcription

1. Linearize plasmid DNA by digesting in appropriate buffer and according to standard molecular biology protocols.
2. Resuspend DNA pellet in DEPC water at a final concentration of 1 µg/µL.
3. Prepare the transcription reaction
   - 10 µL Promega 5X transcription buffer
   - 2.5 µL Dig mix
   - 30 µL DEPC treated water
   - 2.5 µL 1 µg/µL linearized template DNA
   - 0.5 µL 40 U/µL RNase inhibitor
   - 4.5 µL RNA polymerase (T3, T7, or SP6)
   - Incubate for 2 hours at 37°C.
4. Add 1 µL of 1 mg/mL DNase, and incubate for 15 minutes at 37°C.
5. Purify on a Micro Bio-spin column following manufacturers' instructions.
6. Add 100 µL of nuclease-free water, 20 µL of 3 M sodium acetate pH 5.3 and 500 µL of ethanol. Vortex vigorously and incubate at −80°C for at least 1 hour.
7. Wash with 70% ethanol.
8. Air dry pellet and resuspend in 50 µL of DEPC water.
9. Check 1 µL on a freshly made agarose gel.
10. Store the stock probe at −80°C.

2.3 In Situ Day 1

In all steps, samples should be rocked at moderate speed at room temperature, unless otherwise noted.

1. Place a 40-well intavis rack in a bath containing 100 mL of 100% methanol. We have found that Gilson tip box lids work well as baths. The fixed samples are then distributed as appropriate into separate mesh baskets sitting in the rack. We recommend using baskets as they reduce the risk of damaging the embryos when aspirating and allow large-scale in situ screens. For best results, do not place more than 10–15 embryos in the same mesh basket

and use a trimmed and fire-polished glass Pasteur pipette to transfer them.

Note: If embryo genotyping after WMISH is desired, we recommend placing five embryos in a basket.

2. Re-hydrate the embryos with 5-minute 100 mL washes in:
   - 75% methanol, 25% H_2O
   - 50% methanol, 50% H_2O
   - 25% methanol, 75% PBS Tween-20
   - then 2×5 minute washes in 100% PBS Tween-20

3. Use forceps to transfer each basket (containing embryos) from the 40-well intravis racks to 7 mL individual plastic tubes containing 1 mL of proteinase K solution. Care must be taken to record the order of the samples when transferring from the rack to tubes. Extensive proteinase K treatment can result in excessive degradation of tissue, particularly the epidermis. It is therefore recommended to test different incubation lengths for different batches of proteinase K due to variability.
   - *Xenopus laevis*—10 μL 20 mg/mL proteinase K in 50 mL PBS Tween-20
   - *Xenopus tropicalis*—3 μL 20 mg/mL proteinase K in 50 mL PBS Tween-20

4. Transfer the samples back to the rack and wash twice for 5 minutes in 100 mL of 0.1 M triethanolamine pH 7.5.

5. Do not remove the last 5 mL triethanolamine wash; instead add 12.5 μL of acetic anhydride to each basket. Rock for 5 minutes, then add another of 12.5 μL acetic anhydride, and rock for a further 5 minutes.

   Note: If embryo genotyping is required after WMISH, omit acetic anhydride treatments and wash embryos in triethanolamine alone for 5–10 minutes.

6. Wash two times for 5 minutes in PBS Tween-20.

7. Re-fix the samples for 20 minutes with 3.7% formaldehyde in PBS Tween-20. 10 mL of 37% formaldehyde and 90 mL of PBS Tween.

8. Wash five times for 5 minutes in PBS Tween-20.

9. Aliquot 1 mL of hybridization buffer into 7 mL screw cap plastic bijou vials. Transfer the baskets into the buffer and pre-hybridize for 4–6 hours at 60°C in a shaking incubator.

10. Transfer the baskets from the hybridization buffer to 1 mL of probe solution (hybridization solution with 1 μg/mL antisense mRNA probe). Hybridize overnight at 60°C in a shaking incubator. Save the hybridization buffer as it can be used the following day.

Place washing solutions required for Day 2 to warm overnight in the 60°C incubator.

2.4 In Situ DAY 2

1. Remove the baskets from the probe solution and transfer them to the hybridization buffer already saved from the previous day. Incubate at 60°C for 5 minutes. The probe solution can be stored at −20°C and reused a maximum of three times.

2. Wash three times for 20 minutes at 60°C in 2X SSC.

3. Wash for 20 minutes at 37°C in 2X SSC.

4. Incubate for 30 minutes at 37°C with 20 μg/mL of RNase cocktail in 2X SSC.

5. Wash for 10 minutes in 2X SSC at room temperature.

6. Wash twice for 20 minutes at 60°C in 0.2X SSC.

7. Wash twice in MAB for 5 minutes at room temperature.

8. Transfer the samples to 7 mL screw cap plastic tubes and block for 1 hour in 1 mL MAB +2% BR +2% HTLS. Add 2 mL of 10% BR and 2 mL HTLS to 6 mLMAB to make 10 mL of blocking solution.

9. Incubate with 1 mL antibody diluted in MAB +2% BR +2% HTLS. Incubate for 5 hours at room temperature.

10. Transfer the samples back to racks and wash three times for 5 minutes in MAB.

11. Leave in MAB overnight at 4°C rocking.

2.5 In Situ Day 3

Chromogenic reaction

1. Wash twice for 5 minutes at room temperature with AP buffer. AP buffer needs to be made up fresh when needed. A 10X stock can be kept if it is made without $MgCl_2$.

2. Transfer the embryos out of the baskets and into a 24-well plate containing AP buffer.

3. Aspirate the AP buffer from the samples and replace with chromogenic substrate diluted in AP Buffer, 1 mL per well. To make the chromogenic substrate, add 4.5 μL 75 mg/mL of NBT and 3.5 μL 50 mg/mL of BCIP per mL of AP Buffer. Chromogenic substrates are light-sensitive, so cover trays with foil and rock at room temperature. Incubate the samples until a satisfactory signal is obtained. This usually takes between 30 minutes and 24 hours.

4. Once the signal is satisfactory aspirate the chromogenic substrate and replace with PBS Tween.

5. Wash three times for 5 minutes in PBS Tween.

6. Leave overnight in PBS Tween at 4°C rocking.

2.6 In Situ Day 4
Fixation and bleaching
1. Aspirate the PBS Tween and post fix the samples in Bouin solution for 1 hour.
2. If genotyping of the embryos is desired, wash them twice in PBS Tween for one to two minutes and bleach in 5% H_2O_2 (in 1X PBS) for 2–3 hours in the presence of white light. Wash embryos twice in PBS Tween. Ignore steps 3–6 and continue to the section 3 (method for a genotyping analysis).
   If genotyping on embryos will not be performed, please ignore this bleaching step and follow the steps 3–6.
3. Wash for 5 minutes in 70% ethanol.
4. Wash for 5 minutes in 100% methanol.
5. Bleach for 1 hour or the appearance of a satisfactory signal. A satisfactory signal is obtained when pigmentation of the embryos has bleached leaving the contrast between the chromogenic staining and the embryo easily distinguishable. Bleaching is carried out on in the presence of white light, we use a light box.
6. Re-hydrate the embryos with 5-minute washes in:
   – 75% methanol, 25% PBS Tween
   – 50% methanol, 50% PBS Tween
   – 25% methanol, 75% PBS Tween
   – Wash twice for 5 minutes in 100% PBS Tween.

3. Method for genotyping analysis
3.1 Genomic DNA extraction
1. Transfer single embryos into 1.5 mL Eppendorf tubes using either a plastic or glass pipette. Remove liquid and add 50 µL of lysis buffer containing 0.4 µg/µL proteinase K.
2. Incubate tubes at 55°C for at least 1 hour with occasional tapping.
3. Transfer tubes to 95°C for 10 minutes. Make sure embryos are dissolved after this stage.
4. Spin tubes at the maximum speed for 1 minute and transfer supernatant into fresh 1.5 mL Eppendorf tubes.
5. Store extracted DNA at 4°C until usage.

3.2 Polymerase chain reactions (PCRs)
The volume of 20 µL PCR reaction is enough for both gel electrophoresis and sequencing analysis.
1. Design appropriate PCR cycling parameters based on your primer melting temperatures.
2. Amount of DNA lysate added may significantly affect the PCR reaction due to varying DNA and salt concentrations. Therefore, determine the most efficient DNA lysate amounts and primer concentrations for your PCR reactions considering the target fragment and primer sizes.

Recommended final primer concentrations: 0.05–1 μM of each primer.

3. Once the best PCR conditions are identified, prepare a mastermix containing 1X Taq polymerase buffer, 200–250 μM of each dNTPs, 0.5 U of Taq polymerase per 20 μL reaction, and both reverse and forward primers of selected concentrations. Keep the prepared mastermix on ice!

4. Positive and negative controls should be included to identify the success, specificity, and/or contamination of PCR reactions.

5. Mix required volumes of PCR mastermix and DNA in 0.2 mL PCR tubes.

6. Transfer PCR tubes from ice to a PCR thermocycler that is preheated to a required temperature.

7. After PCR cycles are finished, transfer tubes on ice. PCR reactions can be stored at −20°C until usage.

3.3 Gel electrophoresis

The success of DNA extraction followed by PCR should be verified by gel electrophoresis prior to sequencing. This helps to select PCR reactions to be sequenced and also to roughly estimate PCR product concentrations.

1. Prepare 0.5–2.0% agarose gel based on the size of your PCR fragment (use higher agarose gel concentrations to resolve small PCR fragments and vice versa) and visualize the gel with ethidium bromide.

   Other safer alternative stains, such as SYBR Safe (Invitrogen) can be used instead of ethidium bromide.

2. Mix 9 μL of a PCR reaction with 1 μL of 10X DNA loading dye. Load samples in parallel with an appropriate DNA ladder.

3. Load agarose gel in 0.5X TAE buffer for 30–60 minutes at 100 V.

4. Image the gel under UV light.

3.4 PCR product purification prior to sequencing

We do not perform DNA sequencing ourselves and send purified PCR products for sequencing by an external sequencing company. We use Sanger sequencing approach for genotyping analysis.

1. The amount of DNA that is required for sequencing analysis depends on the size of a PCR product (see Table 3.1). Please dilute your PCR product accordingly.

   Preparation of PCR products for sequencing.

2. Prepare a mastermix containing 4–5 U of ExoI nuclease and 0.5 U of TSAP per 10 μL.

3. Mix 10 μL of PCR reaction with 10 μL of ExoI nuclease/TSAP mastermix.

4. Incubate ExoI/TSAP reaction at 37°C for 1 hour.

TABLE 3.1 Required Amounts of DNA for Sanger Sequencing Analysis

PCR Product Size (bp)	DNA Amount for Sequencing (ng)
100–200	1–3
200–500	3–10
500–1000	5–20
1000–2000	10–40

5. Heat inactivate ExoI and TSAP at 75°C for 10 minutes.
6. Perform or send reactions for sequencing. Note that only a forward or reverse primer, which targets a DNA sequence within your PCR product sequence, is needed.

APPLICATIONS

WMISH is an essential technique for characterizing the spatiotemporal expression of genes at a single-cell resolution level. Thus, WMISH is an essential technique used in all model organisms, including invertebrates and vertebrates. Using this technique, one can catalog the expression pattern of hundreds, if not thousands, of genes, resulting in a descriptive atlas of expression patterns throughout development (see, e.g., Ref. 3). In addition, WMISH is a powerful technique to be added to a geneticist's toolkit, as it permits the characterization of phenotypic changes in both forward and reverse genetic approaches, or during large-scale gain of function screens [13]. Importantly, it is an indispensable technique to detect even subtle spatial changes in the expression of genes following targeted genetic manipulations in embryos. Furthermore, WMISH allows assessment of the most immediate effects in gene expression alterations even before any morphological phenotypes become apparent in mutant embryos. As such, WMISH is one of the most versatile and important techniques to master, as clinical scientists build their experimental repertoire.

Scenario

To illustrate how the WMISH combined with genotyping method can be applied, we present a case study that aimed to characterize the phenotype and genotype of heterozygote *X. tropicalis spib* mutants. The West African clawed frog, *X. tropicalis* is an excellent model system for developmental biology, in particularly embryology, because it offers large, thus easily manipulated, oocytes and embryos that develop ex utero [8], as well as the capacity to perform genetic analyses [14,15]. This animal is distinguished from other related amphibians by its diploid genome that facilitated sequencing of its genome, bringing *X. tropicalis* into the genomic era [16,17].

The *spib* gene codes for a homolog of the Spi1 protein that belongs to the family of erythroblast transformation-specific (ETS) motif-containing transcription factors [18]. Our previous study showed that *spib* is one of the earliest genes expressed in the anterior ventral blood islands in Xenopus embryos, and furthermore, we showed that spib is required for the development of primitive myeloid cells, including macrophages and neutrophils [19]. We introduced a *spib* mutation using TALEN-targeted mutagenesis. After they reached sexual maturity, we performed a genetic cross between one of the heterozygous mutant males and a wild-type female frog. This mating would be expected to generate 50% heterozygote mutant progeny and 50% wild type progeny. We performed WMISH for *mpo*, a marker for primitive myeloid cells, to investigate the effect of *spib* mutant heterozygosity on the development of primitive myeloid cells in the resulting F1 embryos. After WMISH, we performed genomic DNA extractions from the post-WMISH embryos. This was followed by PCR method to amplify part of the *spib* gene with primers flanking the TALEN mutagenesis site. We then purified the PCR products and sequenced them using Sanger sequencing.

We successfully distinguished single heterozygote *spib* mutant embryos from their wild-type sibling (Fig. 3.3). Furthermore, we did not observe any phenotypic or gene expression changes between wild-type and heterozygote

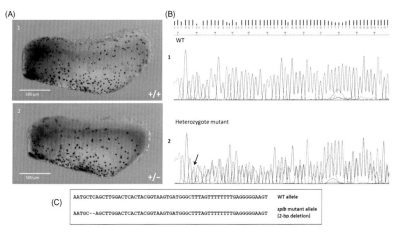

FIGURE 3.3 Genotyping single wild-type and heterozygous spib mutant embryos. (A) Offspring of a wild-type female and heterozygote *spib* mutant male were fixed in MEMFA at the tailbud stage (approximately 24 hours after fertilization). They then were subjected to the newly developed WMISH method, using an *mpo* probe, to visualize the primitive myeloid cells. Note that the purple spots coincide with individual myeloid cells, confirming that the WMISH provides single-cell resolution. (B) Sequencing results showed that Embryo 1 was wild-type, whereas Embryo 2 was heterozygote for the *spib* mutant allele. Black arrow point, where the 2-base pair deletion in the mutant allele is present. (C) Sequence of wild type and 2-base pair deletion mutation.

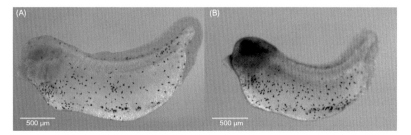

FIGURE 3.4 Comparison between original versus modified WMISH protocol. (A) *Xenopus tropicalis* embryo stained for *mpo*, a marker of primitive myeloid cells, after the standard WMISH protocol. Note that the purple spots coincide with individual myeloid cells, confirming that the WMISH provides single-cell resolution. (B) *Xenopus tropicalis* embryo stained for *mpo*, a marker of primitive myeloid cells, after the modified WMISH protocol, which permits single-embryo genotyping. The required modifications include removal of acetic anhydride and bleaching of embryos with 5% hydrogen peroxide without formamide and methanol.

mutant embryos. Overall, the WMISH-genotyping method offers a unique opportunity to combine reverse genetic approaches with classical molecular phenotyping and genotyping techniques in order to extensively study single *X. tropicalis* embryos.

KEY LIMITATIONS

If embryos are to be genotyped following the WMISH, it is essential to omit the acetic anhydride step from the WMISH protocol. We noticed few associated problems upon acetic anhydride exclusion from the WMISH protocol. As expected, the probe background staining increased to some extent in comparison to the original method (see Fig. 3.4). Also, we noticed that embryos tend to stick to each other and to the walls of baskets during the procedure if acetylation step is omitted. This can be improved by reducing the number of embryos within a single basket. However, some damage of the skin, particularly in the fin area, cannot be avoided most of the time. Given that some genes, such as fibroblast growth factor 7 (*fgf7*), are expressed in these damage-prone areas, the altered WMISH protocol could have major implications on gene expression interpretations [20]. Therefore, these identified problems have to be taken into consideration before deciding to use the altered WMISH protocol designed for genotyping analysis.

CONCLUSION

Here we present a reliable method for performing WMISH in *X. laevis* and *X. tropicalis* embryos. Furthermore, we present key modifications to the standard WMISH protocol, which are necessary for genotyping single *Xenopus* embryos, after they have been submitted to WMISH. Finally, we present a method that permits efficient genotyping of single embryos, after they have been subjected to the WMISH protocol.

REFERENCES

[1] Tautz D, Pfeifle C. A non-radioactive in situ hybridization method for the localization of specific RNAs in *Drosophila* embryos reveals translational control of the segmentation gene hunchback. Chromosoma 1989;98(2):81–5.

[2] Harland RM. In situ hybridization: an improved whole-mount method for *Xenopus* embryos. Methods Cell Biol 1991;36:685–95.

[3] Pollet N, Muncke N, Verbeek B, Li Y, Fenger U, Delius H, et al. An atlas of differential gene expression during early *Xenopus* embryogenesis. Mech Dev 2005;122(3):365–439.

[4] Acloque H, Wilkinson DG, Nieto MA. In situ hybridization analysis of chick embryos in whole-mount and tissue sections. Methods Cell Biol 2008;87:169–85.

[5] Irving C. In-situ hybridization and immunohistochemistry in whole embryos. Methods Mol Biol 2008;461(Chapter 46):687–95. Totowa, NJ: Humana Press.

[6] Broadbent J, Read EM. Wholemount in situ hybridization of *Xenopus* and zebrafish embryos. Methods Mol Biol 1999;127:57–67. New Jersey: Humana Press.

[7] Jin L, Lloyd RV. In situ hybridization: methods and applications. J Clin Lab Anal 1997;11(1):2–9.

[8] Sive HL, Grainger RM, Harland RM. Early development of *Xenopus laevis*. CSHL Press; 2000.

[9] Lea R, Bonev B, Dubaissi E, Vize PD, Papalopulu N. Multicolor fluorescent in situ mRNA hybridization (FISH) on whole mounts and sections. Methods Mol Biol 2012;917(Chapter 24):431–44. Totowa, NJ: Humana Press.

[10] Acton A, Harvey T, Grow MW. An examination of non-formalin-based fixation methods for *Xenopus* embryos. Dev Dyn 2005;233(4):1464–9. Wiley-Liss, Inc.

[11] Hayashi S, Gillam IC, Delaney AD, Tener GM. Acetylation of chromosome squashes of Drosophila melanogaster decreases the background in autoradiographs from hybridization with [125I]-labeled RNA. J Histochem Cytochem 1978;26(8):677–9.

[12] King RS, Newmark PA. In situ hybridization protocol for enhanced detection of gene expression in the planarian *Schmidtea mediterranea*. BMC Dev Biol 2013;13(1):8. BioMed Central Ltd.

[13] Chen JA, Voigt J, Gilchrist M, Papalopulu N, Amaya E. Identification of novel genes affecting mesoderm formation and morphogenesis through an enhanced large scale function screen in *Xenopus*. Mech Dev 2005;122(3):307–31.

[14] Amaya E, Offield MF, Grainger RM. Frog genetics: *Xenopus tropicalis* jumps into the future. Trends Genet 1998;14(7):253–5.

[15] Grainger RM. *Xenopus tropicalis* as a model organism for genetics and genomics: past, present, and future. Methods Mol Biol 2012;917(Chapter 1):3–15. Totowa, NJ: Humana Press.

[16] Hellsten U, Harland RM, Gilchrist MJ, Hendrix D, Jurka J, Kapitonov V, et al. The genome of the Western clawed frog *Xenopus tropicalis*. Science 2010;328(5978):633–6.

[17] Amaya E. Xenomics. Genome Res 2005;15(12):1683–91.

[18] Ray D, Bosselut R, Ghysdael J, Mattei MG, Tavitian A, Moreau-Gachelin F. Characterization of Spi-B, a transcription factor related to the putative oncoprotein Spi-1/PU.1. Mol Cell Biol 1992;12(10):4297–304. American Society for Microbiology (ASM).

[19] Costa RMB, Soto X, Chen Y, Zorn AM, Amaya E. Spib is required for primitive myeloid development in *Xenopus*. Blood 2008;112(6):2287–96.

[20] Lea R, Papalopulu N, Amaya E, Dorey K. Temporal and spatial expression of FGF ligands and receptors during *Xenopus* development. Dev Dyn 2009;238(6):1467–79.

LIST OF ACRONYMS AND ABBREVIATIONS

AP	alkaline phosphatase
BR	blocking reagent
cDNA	complementary deoxyribonucleic acid
DIG	digoxigenin
dNTP	deoxynucleotide
HTLS	heat treated lamb serum
MAB	maleic acid buffer
mRNA	messenger ribonucleic acid
PBS	phosphate-buffered saline
PCR	polymerase chain reaction
SSC	saline-sodium citrate
TALEN	transcription activator-like effector nuclease
WMISH	Whole-mount in situ hybridization

Chapter 4

Microarrays: An Introduction and Guide to Their Use

Frederick D. Park[1,2], Roman Sasik[1] and Tannishtha Reya[1,2]
[1]University of California San Diego School of Medicine, La Jolla, CA, United States; [2]Sanford Consortium for Regenerative Medicine, La Jolla, CA, United States

Chapter Outline

Objectives

In this chapter, we will address the following questions:

- How do microarrays work and what key principles are they based on?
- How do microarrays compare to next generation sequencing approaches?
- What is the workflow of a typical microarray experiment?
- What are the different uses of microarrays and how have they been commonly applied to answer scientific questions?
- What are the main limitations of microarrays?
- What are common problems that might arise in a microarray experiment and how do I address them?

Basic Science Methods for Clinical Researchers. DOI: http://dx.doi.org/10.1016/B978-0-12-803077-6.00004-7

INTRODUCTION

The term "microarray" broadly refers to a solid support such as a glass slide or silicon chip onto which many separate pools of molecules are attached in a regular pattern of microscopic spots in order to analyze multiple genetic or biochemical interactions at the same time. While these pools of attached molecules may consist of antibodies, other proteins, and even whole cells or tissue fragments, they most commonly consist of DNA oligonucleotides (see Glossary). And so, we will focus here on these "DNA microarrays" and their applications. Such microarrays bear many tiny strands of single-stranded DNA called "probes" covalently attached to their surfaces. These probes often consist of gene-specific sequences that collectively span all the genes in the known genome. When fluorescently labeled complementary DNA (cDNA) or RNA (cRNA), known as "targets," are applied to the microarray, the relative amount of target-to-probe hybridization in a particular spot is reflected by the fluorescent signal detected. This signal level then reflects the relative amount of a particular target sequence in the sample. Thus, in a single experiment, microarrays can reveal the genome-wide abundance of nucleic acid sequences in a particular cell or tissue sample of interest.

Microarrays have had a tremendous impact on biomedical research over the past two decades. In combination with whole genome sequencing data, they have helped expand the scope of genetic analyses from one or a few genes to all the known genes in the genome. By enabling rapid genome-wide analyses, microarray users could begin to mine precious functional information from the huge quantities of sequence data emerging from whole genome sequencing. Microarrays have been used to study a variety of genome-wide phenomena including gene expression profiles, single nucleotide polymorphisms (SNPs), DNA copy number variations, DNA methylation, alternative splicing, and transcription factor-binding sites (see Glossary). Each of these phenomena may directly reflect or impact upon when, how, or to what degree genetic information is expressed in cells, in both normal and diseased states. Thus microarrays have played a key role in ushering in the genomic era of biomedical science.

Nevertheless, the recent rise of "next-generation" or massively parallel sequencing technology, in which a large number of sequencing reactions are performed simultaneously, has resulted in a precipitous decline in sequencing cost. For this and other reasons discussed here, it is important to consider both the strengths and weaknesses of microarrays as compared to sequencing when deciding which experimental approach to take for a given study.

IN PRINCIPLE

Microarrays, much like the earlier methods of Southern blotting and bacterial colony blot hybridization, exploit the high specificity under stringent conditions with which complementary nucleic acid strands hybridize [1]. In these earlier blotting methods, DNA samples attached to a membrane are probed with a labeled oligonucleotide in solution. Microarrays function in the reverse-oligonucleotide probes are attached to a supporting structure and labeled sample nucleic acids are applied in solution to the probe-bearing surface. In both cases, labeled nucleic acids that fail to hybridize with attached sequences are washed away, in principle solely due to a lack of perfect complementarity. Targets hybridized to corresponding probes are then detected by measuring signals emitted from persistently bound labels (Fig. 4.1).

There are variations on this theme between different commercially available microarrays, but the same principles apply across platforms. An illustration of this was presented in a large multicenter comparison of microarrays that included popular platforms made by Affymetrix and Illumina [2]. These platforms apply very different technologies—Affymetrix microarrays employ short perfect match and mismatch DNA probes directly synthesized onto a glass slide, while Illumina arrays have long DNA probes attached to silica beads that sit in tiny wells on a slide. When identical samples of RNA were tested, Affymetrix

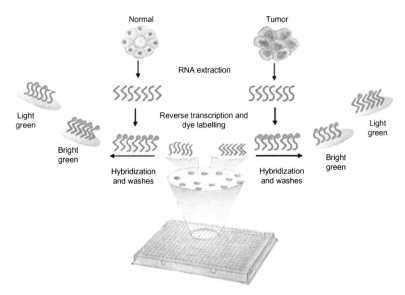

FIGURE 4.1 Schematic representation of microarray hybridization with labeled samples from normal tissue and tumor. Increasing magnified views of the spots on a microarray slide are represented by circles. Each of two representative spots contains bound probes with different sequences, represented by smoothly curving strands ("smooth" gene) and jagged strands ("jagged" gene). RNA extracted from the normal and tumor tissues are reverse transcribed and dye labeled (dye represented by stars). These labeled targets are hybridized to the microarray, and non-annealing targets are washed away. In this schematic example, the normal tissue has more "jagged" gene transcript expression, while the tumor has more "smooth" transcripts. The relative abundance of target-probe hybridization is reflected in the intensity of signal from each spot (bright vs. light green).

and Illumina arrays consistently gave very similar results with a high degree of overlap in differentially expressed genes.

The labels used to visualize target nucleic acids bound to microarrays are fluorophores or dyes, which have the useful ability to re-emit absorbed light at a specific wavelength, or color. Detection of fluorescence can be highly sensitive and provide a fairly wide dynamic range, both important qualities when used to detect targets that can vary in abundance over several logarithmic orders of magnitude. Additionally, two samples labeled with different dyes applied to the same microarray can be distinguished by the color of light they emit. This is the basis for 2-color expression profiling microarrays in which the ratio of the colors produced by bound targets to a certain probe reflects the relative expression between the two samples. In contrast, with 1-color profiling, each sample is labeled with the same color dye but applied to separate microarrays, and relative expression is measured by comparing signal intensity between the two arrays (Fig. 4.1).

A microarray experiment generates large amounts of data, and converting this data into valid and meaningful information is key to unlocking its scientific potential. A single experiment utilizes multiple individual microarrays, and as probe density has increased to provide better genomic coverage, a single array may contain millions of different probes. For instance, the Affymetrix GeneChip Human Transcriptome Array 2.0 currently contains over six million distinct probes. Processing and interpreting such vast amounts of raw data can be challenging, but over the years methods have become well developed. Before the main data analysis, raw fluorescence signal intensities must be preprocessed into signals adjusted for technical sources of variation. This step involves subtraction of background noise by methods that can vary between platforms, and then normalization to correct for nonbiological sources of signal variation. There are different approaches to normalization, but in general when only a small fraction of targets are expected to change between conditions, there are two equally popular normalization methods: lowess normalization and quantile normalization [3]. After the data has been preprocessed, one can proceed to the primary task of identifying statistically significant signals and apply this information in subsequent analyses tailored to address a specific research question.

IN PRACTICE

A typical microarray experiment workflow has the following steps: experimental design planning, sample preparation, hybridization, and washes, scanning, data analysis, and result validation (Fig. 4.2). Here we will give a brief overview of each step in this workflow.

Experimental Design Planning

As with any scientific experiment, careful planning is essential to achieve the goals of a microarray experiment. This is the time to clearly define the goals and to anticipate technical challenges that may arise. Since microarray experiments

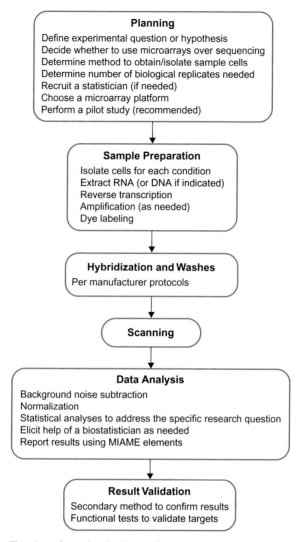

Planning
Define experimental question or hypothesis
Decide whether to use microarrays over sequencing
Determine method to obtain/isolate sample cells
Determine number of biological replicates needed
Recruit a statistician (if needed)
Choose a microarray platform
Perform a pilot study (recommended)

Sample Preparation
Isolate cells for each condition
Extract RNA (or DNA if indicated)
Reverse transcription
Amplification (as needed)
Dye labeling

Hybridization and Washes
Per manufacturer protocols

Scanning

Data Analysis
Background noise subtraction
Normalization
Statistical analyses to address the specific research question
Elicit help of a biostatistician as needed
Report results using MIAME elements

Result Validation
Secondary method to confirm results
Functional tests to validate targets

FIGURE 4.2   Flowchart of steps involved in a typical microarray experiment.

by their nature require a good knowledge of statistics, now is the time to consult a statistician if needed. Also, with the availability of next-generation sequencing approaches, one should decide whether to use microarrays over sequencing (see below for further discussion).

A major objective of planning is to design the experiment to give the best chance of obtaining data with statistical significance. That is, to find expression level differences caused by biological differences, not by chance alone. One aspect of this is to limit, at least initially, the number of variables being tested—in other words, keep it simple. It is easier to test the effect of a drug given at a single dose and single duration than it is to test the effects of several different doses given over several different durations. The latter would require many more microarrays and samples in order to obtain significant data for each of the different combinations of dose and duration. For this reason, a preliminary analysis of the variables being tested, such as a dose optimization study, may be helpful.

Another important aspect in planning is to determine how many biological replicates are needed. Biological replicates are samples from multiple independent sample sources, such as separate tumors, animals, or cell line cultures. It is only through repeated microarray analyses with a sufficient number of biological replicates that statistically significant differences above the level of natural variation across a population can be reliably assessed. To illustrate this, let's compare samples from a highly inbred strain of mice versus samples from human patients. The natural variation of gene expression in the mice is generally lower than in the patient samples. Thus, while a microarray experiment analyzing the mice may only require a handful of samples per condition, the human study may require samples from a large number of patients to account for the greater variability. Note that biological replicates are not technical replicates, which are repeated samples from the same source. Technical replicates are mainly useful for assessing technical variability in the microarray procedure itself, which is generally a smaller factor than biological variability and have largely been dealt with by the microarray manufacturer.

Unfortunately, the question of how many biological replicates are needed can sometimes be difficult to answer with certainty. This is because different types of samples and the methods of obtaining them can yield wide variability in the resulting gene expression profiles. For example, a uniform cultured cell line will usually have less variability from plate to plate than tumors from surgical resection of different patients. To address this problem, it is a good idea to perform a pilot experiment aimed at assessing the range of gene expression variability between samples. One may do this by reverse transcription quantitative polymerase chain reaction (qRT-PCR) for several genes known or suspected to vary in expression across samples. A better approach, particularly for larger studies, may be to perform a pilot microarray analysis on a small number of samples first. This will not only give an estimate of gene expression variance across the whole range of genes on the array, but may also reveal unforeseen technical difficulties that need to be addressed before starting the main experiment. Once variance in gene expression is estimated for the system, one can calculate the number of biological replicates needed to detect the proposed minimum significant change (effect size) at an appropriate level of significance and statistical power.

Naturally, to find real biological expression differences between the conditions being tested, the samples to be compared must represent, as closely as possible, each respective condition. Again, taking the example of cultured cell lines versus patient tumors, this can be relatively straightforward when, for example, assessing the effect of a certain drug on cultured cells. As long as the cells remain similarly viable under each condition, the difference is primarily limited to the treatment administered, experimental or control. Tumors, on the other hand, are composed of heterogeneous populations of cells, which may include stromal cells, epithelial cells, endocrine-type cells, various types of blood cells, and more. If one is only interested in the gene expression differences within a single population of cells in the tumors, then a method to isolate or significantly enrich for this population is needed. Otherwise the gene expression of admixed cells may obscure the true expression pattern of the cells of interest. Thus, one should carefully plan, and ideally practice beforehand, the optimal method for obtaining the desired samples.

Finally in the planning stage, one should select a suitable microarray platform designed to provide the type of information needed for the goals of the experiment. The choice of platform may be dictated largely by which resources are available at your institution, but even then it is important to make sure the available platform is suitable before proceeding further. For example, if one is interested in analyzing variations in splice isoform expression, one should use microarrays with probes covering gene expression across all known individual exons. It would also be desirable for the array to include probes containing known splice junction sites, allowing for analysis of splicing events associated with specific splice isoform expression.

Sample Preparation

Once the desired sample cells or tissues are obtained, one may proceed to RNA or DNA extraction. Given its stability, extraction of high-quality DNA is generally reliable in most hands without taking special precautions. RNA, however, is more labile and prone to degradation. Therefore, among the steps in a typical microarray experiment, RNA extraction and handling probably has the greatest impact on success or failure of the experiment. If the sample RNA has degraded substantially, then expression data derived from such RNA will be correspondingly poor. (See Troubleshooting section below for more on this.)

After RNA isolation, different platforms have differing protocols that lead to the final labeled nucleic acid sample that is to be applied to the microarray. Total RNA may be used directly or enriched for mRNA by affinity purification or depletion of ribosomal RNA. An amplification step can also be performed if the quantity of RNA is limiting. RNA may then be dye-labeled directly or, more commonly, converted to cDNA by reverse transcription. Dyes may be incorporated during cDNA synthesis or in a subsequent transcription step to make cRNA. In any case, the final product of sample preparation is a collection of dye-labeled complementary nucleic acids representing the relative abundance of endogenous transcripts expressed in the original biologic sample.

Hybridization, Washes, and Scanning

Labeled sample is then applied in solution to the microarray for hybridization with complementary probe sequences. This is followed by a series of wash steps to remove unbound sample. If the sample was labeled with biotin (as with the Affymetrix platform) fluorescently labeled streptavidin is then applied. The microarray is now ready for scanning. A laser excites bound fluorophores at each spot on the array, and the emitted fluorescent signal intensity at the appropriate wavelength is measured and recorded. This yields an image of the microarray [typically in Tagged Image File Format (TIFF) format] wherein each gene or target corresponds to a single spot, and the relative expression of that gene or target is reflected in a measurable intensity of hybridization (Fig. 4.3).

Data Analysis

Data analysis begins with processing the raw microarray image (Fig. 4.3). For this step well-developed software is included with all major microarray platforms, which reliably generate high-quality measures of probe hybridization intensities. This software will identify the spots and their boundaries on a microarray and distinguish them from spurious signals, such as dust, scratches, or manufacturing defects. The intensity of each pixel within a spot is measured and mean, median, or integrated pixel intensity is determined for each spot. Final hybridization intensity values for each spot are then calculated by subtraction of average background hybridization intensity in the areas surrounding the spots.

The next step is to translate these signal intensity values into quantities that serve as surrogates for the relative expression levels of each gene or target. This generates a spreadsheet of expression values that can be used for further

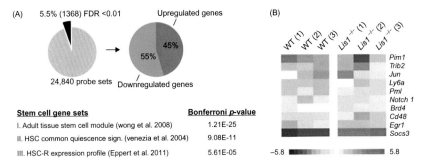

FIGURE 4.3 (A) Pie chart representation of microarray results. Comparative gene set enrichment analysis identified 5.5% of genes as changed with a false discovery rate of <0.01. Of these genes, 55% were downregulated and 45% were upregulated. (B) Heat map showing known regulators of stem and progenitor cell activity significantly affected by the loss of Lis1. *Reproduced with permission from Zimdahl B, Ito T, Blevins A, et al. Lis1 regulates asymmetric division in hematopoietic stem cells and in leukemia. Nat Genet 2014;46(3):245–52.*

statistical analysis. Note that expression values are usually described as a ratio between the signal intensity of a test sample versus that of a reference sample. For example, the expression value of gene X in a certain cancer might be the ratio of gene X signal intensity in the cancer sample to the gene X signal intensity found in surrounding normal tissue. Furthermore, ratios are usually transformed to logarithm base 2 [i.e., $\log_2(\text{ratio})$], in order to generate a continuous spectrum of values across up- or downregulated targets.

The methods for preprocessing, including technical quality control measures and normalization, and subsequent analysis of microarray data can vary by platform, types of samples analyzed, and the specific application or type of information desired. There are numerous commercial and open-source data analysis software packages available. Of the major manufacturer platforms, those of Agilent and Illumina are quite good. Also, Bioconductor, an open-source, open-development analysis platform written in the R programming language is available for free and is widely used. Since a discussion of data analysis methods is beyond the scope of this chapter, the reader is directed to recent focused articles on the use of Bioconductor for microarray data analysis [4,5] and the www.bioconductor.org website that offers a variety of educational materials and courses. It may be best for the novice microarray user to consult with a biostatistician familiar with microarray data analysis. Also, when reporting microarray data, it is helpful to follow the Minimum Information About a Microarray Experiment (MIAME) elements to ensure that the results can be interpreted and potentially reproduced by others [6].

The net result of microarray data analysis is most often a list of target genes with significant differences in expression between conditions of interest. Generally speaking, in order to identify significant expression levels, one must first determine a threshold level above which a probe signal is considered detected. Following this, detected signals can be analyzed for differential expression levels that can be considered significant. In very early microarray studies, an arbitrary fixed cut-off of two-fold change was used, but better statistical methods have since been applied to obtain more meaningful threshold levels. An example of this is assignment of a false discovery rate to each expression value, above which it is considered significant.

This resulting data can be presented in a variety of formats, but most often shown graphically by a heat map in which relative under- or overexpression of a target is represented by a color. Since a microarray experiment typically yields hundreds and even thousands of statistically significant targets, it is useful to group sets of targets by gene ontology—known properties of genes such as molecular function, cellular localization, or involvement in known pathways or biological processes, keeping in mind that a gene may have numerous properties that remain to be discovered. This "gene set enrichment analysis" can be useful when the phenotype of interest results from additive changes across sets of genes that might be missed by looking only at smaller changes in expression of individual genes within a set. In the somewhat reverse application called cluster analysis, genes can be grouped by similar expression level change to explore or mine the data for genes potentially involved in a similar process or phenotype.

Result Validation

Confirmation of expression differences for these genes should be done by an independent method such as qRT-PCR or Northern blot. Analysis on the protein level (e.g., by Western blot) is better still, since transcriptional expression does not always correlate with that of translation. The final and maybe most important question is whether the identified targets are functionally relevant to the process or disease being studied. If feasible, one should directly test their functional importance by knockdown or genetic knockout. For example, if a gene was found in a microarray to be significantly upregulated in metastatic tumors compared to nonmetastatic tumors, one might test whether specifically knocking down the expression of this gene by RNA-mediated interference reduces metastasis formation by tumor cells xenografted into immunodeficient mice.

APPLICATIONS

Thus far, we have focused on using microarrays to compare gene expression profiles between different conditions, which is the most common use of microarrays. Before touching upon other applications, let us briefly look at how gene expression patterns can be used.

Supervised versus Unsupervised Analyses

Broadly speaking, data analysis can be split between "supervised" and "unsupervised" approaches. In supervised analysis, one is typically guided by prior knowledge of genes of interest or a pathway of interest in the context of the experiment, and a limited set of genes with significant expression differences are identified for their potential role in the characteristic that distinguishes the samples.

In contrast, with unsupervised analysis, expression patterns are not constrained by prior expectations and the analysis is more exploratory in nature. The data may be analyzed using machine learning algorithms, the simplest of which is clustering or class discovery in which genes or samples with similar expression profiles are grouped together. This method is useful for discovering new information about a certain sample type or identifying new genes involved in a process or clinical outcome of interest. One application of class discovery is identification of subtypes of a disease, such as cancer, which can be distinguished by their gene expression patterns without prior biological knowledge. Building upon this example, after the specific gene expression profile of a particular cancer type is found, this can be used clinically to identify the primary source of a metastasis of unknown origin or to predict prognosis (e.g., Agendia's Mammaprint). This is class prediction, another application of machine learning.

Speaking of clinical utility, microarrays may be useful when the changes in expression of many genes provide more useful information than that of a few genes or other current markers of disease. Furthermore, clinical use is only warranted when array information will have an impact on patient management decisions, are shown to improve outcomes, and is cost-effective.

Analysis of Genomic DNA Features and DNA-Protein Interactions

In addition to profiling RNA expression, microarrays can be used to analyze a variety of genomic DNA features. Using array comparative genomic hybridization (array CGH), presence of genomic DNA copy number variations including deletions or duplications can be compared between samples. Arrays have often been used in genome-wide association studies (GWAS) to identify single nucleotide polymporphisms (SNPs) associated with certain diseases. Much like the use of gene expression profiles to identify cancer types and predict prognosis, small DNA arrays that probe tumor samples for known DNA mutations are commercially available (e.g., GeneID).

Combining immunoprecipitation (IP) with DNA microarrays has enabled efficient genome-wide analysis of DNA methylation and DNA-protein interactions. When methylated fragments of genomic DNA are purified by IP with an anti-methylcytidine antibody and hybridized to a microarray, positive probe signals identify which genes in the sample are methylated. Likewise, antibodies specific to proteins known or suspected to bind DNA can be used to IP-specific protein-DNA complexes. Bound DNA can then be purified and identified by microarray (ChIP-on-chip). However, it should be noted that chromatin immunoprecipitation-sequencing (ChIP-Seq), in which the bound DNA is identified by sequencing, rather than by hybridization to a microarray, has much better specificity and so has largely replaced microarrays for this application.

SCENARIO

We will now review a typical case in which microarrays were used to answer a scientific research question. Zimdahl and colleagues found that genetic knockout of the gene encoding dynein-binding protein Lis1 in mice resulted in severe anemia and a bloodless phenotype leading to embryonic lethality [7]. This was consistent with a requirement for Lis1 in the maintenance of hematopoetic stem cell function. The question was then asked whether Lis1 knockout caused expression changes in genes of known importance for stem cell maintenance. To answer this question, Affymetrix GeneChip microarrays were used to compare gene expression between stem cell-enriched samples from control and Lis1-knockout mice.

This case provides a useful illustration of steps taken to obtain appropriate experimental samples in order to answer the question at hand. Since the phenotype of interest is a function of hematopoietic stem cells, it was important to enrich samples for this cell population. To do this, bone marrow cells were sorted using a cell surface marker expression pattern that distinguishes hematopoietic stem cells from other bone marrow cells. Next, multiple complementary methods were used to rigorously confirm that Lis1 expression was knocked out in test cells compared to control cells. Genomic PCR demonstrated that the Lis1 gene was deleted in hematopoietic stem cells from Lis1-knockout mice but remained intact in control cells. Likewise, RT-PCR confirmed that Lis1 mRNA expression was absent and immunofluorescence with anti-Lis1 antibody showed that Lis1 protein expression was lost in Lis1-knockout cells. Finally, a functional difference in hematopoietic capacity was demonstrated between Lis1 knockout and control cells. Altogether, this preliminary analysis instills confidence that the appropriate cells were compared using microarrays to find differences in gene expression patterns specifically due to loss of Lis1.

As mentioned earlier, it is important to have sufficient biological replicates to ensure that statistically significant differences in gene expression above the level of natural variation are found. In this study, highly inbred mouse strains that differ primarily in a single well-defined deletion at a particular gene locus were used. Thus, since natural variation in baseline gene expression is expected to be low, samples were obtained from just three test and three control mice. Total RNA of the sorted hematopoetic stem cells from these mice was purified, amplified, labeled, and hybridized to the Affymetrix GeneChip Mouse Genome microarray according to the manufacturer's protocols.

The resulting raw hybridization data was then processed and analyzed by a biostatistician with expertise in microarray data analysis. In brief, expression-level data was normalized and a threshold expression level value was determined using validated methods. Probes whose expression exceeded this threshold value were considered detected. Detected probes were then sorted by the smallest false discovery rate at which the gene was called significant. The sorted list of significant genes was then subjected to Gene Set Enrichment Analysis to identify the genes with significant differential expression between the test and control samples. A subanalysis focusing on genes within published stem cell signature gene sets was also performed (Fig. 4.4).

It was found that 5.5% of genes were differentially expressed, with 746 probe sets downregulated and 622 upregulated (Fig. 4.3A). There were highly significant differences in expression of previously identified stem cell signature gene sets between the two cell samples. Fig. 4.3B shows a typical heat map of expression levels for a subset of these genes; red indicates upregulation and blue indicates downregulation. The loss of expression of multiple core stem cell signature genes in Lis1-knockout cells compared to controls is consistent with a role for Lis1 in maintaining stem cells.

KEY LIMITATIONS

The limitations of microarrays become obvious when they are compared to competing next-generation sequencing (NGS)-based approaches such as whole transcriptome sequencing, or RNA-Seq (Table 4.1).

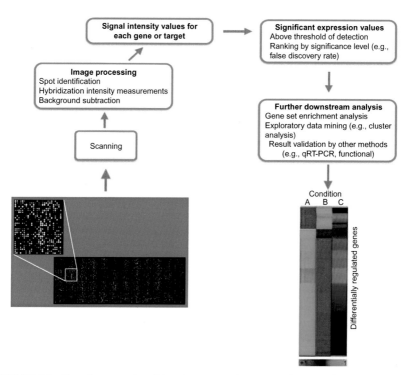

FIGURE 4.4 Steps from scanning of the microarray to data presentation. Scanning of a two-color Agilent microarray slide to which two samples labeled with different color fluorophores, red and green, has been hybridized. Note that yellow spots reflect binding of both red and green labeled samples. The scanned image is then processed to yield signal intensity values for each probe-target hybridization. This is followed by identification of significant expression values. Differentially expressed gene expression can be presented in a heat map format in which the degree of expression change is represented by colors; green and red for relative over- and under-expression, respectively. In addition, the data can be subjected to further downstream analysis, and additional experiments to validate the results may be performed.

Sensitivity, Specificity, and Dependence on Known Sequences

Microarrays are both less sensitive and less specific than RNA-Seq. The range of microarray probe coverage is entirely dependent on prior sequence knowledge, whereas RNA-Seq is not biased or limited by such information since it provides a comprehensive view of the entire transcriptome. And so, even though high-density microarrays are narrowing the gap, microarrays may still miss novel transcripts or isoforms that can be detected by RNA-Seq. In addition, SNPs found by RNA-Seq may be both novel and more informative since they are expressed in RNA transcripts, whereas microarrays only detect known SNPs in genomic DNA. Another disadvantage of microarrays is that as new sequence data becomes available, earlier microarray data may become obsolete. A set of

TABLE 4.1 The Limitations and Advantages of Using Microarrays Compared to Next-Generation Sequencing

Limitations	Advantages
Depends on prior sequence knowledge • Will not detect novel transcripts • Samples need to be re-run as new sequence data emerges	Mature technology • Well-established protocols and tools • Well-developed infrastructure
Based on hybridization • Less sensitive for low abundance targets • Lower specificity than deep sequencing, especially for closely homologous gene families	Relatively fast • Typically <1 week per experiment • Scalable for automated high-throughput analyses
Signal levels are an indirect readout • Only gives relative not absolute abundance of targets	Cost-effective • Still significantly less expensive at this time
	Lower computing requirements • Data analysis can be done on a standard PC • Requires less computing expertise • Much smaller data storage requirements
	Simpler data analysis • Consensus methods for data analysis • Less statistical data analysis expertise required

samples would need to be run anew with a newer microarray that covers the new sequence, whereas the more comprehensive RNA-Seq data would remain relevant. RNA-Seq can also distinguish host from parasite transcripts and examine transcripts of any organism without prior sequence knowledge. This is especially useful for bacteria, which have large variations in genomic sequence even within a single strain.

Microarrays are also inherently less sensitive and specific than RNA-Seq because they rely upon nucleic acid hybridization. Although probe sequences are selected for good specificity to a target sequence, a probe may not be completely specific to a single target. Also, fluorescent dyes can differentially affect hybridization. Given this fundamental difference, one might say that microarray data versus RNA-Seq data are "analog" versus "digital," respectively. Microarrays can only provide relative expression by comparing probe intensities. They do not provide absolute quantitation of transcript expression such as RNA-Seq. In addition, microarrays have less potential dynamic range due to nonlinear hybridization kinetics at either extreme of abundance—probes can be saturated by high abundance targets and give little signal above background noise with very low abundance targets. With increased depth of sequencing, RNA-Seq offers much better sensitivity and signal-to-noise ratio for low abundance transcripts.

Limitations Shared with RNA-Seq

Some limitations, however, are shared by microarrays and RNA-Seq. Both rely on high-quality RNA from samples that accurately reflect the conditions being tested. Also, both report gene expression on the transcriptional level, which may not reflect the true biological activity of the gene on the protein level. An important problem inherent to genome-wide expression studies that should be understood is the "multiple comparisons problem." Given that microarrays (and RNA-Seq) compare the level of tens of thousands of signals simultaneously, one must guard against a potentially large number of false positives. For example, if one selects a standard p-value cut-off for significance of <0.05, and 10,000 tests are simultaneously performed by microarray, then we can expect 500 targets to be falsely identified as differentially expressed. A conservative method to address this problem is the Bonferroni correction, which essentially divides the p-value by the total number of tests performed. In our example, this would adjust the p-value cut-off to $<0.05/10,000$, or $p<0.000005$, which can be overly stringent, resulting in detection of few, if any, true positives. For this reason, alternative statistical concepts with more power have been developed. One is false discovery rate, which is the expected fraction of false positives among discoveries [8], later generalized in the Bayesian framework into local false discovery rate [9]. A popular tool called Significance Analysis of Microarrays calculates a quantity similar to false discovery rate, which is defined as the ratio of expected number of false positives and the number of discoveries [10].

Practical Advantages That Remain

Despite their limitations, microarrays maintain some practical advantages over RNA-Seq. Microarrays represent a more mature and reliable technology that has been improved and validated over decades of use. A microarray experiment is relatively fast, typically taking less than a week and can be automated for high throughput, with the potential to analyze thousands of samples in a month. They also have a well-established infrastructure with widely accessible institutional cores and companies to support their use. Data analysis is now relatively straightforward with established and refined protocols that can be performed on any PC. In contrast, data analysis for RNA-Seq is more complex, typically requiring special expertise and computing capability. As such, there is not yet full consensus on the protocols for RNA-Seq data analysis. Also, RNA-Seq produces far more data than microarrays and data storage requirements can be very large—typically thousands of times more than microarrays require. Finally, costs for NGS are dropping, but at this time microarrays are still significantly cheaper per sample than NGS. For these reasons, even when RNA-Seq is used to obtain initial study findings, microarrays can complement RNA-Seq with a rapid and more cost-effective method for follow-up analyses based on the RNA-Seq data.

TROUBLESHOOTING

Problems may occur in any of the steps of a microarray experiment described above. In the following table, some potential problems are listed along with comments and suggestions to address them. Note that microarray platforms utilize built-in controls, broadly split into sample-dependent and sample-independent controls that may help to detect the source of problems.

Problem	Comments and Suggestions
Poor sample RNA quality	• Isolation of high-quality RNA is essential to good quality microarray results, and so problems often arise during sample preparation steps, including RNA isolation, reverse transcription, and labeling.
	• Sample dependent controls can help to assess problems with sample quality. These include housekeeping control probes that give the signal of common housekeeping genes expressed in all cells. These positive controls should be highly expressed if the sample is not overly degraded and the array is working properly.
	• A set of previously validated commercially available RNA samples, such as those used in the MicroArray Quality Control (MAQC) project [2], may also be used as a positive sample control to ensure that microarray results of comparable quality are obtained in your hands using your array platform.
	• Work quickly to minimize handling time of RNA.
	• Precautions should be taken to minimize degradation by RNA endonucleases (RNAses). Sources of RNases to consider are those in the external environment as well as endogenous nucleases within cells and tissues.
	• To reduce external RNase contamination, a dedicated RNA work area should be treated with RNase inhibitor solution before beginning, dedicated pipettes and filter tips for RNA work should be used, and all reagents and plasticware should be certified as RNase-free or treated with diethylpyrocarbonate (DEPC) or autoclaved to inactivate RNases.

Problem	Comments and Suggestions
	• Different cells and tissues contain widely varying levels of endogenous RNases, but it is a good idea to take precautions to minimize their activity regardless of the cell or tissue source. Samples should be kept on ice and processed or frozen as soon as possible. If samples are to be frozen, it is preferable to flash freeze using liquid nitrogen or dry ice. Also, thawing of frozen samples should be avoided before lysing for RNA extraction. The addition of beta-mercaptoethanol to the lysis solution can further limit endogenous RNase activity. • Once RNA is isolated, one should check RNA integrity and quantity before proceeding further. RNA integrity can be roughly assessed by agarose gel electrophoresis; sharp, well-defined rRNA bands with upper band intensity roughly twice that of the lower band is consistent with good RNA integrity. Alternatively, integrity can be checked by 260/280 wavelength absorbance ratio measurement, or more accurately using a bioanalyzer such as the Agilent bioanalyzer, if available.
Poor hybridization	• Given the maturity of current microarray platforms, most problems arise from the steps outside the standardized procedures of hybridization, washes, and microarray scanning. Nevertheless, one should strive to minimize variation—use the same equipment, operator, batch of reagents, and incubation times, as well as microarrays preferably from the same lot throughout the experiment. • Sample-independent controls can help identify problems in hybridization. Hybridization controls are fluorescently labeled oligonucleotides spiked into the sample solution at graded concentrations. These should give a corresponding gradient of signal level responses if the hybridization is working properly. These may also include low stringency control oligonucleotides with mismatch bases in their sequences. The signal produced by these should be low in the high-stringency hybridization conditions desired for microarrays compared to positive controls with no mismatches.
High natural variation in gene expression across samples	• As discussed above, this may obscure significant differences in gene expression. • In general, a larger number of biological replicate samples will be needed to increase the statistical power of the study. • A preliminary analysis by qRT-PCR for selected genes or a small pilot microarray study can be performed to estimate variance of gene expression across samples.
Difficulty with data analysis	• The www.bioconductor.org website is a useful resource for educational materials. • Consultation with a biostatistician familiar with microarray data analysis is recommended for the novice microarray user.

Problem	Comments and Suggestions
General workflow problems	• It is a good idea to start with a pilot experiment in which a trial run is performed using a small number of samples and arrays. This is probably the best way to avoid unforeseen problems in the main microarray experiment. It will also give the opportunity for the novice user to practice hands-on protocol steps from sample isolation and preparation through microarray scanning.

CONCLUSION

The microarray has proven to be a valuable tool in the effort to translate the large amounts of available genomic sequence data into meaningful insights about biological function. Since microarrays have been in wide use for a long period of time, the technology is now well-developed, cost-effective, and highly reliable. Applications for microarrays are numerous and better microarrays continue to be developed. However, microarrays have inherent weaknesses that prevent them from matching the performance of next-generation sequencing-based approaches such as RNA-Seq. It is likely that as the cost of sequencing falls to match that of microarrays and as consensus protocols for RNA-Seq data analysis emerge, microarrays will be largely supplanted. In the meantime, however, microarrays continue to offer real practical advantages.

REFERENCES

[1] Pirrung MC, Southern EM. The genesis of microarrays. Biochem Mol Biol Educ 2014;42(2):106–13.
[2] Consortium M, Shi L, Reid LH, et al. The MicroArray Quality Control (MAQC) project shows inter- and intraplatform reproducibility of gene expression measurements. Nat Biotechnol 2006;24(9):1151–61.
[3] Bolstad BM, Irizarry RA, Astrand M, Speed TP. A comparison of normalization methods for high density oligonucleotide array data based on variance and bias. Bioinformatics 2003;19(2):185–93.
[4] Rodrigo-Domingo M, Waagepetersen R, Bodker JS, et al. Reproducible probe-level analysis of the Affymetrix Exon 1.0 ST array with R/Bioconductor. Brief Bioinform 2014;15(4):519–33.
[5] Mohapatra SK, Krishnan A. Microarray data analysis. Methods Mol Biol 2011;678:27–43.
[6] Brazma A, Hingamp P, Quackenbush J, et al. Minimum information about a microarray experiment (MIAME)-toward standards for microarray data. Nat Genet 2001;29(4):365–71.
[7] Zimdahl B, Ito T, Blevins A, et al. Lis1 regulates asymmetric division in hematopoietic stem cells and in leukemia. Nat Genet 2014;46(3):245–52.
[8] Benjamini Y, Hochberg Y. Controlling the false discovery rate: a practical and powerful approach to multiple testing. J R Stat Soc B 1995;57:289–300.
[9] Efron B. Microarrays, Empirical Bayes and the Two-Groups Model. Stat Sci 2008;23(1):1–22.
[10] Tusher VG, Tibshirani R, Chu G. Significance analysis of microarrays applied to the ionizing radiation response. Proc Natl Acad Sci USA 2001;98(9):5116–21.

SUGGESTED FURTHER READING

[1] Wang C, Gong B, Bushel PR, et al. The concordance between RNA-seq and microarray data depends on chemical treatment and transcript abundance. Nat Biotech 2014;32:926–32.
[2] Quackenbush J. Microarray data normalization and transformation. Nat Genet 2002; 32(Supplement):496–501.
[3] Canales RD. Evaluation of DNA microarray results with quantitative gene expression platforms. Nat Biotechnol 2006;24(9):1115–22.

GLOSSARY

Alternative splicing   The process by which particular exons can be included or excluded from the final processed messenger RNA (mRNA) produced by a gene. This allows for a single gene to express multiple different alternatively spliced mRNA that may be translated into different proteins with different biological functions.

DNA copy number variation   A structural alteration of the genomic DNA resulting in a cell with a variant number of one or more sections of DNA. For example, a deletion may result in a lower DNA copy number than normal for the deleted section of DNA, and conversely, a duplication event may result in a greater DNA copy number of the duplicated section of DNA.

DNA methylation   Addition of methyl groups to DNA, which typically results in suppression of gene transcription.

Fluorophore   A fluorescent chemical compound that can re-emit light upon light excitation

Genome   The complete set of genes or genetic material present in a cell or organism.

Gene expression profile   A measurement of the activity or "expression" of thousands of genes at once to create a global picture of cellular function.

Housekeeping genes   Constitutive genes that are required for the maintenance of basic cellular function and are expressed in all cells of an organism under normal and pathophysiological conditions.

Hybridization   The process of combining two complementary single-stranded DNA or RNA molecules and allowing them to form a single double-stranded molecule through base pairing.

Isoform   Any of two or more functionally similar proteins that have a similar but not identical amino acid sequence and are either encoded by different genes or by RNA transcripts from the same gene, which have had different exons removed.

Microarray   A solid support such as a glass slide or silicon chip onto which many separate pools of molecules are attached in a regular pattern of microscopic spots in order to analyze multiple genetic or biochemical interactions at the same time.

Oligonucleotide   A polynucleotide whose molecules contain a relatively small number of nucleotides.

***p*-Value**   Estimated probability of rejecting the null hypothesis of a study question when that hypothesis is true.

Pilot study   A small-scale preliminary study conducted in order to evaluate feasibility, time, cost, adverse events, and effect size (statistical variability) in an attempt to predict an appropriate sample size and improve upon the study design prior to performance of a full-scale research project.

Single nucleotide polypmorphism (SNP)   A variation in a single base pair in a DNA sequence.

Transcription factor A protein that binds to specific DNA sequences, thereby controlling the rate of transcription of DNA to messenger RNA.

Transcriptome The sum total of all the messenger RNA molecules expressed from the genes of an organism.

LIST OF ACRONYMS AND ABBREVIATIONS

cDNA	Complementary deoxyribonucleic acid
ChIP-on-chip	Chromatin immunoprecipitation-microarray
ChIP-Seq	Chromatin immunoprecipitation-sequencing
cRNA	Complementary ribonucleic acid
DEPC	Diethylpyrocarbonate
GWAS	Genome wide association studies
HSC	Hematopoetic stem cell
IP	Immunoprecipitation
MAQC	Microarray quality control
MIAME	Minimum Information About a Microarray Experiment
NGS	Next-generation sequencing
qRT-PCR	Quantitative reverse transcription—polymerase chain reaction
RNase	Ribonucleic acid endonuclease
RNA-Seq	Ribonucleic acid—sequencing
SNP	Single nucleotide polypmorphism

Chapter 5

Analysis of Human Genetic Variations Using DNA Sequencing

Gregory A. Hawkins

Wake Forest School of Medicine, Winston-Salem, NC, United States

Chapter Outline

Objectives

The primary objective of this chapter is to understand how DNA (deoxyribonucleic acid) sequencing is used to identify genetic variations that can contribute to disease. Two methodologies will be discussed: Sanger sequencing and next-generation DNA sequencing (NGS).

INTRODUCTION

DNA is the molecule contained in every cell of the body, except for red blood cells, and contains the genetic code utilized for regulating the production of

Basic Science Methods for Clinical Researchers. DOI: http://dx.doi.org/10.1016/B978-0-12-803077-6.00005-9

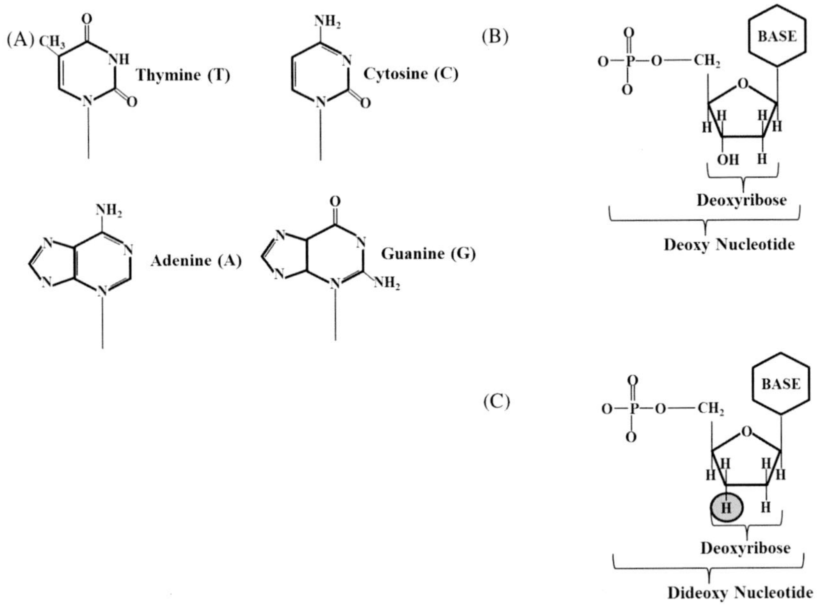

FIGURE 5.1 Nucleotide structure. (A) Chemical structure of base attached to ribose ring in DNA. (B) Chemical structure of deoxy nucleotide. (C) Chemical structure of dideoxy nucleotide used in Sanger sequencing termination reaction. Note the loss of the hydroxy group on the ribose ring. Loss of the hydroxyl group does not allow the addition of additional nucleotides in the polymerase reaction, thus terminating DNA synthesis in Sanger sequencing.

proteins. The basic unit of the DNA genetic code consists of four nucleotides distinguished by the attached bases adenine (A), cytosine (C), guanine (G), and thymine (T) attached to a deoxyribose molecule (Fig. 5.1A and 5.1B). The human genome contains approximately 3 billion nucleotides, and each cell in the body contains 2 genome copies: one copy inherited from the mother and one copy inherited from the father. The DNA of genes is copied from DNA into RNA (ribonucleic acid) during a process called gene transcription (Fig. 5.2A). DNA sequences that promote and enhance transcription lie in the 5' region of the gene. During transcription, the coding regions, termed exons, are spliced together to form the mature messenger RNA (mRNA) (Fig. 5.2B). The messenger is then translated into a protein by a process called translation (Fig. 5.2C). Translation of the RNA is performed by ribosomal RNA and tRNA (transfer RNA). During translation, the three-letter DNA code, termed a codon, is used to produce a protein by sequentially attaching amino acids base on the end of a growing peptide chain. There are 64 codons represented by a three-letter genetic code (Figure 5.3). Only 61 codons code for an amino acid, while the remaining three codons signal the translation process to stop and thus terminate protein synthesis. Mutations that occur in the codons are the source for many genetic variations, which contribute to genetic diseases.

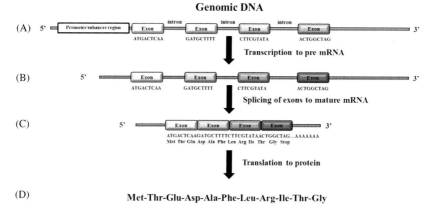

FIGURE 5.2   Basic overview of transcription and translation. (A) Basic gene structure in genomic DNA. (B) Pre-mRNA before exon splicing. (C) Mature mRNA ready for translation. (D) Protein synthesized by ribosomal machinery during translation of mature mRNA.

ATT		ACT		AAT	Asparagine (Asn/N)	AGT	Serine (Ser/S)
ATC	Isoleucine (Ile/I)	ACC	Threonine (Thr/T)	AAC		AGC	
ATA		ACA		AAA	Lysine (Lys/K)	AGA	Arginine (Arg/R)
ATG	Methionine (Met/M)	ACG		AAG		AGG	
CTT		CCT		CAT	Histidine (His/H)	CGT	
CTC	Leucine (Leu/L)	CCC	Proline (Pro/P)	CAC		CGC	Arginine (Arg/R)
CTA		CCA		CAA	Glutamine (Gln/Q)	CGA	
CTG		CCG		CAG		CGG	
GTT		GCT		GAT	Aspartic Acid (Asp/D)	GGT	
GTC	Valine (Val/V)	GCC	Alanine (Ala/A)	GAC		GGC	Glycine (Gly/G)
GTA		GCA		GAA	Glutamic Acid (Glu/E)	GGA	
GTG		GCG		GAG		GGG	
TTT	Phenylalanine (Phe/F)	TCT		TAT	Tyrosine (Tyr/Y)	TGT	Cysteine (Cys/C)
TTC		TCC	Serine (Ser/S)	TAC		TGC	
TTA	Leucine (Leu/L)	TCA		TAA	Stop (Ter)	TGA	Stop (Ter)
TTG		TCG		TAG		TGG	Tryptophan (Trp/W)

FIGURE 5.3   Condon usage chart.

The first draft of a human genome was completed in 2001 and took over a decade and ~$2.7 billion to complete [1,2]. Approximately 25,000 genes were identified in the human genome, far fewer than what was predicted given the complexity of the human cell. The coding region of the genome, the region containing genes, only occupies about 1–2% of the human genome. Millions of genetic variations have been identified in the human genome, primarily consisting of single nucleotide polymorphisms (SNPs) and small insertions or deletions (InDels) of one or more nucleotides (Fig. 5.4). Each individual's genome contains 2–3 million of these polymorphisms, a small portion of which may be de novo(new) variations created during development of the embryo and not inherited from either parent.

(A) **Synonymous Change**

ATG GAC TCG ATT CAG CAT >> ATG GAC TCG ATC CAG CAT
Met--Asp--Ser--- Ile---Glu --His   Met---Asp---Ser--Ile—Glu--His

(B) **Missense/Non-synonymous**

ATG GAC TCG ATT CAG CAT >> ATG GAC TCG AAT CAG CAT
Met--Asp--Ser---Ile---Glu ---His   Met---Asp---Ser--Asn—Glu--His

(C) **Nonsense**

ATG GAC TCG ATT CAG CAT >> ATG GAC TCG ATT TAG CAT
Met--Asp---Ser---Ile---Glu --His   Met--Asp--Ser---Ile---Stop

(D) **Insertion/Deletion**

ATG GAC TCG ATT CAG CAT >> ATG GAC TCG AAT TTC AGC AT
Met--Asp---Ser---Ile---Glu --His   Met--Asp---Ser---Ile---Phe—Ser-----

FIGURE 5.4 Basic classes of genetic variations in DNA coding regions. (A) Synonymous change changes the DNA sequence but not the protein sequence. (B) Missense or nonsynonymous change changes the DNA and protein sequence. (C) Nonsense changes introduces a premature stop codon which results in premature termination of protein synthesis. (D) Insertion/Deletion changes the DNA sequence by adding or removing nucleotides thus change the peptide sequence following the insertion or deletion point.

The methodology called Sanger sequencing was used to sequence the first human genome. A new DNA sequencing methodology, NGS, was introduced in 2007. NGS is defined as a "mass array" methodology that allows millions of unique DNA fragments to be sequenced simultaneously. NGS has advanced significantly over the past few years and instruments are now available that can sequence a whole human genome in a few days at a cost approaching $1000 per genome. The principles of Sanger sequencing and NGS will be described in more detail later.

IN PRINCIPLE

Sanger Sequencing

Sanger sequencing was developed by Dr. Frederick Sanger in 1977 [3] and has been the most widely used DNA sequencing technique over the past 40 years. Sanger sequencingutilizes a mixture of a DNA polymerase, an enzyme that copies of DNA strands, and the four deoxynucleotides dATP, dCTP, dGTP, and dTTP (the basic chemical units of DNA) to synthesize a copy of a small region of the genome (Fig. 5.5). The DNA polymerase step is initialized by a short synthetic DNA molecule (15–25 bases), termed the sequencing primer, that is 100% complementary to a DNA region flanking the desired sequence target.

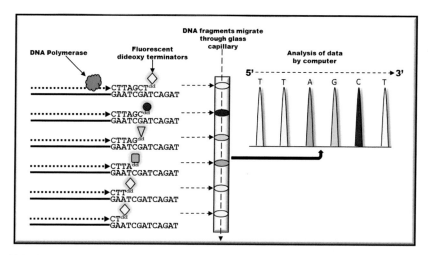

FIGURE 5.5 Basic diagram of Sanger sequencing performed on a capillary DNA sequencer.

During the polymerase extension step, the DNA strand synthesis is randomly terminated by incorporating dideoxynucleotides (ddNTPs), a form of nucleotide that cannot participate in the DNA extension reaction (Fig. 5.2C). The random termination of DNA synthesis creates a ladder of DNA fragments differing by 1 nucleotide, or 1 base pair (bp), in length (Fig. 5.5).

Analysis of Sanger Sequencing by Capillary Electrophoresis

State-of-the-art Sanger sequencing technology uses capillary electrophoresis (a small hollow glass fiber filled with a resin) to separate the DNA fragments at a single bp resolution. The ddNTP terminators are labeled with one of four fluorescent dyes, which emit light at a specific wavelength when illuminated with a laser as the DNA fragments pass through a glass capillary. Each form of ddNTP is labeled with a different fluorescent dye, which permits all four dideoxy terminations to be performed in a single reaction and the fragments to be resolved in a single glass fiber. The DNA sequence from DNA fragments as small as 50–100 bp and up to 1000 bp can be resolved in a single capillary. As the light detectors collect the color information from the DNA passing through the capillary, it is translated into a chromatogram that can be visualized using DNA analysis software. This data is also further processed into the genetic code represented by the letters A, C, G, and T.

Generation of Polymerase Chain Reaction (PCR) Templates for Sanger Sequencing

In clinical genetics studies, almost all Sanger sequencing is performed on short homogeneous DNA fragments generated by the polymerase chain reaction (PCR) [4] (Fig. 5.6). As discussed in Chapter 1, PCR has greatly simplified

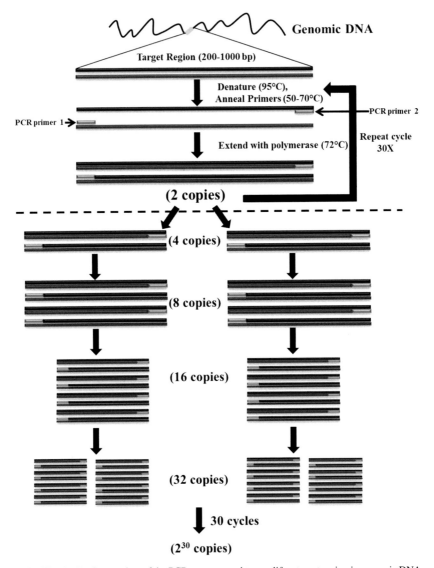

FIGURE 5.6 Basic overview of the PCR process used to amplify a target region in genomic DNA.

the generation of homogenous DNA fragments necessary for DNA sequencing. PCR-generated DNA fragments ranging in size from ~200 to 1000 bp are the most optimal for Sanger sequencing. In PCR, a DNA template (i.e., human genomic DNA), is mixed with two synthetic oligonucleotides (PCR primers), which anneal to regions on opposing DNA strands on each side of the targeted region of the genome. The DNA template and primers are mixed with dNTPs, a reaction buffer containing Mg^{+2}, and a thermal stable DNA polymerase able to withstand temperatures >100°C without becoming inactivated. The reaction is heated to >95°C (denaturing step) and then rapidly cooled to allow the PCR primers to anneal to the flanking ends of the genomic DNA (annealing step). The annealing temperature (50–70°C) will be different for each PCR reaction and must be determined by experimental optimization prior to running the final PCR. After annealing of the PCR primers, the reaction is then heated to 72°C, where the thermostable polymerase begins to process and copy each strand (extension step). The denaturing, annealing, and extension steps are generally repeated sequentially for 30–35 cycles. During each cycle, the target region is copied, and the number of copies doubles each cycle. After 30 cycles, the target region has increased ~1 billion-fold (2^{30}), generating an enriched homogenous fragment for sequencing.

Next-Generation DNA Sequencing

NGS is a massive high-throughput DNA sequencing [5]. Compared to low-throughput Sanger sequencing, which is only capable of analyzing 96–384 sequencing templates per run and may be a few thousand templates in a day, NGS utilizes nanoscale technology to sequence millions of DNA templates in a single sequencing run (~12–36 hours). To accomplish this massive level of sequencing, sequencing reactions are performed on the surface of small silicon chips or glass plates and imaging of individual sequencing reactions are performed by exciting a fluorescent dye incorporated during the NGS reaction (see Fig. 5.7).

Generation of NGS Templates Using DNA Shearing

There are several ways to generate NGS templates; however the simplest approach is mechanical shearing (Fig. 5.7). DNA shearing fragments the target DNA, such a whole human genome, into a random pool of DNA fragments in the range of 150–600 bp, depending on the NGS application. Short synthetic pieces of DNA called adapters are then attached to the ends of each DNA fragment. The adapters contain unique DNA sequences that are used to further prime downstream PCR that will generate additional sequences on the ends of the fragments. The additional sequences are used to attach the DNA fragments to the DNA sequencing matrix (glass plate or silicon chip), act as the priming sites for additional PCR amplification, and to prime the DNA sequencing reaction.

Generation of NGS Templates Using PCR

An alternative method for generating a pool of DNA fragments for NGS is to combine multiple PCR products (Fig. 5.7). These PCR products can be generated independently, or can be generated by performing PCR on numerous genomic targets in the same reaction. Hundreds or thousands of PCR products can be generated in a multiplex reaction. An advantage to generating the sequencing pools by PCR is that the oligonucleotide primers not only amplify the target sequence, but also generate the adapter sequences.

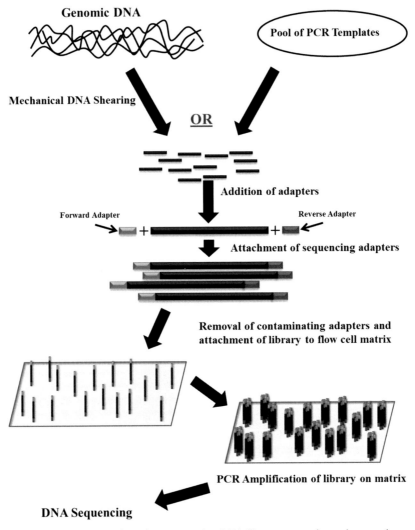

FIGURE 5.7   Basic overview of next-generation DNA library construction and sequencing on flow cell.

Cleaning and Pooling NGS Templates

Before NGS is performed, the PCR-amplified pool of fragments is cleaned to remove excess adapters or other contaminants. This pool of PCR-amplified DNA fragments is now called a DNA sequencing library. The DNA sequencing library is then injected into a flow cell, a module that holds the sequencing matrix (glass plate or silicon chip). The DNA library concentration is carefully controlled as to not saturate the surface area of the sequencing matrix. In a

Single Index

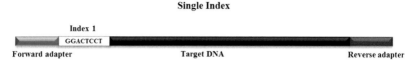

* Up to 12 forward indices allows 12 NGS libraries to be pooled and de-multiplexed

Dual Indices

* Up to 12 forward indices and 8 reverse indices allows up to 96 NGS libraries to be
pooled and de-multiplexed

FIGURE 5.8 Illustration of single index and dual indices incorporated into NGS sequencing adapters.

process similar to PCR, multiple rounds of DNA amplification increase the local concentration of the template on the matrix surface, creating a colony of the homozygous-enriched fragment, now termed a DNA cluster. Millions of these DNA clusters are generated on the matrix surface of a flow cell.

Barcoding of NGS Sequencing Libraries

A very important feature of the NGS sequencing adapters is the ability to incorporate a "DNA barcode," or sequencing index, on the end of each DNA library fragment (Fig. 5.8). The indexing sequence consists of a 6–8 bp sequence that is unique to one DNA sequencing library. A second index can be incorporated onto the opposite end of each DNA fragment, thus allowing "dual indexing." These indices can be incorporated into the adapters originally ligated to the fragments, or generated during the library amplification step. The purpose of the index is to allow multiple DNA libraries to be combined into sequencing pools. Each DNA sequencing cluster will have a unique sequencing index (or indices) that identifies from which DNA sequencing library the data were derived. A separate DNA sequencing step is included in the sequencing protocol to read the 6–8 bp code of each sequencing index assigned to each DNA sequencing cluster. The index sequence assigned allows computer software to "de-multiplex" the sequencing reads and combines DNA sequencing data containing the same DNA sequencing index. Indexing libraries allows simultaneous sequencing of pools of up to 12 libraries (single index) and up to 96 (or more) libraries using dual indexing. Indexing is more cost-effective than sequencing individual DNA sequencing libraries and also reduces potential experimental variation in data when matched data sets are analyzed.

NGS Sequencing Process

There are several DNA sequencing chemistries used to sequence the DNA clusters, with the most common termed "sequencing by synthesis" (SBS) (Fig. 5.9). In SBS, dNTPs modified with a fluorescent dye, are attached during a single bp

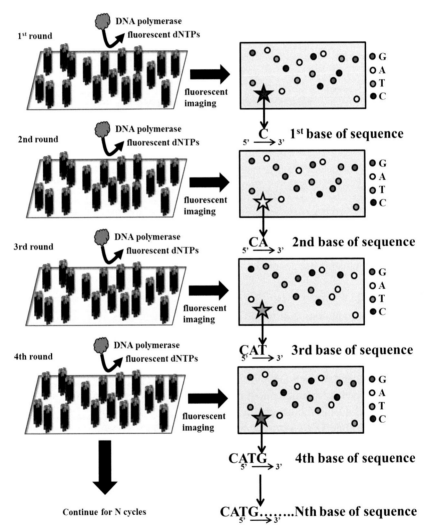

FIGURE 5.9 Overview of NGS sequencing reaction. The NGS sequencing reaction is performed on each cluster on the flow cell. After each round of sequencing, the fluorescent nucleotide incorporated at the end of the sequencing template is illuminated with a laser. Each base is incorporation is indicated by the a different fluorescent color. Sequential sequencing reactions on each cluster indicates the nucleotide in the growing DNA sequence. The colors are converted into the corresponding DNA code with additional quality data.

extension reaction, with each of the four nucleotides A, C, G, and T, having a different fluorescent dye attached. Each DNA cluster is extended by a DNA polymerase, incorporating a single modified fluorescently labeled nucleotide per reaction. The flow cell is washed after each single base extension reaction and each DNA cluster is imaged by a high-resolution camera after laser excitation of each fluorescent dye. The fluorescent dye is removed and the next sequential single nucleotide extension is performed. This process is repeated until the desired sequencing length is completed. Since the adapters are attached to both ends of the DNA fragments, the DNA sequencing reaction can be primed from each end during the sequencing process, thus allowing "paired-end" sequencing to be performed. Paired-end sequencing can double the amount of DNA sequence obtained during NGS, thus increasing the depth and coverage of DNA sequencing. If multiple NGS libraries have been pooled, then additional sequencing cycles are included to read the 6–8 base barcodes incorporated into the sequencing adapters.

IN PRACTICE

Design of Sanger Sequencing Experiment

Use of Bioinformatic Databases to Identify Reference Sequence

Before a Sanger sequencing experiment is designed, the reference sequence must first be identified from a bioinformatics database. The reference sequence is used to identify the target region to be sequenced and used to design the PCR primers that will be used to amplify the target region. One of the most flexible and comprehensive bioinformatics databases to find DNA sequence is supported by the University of California at Santa Cruz (http://genome.ucsc.edu). This UCSC website allows searches for genomic DNA sequences by gene name, chromosome position, protein name, or using a "Blast" like alignment tool (BLAT) that permits entering a DNA sequence and searching the genome for similar or identical sequences. The UCSC website also links multiple bioinformatics databases for the human genome such as the National Center for Biotechnology Information (NCBI) databases dbSNP (database of small genetic variations), dbVar (database for large structural DNA variations), and CinVar (database of genetic variations of clinical importance). The UCSC site also links to other genome databases such as primates and mouse in order to perform comparative sequence analysis.

Design of PCR and Sequencing Primers

Most clinical Sanger sequencing protocols use PCR-generated products as sequencing templates. It is critical to design proper PCR primers to the reference sequence or the PCR will fail or could amplify a nonspecific region from the genome. A PCR product of 500–800 bp is the optimal length for Sanger sequencing. Computer software for primer design software is highly recommended, and biotechnology companies that synthesize and sell oligonucleotide primers have online tools for PCR primer design. Primer design tools increase the chance of obtaining optimized primer pairs for PCR. The target sequence, or genetic variation in the sequence, should be located approximately in the middle of the target sequence in order to increase the chance of capturing the target region where the Sanger sequencing quality will be the highest.

Performing PCR and Analysis of PCR Products

The basic components for PCR (dNTPs, enzymes, and reaction buffers) can be purchased separately or prepared in the lab; however, prepared PCR kits can be purchased and will greatly increase the chance that the PCR will be the highest quality. Only the PCR primes need to be purchased separately. Some of the kits will contain PCR additives (see Trouble Shooting section) that increase PCR efficiency, especially through high G/C (guanine/cytosine) content regions. Preanalysis of the PCR products using gel electrophoresis should be performed to validate that the reaction indeed works, that the size of the PCR product is correct (size standards must be included in the gel analysis), and that the PCR produces a single clean PCR product. If more than one PCR product appears or secondary banding occurs, reoptimization of the PCR should be performed or the PCR primers should be redesigned. It should be noted that clinical sequencing performed for patient diagnosis might require processing in a Clinical Laboratory Improvement Amendments (CLIA) approved laboratory, and the PCR kits and components may require additional validation to be used for clinical purposes.

Interpretation of Capillary-Based Sanger Sequencing Chromatograms

Interpretation of Sanger sequencing can be performed using numerous sequence analysis software packages that can be purchased or found online for free. The goal of clinical Sanger sequencing is to identify genetic variations in the targeted genomic region. PCR amplification of a target genomic region generates DNA products from both chromosomes. When a genomic region does not contain genetic variations, the DNA sequencing data will be the same from both chromosomes and no differences will be observed in the DNA sequence chromatogram (Fig. 5.10A and B). When Sanger sequencing is performed on a PCR-generated product from a targeted region of DNA where the genetic code is different between the paternal and maternal genomes, these differences will be visible as overlapping chromatograms of different colors representing the sequence differences that occur between the two chromosomes (Fig. 5.10C), and that the subject is heterozygous for the polymorphism. Genetic variations can be noncoding (not found in a gene) or coding (found within a gene). Coding variants occur in the codon of a gene and have several designations (see Fig. 5.4). Synonymous changes in the coding region change the DNA sequence in a gene, but do not alter the protein sequence. A nonsynonymous change in the

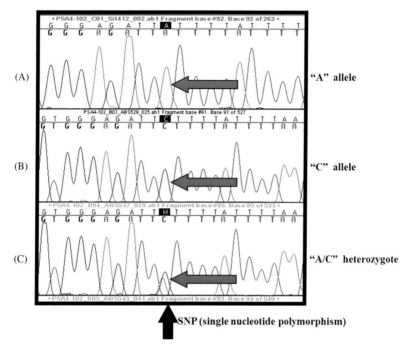

FIGURE 5.10 Analysis of single nucleotide polymorphism (SNP) using Sanger sequencing. (A) Single "A" peak indicative of a A homozygous allele. (B) Single "C" peak indicative of a C homozygous allele. (C) Presence of both the "A" and "C" peak indicative of a A/C heterozygous allele.

coding region will change the DNA sequence and alter the protein sequence. In subclassifications of nonsynonymous coding changes include missense changes (a change in a codon that alters the amino acid incorporated), nonsense changes (change in a codon that creates a new termination signal), or an insertion/deletion change (change in a peptide sequence following the mutation).

Next-Generation DNA Sequencing

Bioinformatic Analysis of NGS Data

The strength of NGS compared to Sanger sequencing is the ability of NGS to analyze a large panel of genes, or a complete genome, in a single assay. Unlike Sanger sequencing, however, NGS requires powerful computing and data storage resources to handle the large data files produced during a single analysis. There are numerous commercial software programs that can be purchased for analysis of NGS data; however, many of the best bioinformatics programs can be acquired for free. Much of the free NGS analysis software do not have graphical user interfaces (GUI) and thus experience with "command line" programing may be necessary. It is advisable to have a strong bioinformatics resource before attempting NGS data analysis.

Example of NGS Data Formats

The typical work flow for producing and analyzing NGS data is shown in Fig. 5.11. The biggest consideration in NGS is data analysis and data storage. It is advisable to prepare for very large data storage exceeding 1 Tb for processing a single NGS project. The raw data file format generated from NGS is a .fastq file. This file stores the DNA sequence data and assigns a quality score to each nucleotide in the sequence. The .fastq files hold millions of DNA sequencing reads, and thus the files can be very large (10–100 Gb). The .fastq data can be filtered based on Q scores (quality scores), with Q scores >30 (1 in 1000 chance of error) being a desired quality threshold. The .fastq data can be also merged and edited. The editing removes the adapter sequences and indexing barcodes sequences that will not align to the reference DNA sequence. The curated .fastq data can then be mapped to the reference genome using one of many DNA sequence "aligner" programs. Not all aligners are appropriate for every project, thus several aligners might have to be used to obtain the best quality data. The aligned data is then processed into a BAM (binary alignment). Visualization of NGS data in the BAM file can be performed using the integrated genome viewer (IGV) program (http://www.broadinstitute.org/igv/). The BAM file is subsequently used in downstream computer programs for DNA variant calling. The variant calls are reported in a variant call file (.VCF). Some of the information found in a VCF include chromosomal location of the variant, several

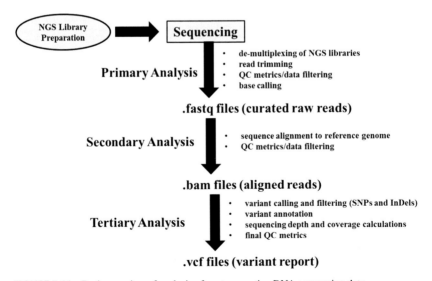

FIGURE 5.11   Basic overview of analysis of next-generation DNA sequencing data.

parameters to evaluate the DNA sequence quality around the variant, the variant genotype, the reference sequence allele, the rs# from the dbSNP database (if known), and the type of variant (coding or noncoding). Other data, such as amino acid changes and the potential effects of the mutation changes may also be listed in the VCF depending on preferences and DNA filtering choices. The data in the VCF will be the primary source for any clinical diagnosis made from the sequencing data.

APPLICATIONS

In this section, the focus will be on the primary applications used to identify genetic variations, and will be based on a global approach [whole genome sequencing (WGS)] and targeted approaches (exome sequencing and DNA sequencing panels).

Whole Genome Sequencing

Overview of Whole Genome Sequencing (WGS)

WGS is the most global approach to identifying genetic variations. Although Sanger sequencing was used to analyze the first human genome, Sanger sequencing has not developed in scale during the last decade, and thus Sanger sequencing is cost- and time-prohibitive for WGS. NGS, on the other hand, was developed with WGS as the primary focus. The biotechnology company Illumina sells multiple NGS platforms capable of WGS, some of which can sequence a whole genome in less than 30 hours. As described in the *In Principle* section, DNA from a single whole genome is fragmented and assembled into a single sequencing library, and then sequenced in single run. Approximately 2–3 million SNPs and small InDels can be expected in a typical human genome, of which only 15–20,000 will occur in the coding region of the genome. Thus if identification of coding changes is the primary goal of sequencing, alternative sequencing strategies (see Exome Sequencing below) should be utilized to limit the bioinformatics burden and reduce sequencing costs.

Important Factors for WGS

Important metrics to consider for WGS are the depth of sequencing and sequencing coverage. Sequencing coverage is the percentage of the genome actually sequenced, while sequencing depth is a measure of how many times a region of DNA has been sequenced. Long DNA sequencing reads (>100 bp) significantly improves the percent coverage of the genome and leave fewer sequencing gaps. However, it is nearly impossible to obtain 100% coverage due to regions that are difficult to sequence, such as long repetitive sequences and DNA sequences with high G and C content. Sequencing depth is reported as average (mean) sequencing depth or as a minimum sequencing depth. For WGS, an average sequencing depth of 30–50 fold (also termed 30–50X) is considered a good baseline sequencing depth to detect the majority of genetic variations in germline (inherited DNA). In contrast, minimum sequencing depth is important on a region by region basis. The minimum sequencing depth may need to be very high in order to find difficult-to-detect variants. For example, when performing WGS sequencing on DNA from cancer cells, mutations in the DNA from cancer cells may be de novo (termed somatic mutations) and not present in the germline (inherited) genome. In addition, the somatic mutation may not be present in every cancer cell. This heterogeneous mixture of cells will generate a heterogeneous mixture of DNA, where the mutation may occur in only 5–10% of the DNA. Sequencing depth as high as 100–500X may be necessary to identify DNA mutations in a mixed DNA pool.

Targeted DNA Sequencing

Targeted Sequencing of Small Genomic Regions

Both Sanger sequencing and NGS can be applied in targeted DNA sequencing. Targeted sequencing focuses on sequencing a single gene or set of genes. The choice of Sanger sequencing or NGS for targeted sequencing is determined by the number of samples to be sequenced and the size of the sequencing target. If performing sequencing on a few PCR products, Sanger sequencing is cost-effective and rapid. A single PCR product can be sequenced on 96 DNA samples in approximately 1 week at a cost <$10 per sample. Multiple PCR products can be designed across a genetic region creating a tiling path. An example of using multiple PCR products to analyze DNA sequence across complete gene is shown in Fig. 5.12 for the gene *ADRB2*, which will be discussed later. PCR products should be designed in the range of 600–800 bp, and overlap by 100–150 bases to complete a contiguous sequence. Primer design is critical. PCR and sequencing primers should not be designed across sequence where a known genetic polymorphism is expected and should not be designed in genomic regions for which there are multiple copies in the genome, such as the 180–200 bp Alu repeat, for which there are $\sim 10^6$ copies in the human genome. The location of known genetic polymorphisms and repetitive sequences can be found using the SNP and RepeatMasker windows in the UCSC Genome Browser. Using this targeting strategy, expect to find one SNP or small InDel every 1000 bp on average. An example of this type of targeted sequencing is described in the next section.

Targeted Sequencing of Large Genomic Regions

When the target region exceeds 10,000 bases or analysis of multiple genes is required, NGS-targeted sequencing should be used. NGS-targeted sequencing can be performed by pooling hundreds to thousands of PCR products, or can be performed on DNA regions that have been selectively removed from the whole genome using capture arrays or capture baits. Predesigned disease-specific sequencing panels (i.e., cancer genes panels) can be purchased from biotechnology companies that sell NGS products. The most common form of

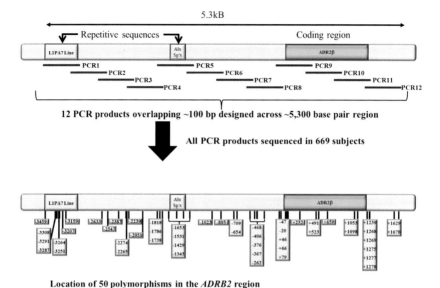

Location of 50 polymorphisms in the *ADRB2* region

FIGURE 5.12 Targeted sequencing of the human ADRB2 gene using PCR. The SNP positions are numbered in respect to the start codon ATG, with the A being the +1 position and the base preceding the start codon A being the −1 position. LIPA7 LINE = Long Interspersed Nuclear Element form LIPA7; Alu Sg/x = Alu repeat of the form Sg/x.

NGS-targeted sequencing is called "exome" sequencing. The process of exome sequencing, shown in Fig. 5.13, targets only the coding region, or exons, in the genome. In brief, exome sequencing requires the use of "capture baits," which form a binding complex with only the exonic region of the genome. These capture baits are synthesized and modified with an affinity molecule, such as biotin, on one end of the capture bait. After a NGS sequencing library is created, the capture baits are mixed with the DNA library, and only the DNA molecules in the library containing the exons bind to the capture baits. Streptavidin, which forms a strong bond with the biotin on the capture baits, is then used to enrich the fragments attached to the capture baits. This enriched pool is purified, amplified using PCR, and then submitted to NGS. Custom capture baits can also be designed to capture a select set of genes or, in an example highlighted in the next section, designed to capture a large contiguous genomic region.

SCENARIO

Case Study: Targeted Sanger Sequencing

Beta-agonists are one of the most common medications used to treat airway constriction in asthma. Beta-agonists target the beta-2 adrenergic receptor, that when activated, relaxes airway smooth muscle cells. The gene for the beta-2

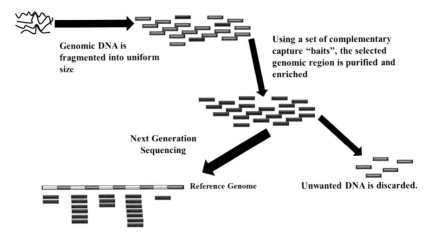

Genomic DNA is fragmented into uniform size

Using a set of complementary capture "baits", the selected genomic region is purified and enriched

Next Generation Sequencing

Reference Genome

Unwanted DNA is discarded.

Sequenced DNA fragments are mapped back to the reference gene aligning only in the regions selected for capture.

In exome sequencing, all of the region of the genome that encode proteins (exons) are captured (38-50 Mbases)

FIGURE 5.13   Basic overview of exome target capture sequencing.

adrenergic receptor, *ADRB2*, contains only one exon and is thus small. Two common genetic variation creating the amino acid changes Gly^{16}Arg and Gln^{27}Glu, occur in the *ADRB2* gene and have been implicated in affecting a person's responsiveness to beta-2 adrenergic agonists (a pharmacogenetics effect) [6]. In order to identify additional *ADRB2* variations with potential effects on gene function, a ~5300 (5.3 kb) genomic region centered on *ADRB2* was sequenced in 669 subjects [7]. Twelve PCR products overlapping ~100–150 bp were designed across the region, and each PCR product was sequenced (Fig. 5.12). Fifty genetic variations were identified in the 5.3 kb region, ranging in frequency of <1% to 45%. No new coding changes were found. However, some of the newly discovered rare variants have been further studied and implicated in increasing the risk of severe asthma exacerbations in subjects taking long-acting beta-agonists [8].

Case Study: Targeted NGS

A large 275,000 (275 kb) bp region of human chromosome 22 encodes the genes *APOL1*, *APOL2*, *APOL3*, *APOL4*, and *MYH9*, and several coding genetic variations found in *APOL1* are highly correlated with increased risk of kidney disease in subjects of African ancestry [9]. This 275 kb region has been further sequenced using targeted NGS to determine if additional genetic variations in the other four genes may contribute to kidney disease risk [10]. DNA capture baits were designed and synthesized to capture the complete 275 kb region. Repetitive DNA sequencing, such as Alu repeats, were filtered out from the bait design using the algorithm RepeatMasker (Institute for Systems Biology) and were not sequenced. In a process identical to exome sequencing, the capture baits were used to enrich for the 275 kb region in 192 subjects, and NGS sequencing libraries constructed for each subject. DNA barcode indices were incorporated into each 192 libraries, and the pools of 48 libraries were sequenced on a single NGS flow cell. This NGS-targeted sequencing strategy identified 3547 SNPs and InDels; however, no new coding variations were identified in the any of the 5 genes in the capture region. Of the 3547 variants found, only a single rare nonsense coding change (Gln^{58}Ter) in the gene *APOL3* was found to be nominally associated with kidney disease risk.

KEY LIMITATIONS

Primary Limitations of Sanger Sequencing

The primary limitation of Sanger sequencing is low-throughput capacity. The largest capillary DNA sequencer can only process 96–384 samples per run (~1–2 hours per run) at fragment lengths <1000 bases. The cost per bp is thus high (>$0.01) and sequencing long contiguous DNA regions becomes time- and labor-intensive. Sequencing depth is thus limited and identifying low-frequency somatic mutations in mixed pools of DNA is very difficult. Since PCR products are now the primary source for Sanger sequencing, PCR products contaminated with nonspecific amplified products can make Sanger sequencing almost uninterpretable and increase the risk of identifying false polymorphisms. Finally, Sanger sequencing hasn't been improved much over the past decade since the focus is now on utilizing NGS in research. Thus there are few new products supporting or improving Sanger sequencing methodologies.

Primary Limitations of NGS

The primary limitations of NGS are not based on methodologies, but in implementing NGS in a lab. First, NGS instruments can be expensive, with some instruments costing >$1 million. Second, large computing and bioinformatics resources are critical to store and process NGS data, and access to an experienced bioinformaticist is usually necessary. Other limitations of NGS are capacity and cost issues. For example, in order to fill the large capacity flow cells, flow cells may have to be shared by investigators. Otherwise, the total cost of running small numbers of samples by NGS can exceed $1000 per sample. It isn't uncommon for investigators to wait weeks to find another investigator to share flow cell time to keep sample costs reasonable.

COMMON TROUBLE SHOOTING FOR SANGER SEQUENCING AND NGS

Common Problem	Causes	Solution
Difficulty constructing DNA library for NGS or difficulty performing PCR to generate Sanger sequencing templates	–Degraded or low-quality genomic DNA	Check DNA quality on gel to determine if degraded. If degraded, new DNA source DNA must be purified
	–Low quantity of genomic DNA	Check DNA quantity by spectrophotometer. Concentrate DNA using speed vacuum or by ethanol precipitation
	–RNA contamination	Check DNA by spectrophotometer. If A260/A280 readings are >2.2, re-purify DNA with high-quality DNA purification kit that removes RNA
	–Protein contamination	Check DNA by spectrophotometer. If A260/A280 readings are <1.7, re-purify DNA with high-quality DNA purification kit that removes protein
Good quality genomic DNA, but PCR does not work	–Poor design of PCR primers	Redesign PCR primers
	–PCR reaction is not optimized	Adjust annealing temperature for PCR. Adjust $MgCl_2$ concentration used in PCR
	–High G/C content in target region	Using additives such as betaine or dimethylsulfoxide in PCR
Poor Sanger sequencing results	–PCR template concentration too high or too low in sequencing reaction	Re-quantify PCR template using spectrophotometer and adjust template concentration in Sanger sequencing reaction

Common Problem	Causes	Solution
	–Low sequencing primer concentration	Verify that sequencing primer concentration correct in Sanger sequencing reaction
Poor NGS sequencing results	–Poor library construction	Verify that library is of sufficient concentration and has proper average base pair size
	–Improper loading of NGS sequencer	Verify that the flow cell is not under or overloaded based on NGS flow cell specs
	–Poor pooling of indexed libraries	Carefully re-quantify each library before pooling to ensure that each library has equimolar quantities in the pool
	–Poor DNA capture during capture library construction	Verify that the quantity of NGS library used in the capture reaction is not too high or too low

CONCLUSION

Sanger sequencing and NGS are both robust methods for performing the most comprehensive level of genetic analysis. While Sanger sequencing has been a staple in clinical sequencing over the past decade, NGS is rapidly overtaking Sanger sequencing as the primary method for identifying genetic variations with clinical implications.

REFERENCES

[1] Adekoya E, Ait-Zahra M, Allen N, Anderson M, Anderson S, Anufriev F, et al. Initial sequencing and analysis of the human genome. Nature 2001;409(6822):860–921.
[2] Venter JC, Adams MD, Myers EW, Li PW, Mural RJ, Sutton GG, et al. The sequence of the human genome. Science 2001;291(5507):1304–51.
[3] Sanger F, Coulson AR. A rapid method for determining sequences in DNA by primed synthesis with DNA polymerase. J Mol Biol 1975;94(3):441–8.
[4] Mullis KB, Faloona FA. Specific synthesis of DNA in vitro via a polymerase-catalyzed chain reaction. Methods Enzymol 1987;155:335–50.
[5] Moorcraft SY, Gonzalez D, Walker BA. Understanding next generation sequencing in oncology: a guide for oncologists. Crit Rev Oncol Hematol 2015;96:463–74.

[6] Liggett SB. The pharmacogenetics of beta2-adrenergic receptors: relevance to asthma. J Allergy Clin Immunol 2000;105(2 Pt 2):S487–92.

[7] Hawkins GA, Tantisira K, Meyers DA, Ampleford EJ, Moore WC, Klanderman B, et al. Sequence, haplotype, and association analysis of *ADRB2* in a multiethnic asthma case-control study. Am J Respir Crit Care Med 2006;174(10):1101–9.

[8] Ortega VE, Hawkins GA, Moore WC, Hastie AT, Ampleford EJ, Busse WW, et al. Effect of rare variants in *ADRB2* on risk of severe exacerbations and symptom control during longact-ing beta agonist treatment in a multiethnic asthma population: a genetic study.. Lancet Respir Med 2014;2(3):204–13.

[9] Freedman BI, Langefeld CD, Turner J, Nunez M, High KP, Spainhour M, et al. Association of *APOL1* variants with mild kidney disease in the first-degree relatives of African American patients with non-diabetic end-stage renal disease. Kidney Int 2012;82(7):805–11.

[10] Hawkins GA, Friedman DJ, Lu L, McWilliams DR, Chou JW, Sajuthi S, et al. Re-Sequencing of the *APOL1-APOL4* and *MYH9* gene regions in African Americans does not identify additional risks for CKD progression. Am J Nephrol 2015;42(2):99–106.

Chapter 6

Western Blot

Tomasz Gwozdz and Karel Dorey
University of Manchester, Manchester, United Kingdom

Chapter Outline

Objectives

- Get familiar with the principles of protein detection using Western blotting techniques.
- Provide a "ready-to-use" laboratory protocol.
- Highlight the main limitation of Western blotting and its interpretation.
- Describe the main applications of Western blotting in biomedical research and diagnosis.

INTRODUCTION

Western blotting (WB) or protein immunoblotting is a popular laboratory technique to detect specific proteins from a cell or tissue sample. The technique was initially described by Towbin et al. in 1979 [1] and the name coined by Burnette in 1981 [2] to match similar techniques used for detection of DNA, Southern blotting [3], and RNA, Northern blotting [4]. WB requires to separate proteins

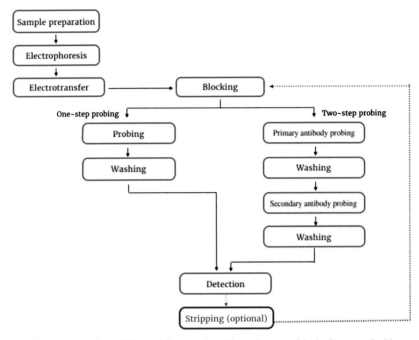

FIGURE 6.1 Flowchart of Western blot procedure. The main steps of the basic protocol with one- and two-step probing are shown.

according to their size by gel electrophoresis. It is then followed by electrotransfer of the proteins to a membrane that is probed with an antibody against the protein of interest. This technique is very powerful to detect the presence of a given protein in a sample, to assess posttranslational modifications such as phosphorylation and to characterize protein complexes. Fig. 6.1 shows the general workflow of typical WB. Although there are variations of the WB and one can optimize the protocol for specific application, in the majority of cases WB is still widely used in its basic form with only subtle adjustments mainly in the generation and characterization of the antibodies used. In this chapter, we will provide the principles underlying WB together with a general protocol with specific optimization points as well as the main applications and limitations.

IN PRINCIPLE

Sample Preparation

The first step of WB is sample preparation. It requires to lyse the cells, which in case of tissues is preceded by mechanical or enzymatic homogenization, to extract and solubilize proteins. Depending on the cellular structures (nucleus, cytoplasm, plasma membrane, etc.) and cell type (e.g., mammalian cells, bacteria, yeast), different lysis buffer might be used. Typically, it is composed of a buffering agent (e.g., Tris–HCl), salts such as sodium chloride, detergent disrupting lipids, and protease inhibitors to prevent protein degradation.

Ionic detergents such as sodium dodecyl sulphate (SDS) are strong solubilizing agents but they denature proteins impairing their activity and their interactions. Ionic detergents also cause an increase in the viscosity of the sample due to the release of the chromatin which then needs to be fragmented by sonication. In contrast, nonionic (NP-40, Triton X100) and zwitterionic (CHAPS) detergents are milder and less denaturing, keeping proteins relatively intact but are less efficient in extracting proteins from subcellular organelles and membranes.

Cell lysis leads to the release of proteases and phosphatases, therefore the samples must be kept at 4°C and the appropriate inhibitors need to be used: ethylenediaminetetraacetic acid/ethylene glycol tetraacetic acid EDTA/EGTA (bivalent cation chelators inhibiting metalloproteases), aprotinin (trypsin, plasmin), leupeptin (lysosomal proteases), pepstatin A (aspartic proteases), PMSF (phenylmethylsulfonyl fluoride, serine, cysteine proteases), NaF (sodium fluoride, serine/threonine phosphatases) and Na3VO4 (sodium orthovanadate, tyrosine phosphatases).

It is useful to know the protein concentration in the lysates to control the amount loaded on the gel and to allow comparison between experiments. Protein concentration can be measured using Bradford [5], Lowry [6], or bicinchoninic acid (BCA) [7] assays, but some detergents above a certain concentration are incompatible with some of those methods, this should be checked before use.

The final step is to mix the samples with a loading buffer composed of (1) glycerol to increase the density of the sample facilitating its loading on the gel, (2) a dye making the sample visible (bromophenol or Coomassie Blue), (3) agents reducing disulfide bonds (β-mercaptoethanol or dithiothreitol, DTT), and in case of denaturing conditions (4) SDS (or LDS). To ensure complete denaturation of the proteins, samples are usually boiled at 95°C for 5–10 minutes but prolonged boiling may lead to protein degradation or aggregation.

Electrophoresis

Samples are then subjected to polyacrylamide gel electrophoresis (PAGE), which separates proteins either based on their structure and isoelectric point (native-PAGE) or their size (SDS-PAGE). Proteins need to possess a negative charge to migrate through the gel pores when subjected to an electromagnetic field. In native-PAGE either the intrinsic charge of a target protein or Coomasie G-250 give proteins a negative charge. In SDS-PAGE, all proteins are negatively charged due to the binding of SDS (Fig. 6.2).

Polyacrylamide gels are composed of a mix of polyacrylamide and bis-acrylamide, which form a crosslinked polymer by the polymerizing agent ammonium persulfate (APS) and the catalyst TEMED (N,N,N,N′-tetramethylenediamine). Gel pore size is regulated by the ratio and concentration of polyacrylamide and

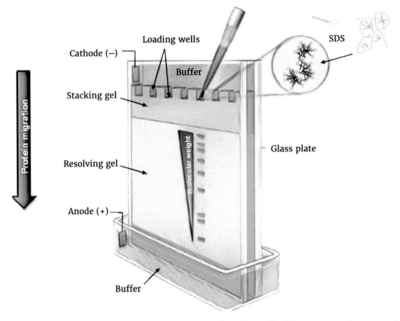

FIGURE 6.2 Schematic of SDS-Page electrophoresis. Polyacrylamide two-part gel composed of a stacking (top) and a resolving (bottom) gel is enclosed between two glass plates and submerged in running buffer. Protein samples denatured with SDS are pipetted into wells on top of the gel. Application of electromagnetic field causes the migration of proteins toward the anode (+) resulting in their separation based on the molecular weight.

bis-polyacrylamide and needs to be adjusted to the size of the protein of interest (see "In Practice" for details). Higher concentrations and increased amount of bis-acrylamide give smaller pores separating smaller proteins and vice versa. There are two main types of gels commonly used in WB: (1) two-part gels, which are composed of a stacking and resolving gels. Stacking gels have a lower percentage of acrylamide, a lower pH, and a different ionic composition that helps to "compress" proteins and (2) gradient gels where the concentration of acrylamide progressively increases.

Electrotransfer

Proteins separated by PAGE need to be transferred from the gel to a membrane by electrophoretic transfer (or electrotransfer) for further processing. A blot sandwich is prepared where a membrane is tightly touching the gel and sandwiched between filter paper (or similar support) in a transfer buffer. It is important to ensure a good contact between the gel and the membrane as air bubbles would prevent the transfer. An electromagnetic field is applied perpendicularly to the gel surface to transfer the proteins from the gel to the membrane (Fig. 6.3).

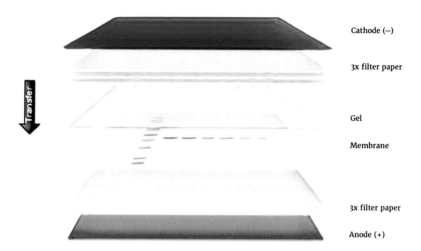

Cathode (−)

3x filter paper

Gel

Membrane

3x filter paper

Anode (+)

FIGURE 6.3 Schematic of a blot sandwich in semidry transfer. Blot sandwich is composed of a polyacrylamide gel placed on top of a membrane and enclosed between three layers of filter paper soaked in transfer buffer. Since, upon electromagnetic field, proteins migrate from cathode (−) toward anode (+), the membrane needs to be placed between the gel and the anode (+).

The efficacy of the transfer may vary significantly depending on the protein (e.g., size, composition), the protocol (e.g., transfer time, buffer) and equipment (e.g., the type and position of the electrodes). There are several options to perform an electrotransfer with a choice of: (A) method, (B) membrane, and (C) buffer.

Method

There are two main methods of electrotransfer: wet and semidry. For a wet transfer, the tight contact between the gel and the membrane during the electrotransfer is ensured by solid support and the sandwich is entirely submerged in the transfer buffer. Wet transfer is more flexible to optimization (e.g., time, voltage) and with proper cooling may be run overnight resulting in more complete transfer. It allows the transfer of broader range of protein size and is favorable for big proteins (>100 kDa). Its main disadvantage is that it takes much longer (at least 1 hour) and requires greater buffer volumes. In contrast, in a semidry transfer, the sandwich is placed directly between electrodes and only the filter paper is soaked with the transfer buffer (Fig. 6.2). It is very quick (10 minutes to an hour) and many membranes can be processed at the same time with minimal use of buffer. However semidry method has low buffering capacity and cannot be run for prolonged time due to the risk of drying. As a result, the transfer is less effective and may result in difficulties of detecting the target, especially if the protein has a high molecular weight and/or is not abundant. Nevertheless semidry transfer is sufficient in most cases and is often preferred due to its convenience.

Membrane

There are two main types of membranes commonly used: nitrocellulose and PVDF (polyvinylidene difluoride). Nitrocellulose is charged and binds proteins through their hydrophilic part very quickly and efficiently. However, it is mechanically weaker and not really suitable for re-probing when stripping is required. PVDF is uncharged and binds hydrophobic parts of proteins. It has better binding capabilities (150 to160 $\mu g/cm^2$ vs. $80\,\mu g/cm^2$) but tends to give a higher background than nitrocellulose. Since PDVF is hydrophobic it requires short "activation" with methanol before use. Since it has high mechanical strength and chemical stability it is a preferable choice when re-probing is required.

Buffer Composition

The most common transfer buffers are based on Tris and glycine solutions supplemented with methanol and SDS [1,8]. The balance between SDS and methanol is important as it may affect the transfer efficiency and should be adjusted to the protein size. SDS improves the solubility and the migration of proteins but reduces their binding to membranes, whereas methanol precipitates proteins reducing their migration but improves protein-binding to membranes (see Note 5 for details).

Blocking

Membranes used in electrotransfer are characterized by high protein-binding capabilities and therefore the surface unoccupied by transferred proteins requires blocking to reduce background. Ideal blocking solution needs to be optimized but in general 1–5% nonfat dry milk (NFDM) or bovine serum albumin (BSA) solutions have proved to be successful. NFDM is widely used but it is not recommended for the detection of phosphorylated proteins as it is rich in casein, which as a phosphoprotein may increase the background.

Probing

After blocking, the membrane needs to be probed with an antibody recognizing the protein of interest. This step is crucial and may require extensive optimization as each antibody will have different strength and specificity.

Probing can be performed in one or two steps. In the one-step probing, the antibody against the protein of interest is conjugated with an enzyme (such as horseradish peroxidase, HRP, or alkaline phosphatase, AP) or a fluorophore allowing its detections. The one-step probing is fast and may be useful for high-throughput applications. In the two-step probing, two different antibodies are used: the primary antibody against the protein of interest and a secondary, conjugated antibody against the primary antibody. In this case, there are two steps of amplification of the signal increasing the sensitivity of the assay. In some cases, an antibody can be replaced by another protein that recognizes the target. The most popular is avidin or streptavidin, which similarly to an antibody can be conjugated and binds biotin hence recognizes all biotinylated proteins.

Washings

Unbound antibodies will cause increased background and should be removed without affecting the specific binding of the antibody to its target. Therefore a balance should be achieved between removal of the background and ability to detect the specific signal. Usually a very mild detergent solution is used for washings (see "In Practice" for details).

Detection

The most popular method is using antibodies conjugated with HRP and AP enzymes. Depending on the substrate the chemical reaction catalyzed by the enzyme releases light (chemiluminescence), which can be detected using X-ray films or imaging equipment (e.g., GelDoc) or gives a visible end-product, which precipitate on the membrane (chromogenic/colorimetric detection). The latter can be monitored in real-time and does not require any additional equipment. It is however less sensitive in comparison to chemiluminescense and it is not suitable for re-probing.

The antibody can also be conjugated with fluorophores such as fluorescein (FITC) or rhodamine (TRITC) allowing direct visualization of the antibody using fluorescence. This detection method is usually 2–4 times less sensitive in comparison to chemiluminescence but it gives a much better linear dynamic range (upto 10-fold) allowing better quantification. It is also suitable for detection of several targets in a single run when different fluorophores are used. The development of infrared and near-infrared fluorophores have improved the signal-to-noise ratio and increased the sensitivity of this technique.

Stripping (optional)

It can be useful to reprobe the same membrane for a different target with a different antibody. First, all antibodies already bound to the membrane need to be removed and several stripping methods were developed: a mild stripping using low pH buffers (e.g., pH = 2.2) with low concentration of detergent (e.g., 0.1% SDS) or harsh stripping, using high incubation temperature (e.g., 50°C) with high concentration of detergent (e.g., 2% SDS). After stripping, the membrane can be again blocked and probed as described above. The stripping may not be consistent throughout the whole membrane, therefore a re-probed blot should not be used for semiquantification and the result should be interpreted with caution.

IN PRACTICE

Having explained the general principles of Western blot, we are now providing a detailed protocol for the detection of a cytoplasmic protein in cell culture. This protocol will describe the preparation of the sample, setting up, and running of the SDS-PAGE, the transfer of the proteins onto a membrane and finally the detection of the protein of interest. This protocol will be sufficient for the vast majority of cases but we have included some advices in the troubleshooting section if needed.

Sample preparation:

1. Place a 35 mM dish with cells on ice. Decreasing the temperature of the culture slows down protease activity thereby protecting the target proteins during the lysis process.
2. Discard media.
3. Wash cells with 2 mL ice-cold phosphate-buffered saline (PBS) and discard the PBS. Washing with cold PBS is important to remove the proteins from the culture media, which interfere with the assay and increase the background.
4. Add 100–500 μL NP-40 lysis buffer freshly supplemented with protease inhibitors. NP-40 is a mild detergent suitable for most cytoplasmic proteins. For other types of proteins see Note1.
5. Lift cells using a cell scraper and transfer to 1.5 mL microcentrifugation tubes.
6. Incubate on ice for 5–30 minutes. During the incubation the plasma membrane is being disrupted and the target proteins are released into the lysate. Depending on the target protein the incubation time and temperature may be adjusted to assure the maximum yield. If the target protein is known to be difficult to release from the cellular structures the incubation can be performed at room temperature (RT) instead of ice.
7. Clear the lysate by centrifugation (10 minutes, 4°C, $\geq$12,000xg). Depending on the used lysis buffer the resulted pellet may contain cell membranes, precipitated DNA, nuclei, and denatured insoluble proteins.
8. Transfer the supernatant to a fresh tube.
9. Measure the concentration of total proteins in the lysate (with commercial assays such as BCA, Bradford, Lowry). This ensures equal loading of proteins from different samples as well as determining the optimal amount of proteins needed for detection of a particular protein.
10. Prepare a sample by mixing the lysate with 2x sample buffer freshly supplemented with the reduction agent beta-mercaptoethanol (1–5% final concentration) or DTT (100 mM). Reduction agent disrupts disulfide bonds, whereas glycerol present in the sample buffer facilitates gel loading.

11. Boil the sample for 5 minutes at 95°C. This step facilitates the denaturation of proteins but in some cases may cause their precipitation or aggregation (see Note 2).

12. Cool down the sample on ice for 1 minutes.

13. **Centrifuge for 5 minutes at RT.** ($\geq$12,000xg) to remove all precipitants. The sample is ready for loading on the gel.

    Note 1: Adjust the lysis buffer composition depending on the protein of interest. Use lysis buffer with 0.25–0.5% CHAPS for membrane proteins and buffer with 0.1% SDS for nuclear proteins. Sample buffer with SDS may be used as a lysis buffer but this prevents the quantification of protein concentration. When SDS is used subsequent fragmentation of genomic DNA by sonication is required.

    Note 2: Some plasma membrane proteins tend to aggregate when boiled. If this is the case incubate sample at 37°C for 30 minutes – 1 hour instead.

    SDS-Page:

14. Cast the two-part gel in advance. Pour 10% resolving gel followed by 4% stacking gel giving enough time for polymerization in between (see Note 3). It may be stored at 4°C for a week.

15. **Load 20–50** μg of total proteins per lane. Overloading the gel results in poor separation of proteins and bulky bands. Try to load the minimum amount which still gives a good detection.

16. Load a protein ladder, preferably prestained as this allows real-time monitoring of the electrophoresis progress as well as the efficiency of subsequent electrotransfer. When molecular weight needs to be determined accurately, avoid prestained protein ladders as their migration is not precisely correlated with the molecular weight of the constituent proteins.

17. **Run the gel in 1**x TGS buffer under constant 50 V until the migration front of electrophoresis enters the resolving gel. Then increase the voltage to 100–150 V and run until the front leaves the gel. The run can be stopped at any point depending on the size of the target protein, for example for big proteins the run can be prolonged as long as the corresponding band from the prestained ladder is still visible. If necessary, decrease the voltage to prevent overheating the gel as it may cause uneven migration and distortion of the bands.

    Note 3. Contact with air prevents acrylamide polymerization. Therefore cover the resolving gel with isopropanol during its polymerization. Remove isopropanol and rinse with double distilled water before pouring the stacking gel.

    Note 4. Adjust percentage of the gel to size of the target protein accordingly to Table 6.1. If proteins of different size need to be detected on the same gel consider using a gradient gel.

    Electrotransfer (semidry):

18. Incubate the PVDF membrane for 5 minutes in methanol and then in transfer buffer. When a nitrocellulose membrane is used the methanol activation step is not required.

TABLE 6.1 Recommended Percentage of Polyacrylamide Gel

Protein Size [kDa]	Gel Percentage [% of Acrylamide]
4–40	20
12–45	15
10–70	12.5
15–100	10
25–200	8

19. Equilibrate the gel in transfer buffer for 5 minutes.
20. Assemble the blot sandwich according to Fig. 6.3. Make sure that the membrane is between the anode (+) and the gel. Use the transfer buffer appropriate to the target protein (see Note 5).
21. Run the electrotransfer according to the manufacturer's recommendation, in general 30 to 60 minutes under constant current (1 mA per 1 cm^2 of membrane).

   Note 5. A good starting point for mid-size proteins (20–100 kDa) is 20% of methanol and 0.025% SDS. For large proteins (>100 kDa) 0.1% SDS and 10% methanol may improve the transfer. For small proteins (<20 kDa) complete removal of SDS and 20% methanol is recommended.

   Note 6. It is useful to determine the efficacy of the transfer by (i) running a prestained protein ladder, (ii) staining the membrane using Ponceau S, and (iii) staining the gel with Coomassie Blue to detect proteins that have not transferred.

   Note 7. If you notice insufficient transfer or your protein cannot be detected consider using wet transfer.

   Blocking:
22. Block the membrane in a 5% NFDM solution in Tris Buffered Saline with Tween-20 (TBST) for 15–30 minutes at RT. Use 50 mL falcon tube or other appropriate tublike vessel. Keep the "gel side" of the membrane exposed to the solution. Rotate or shake the membrane to ensure even distribution of the blocking solution.

   Note 8. As most of the proteins will be immobilized on the "gel side" of the membrane it is good to mark this side before or after the transfer with a pencil.

   Note 9. For detection of phospho-proteins 5% BSA solution in TBST is recommended.

Primary antibody incubation:

23. Prepare dilution of a primary antibody in a blocking solution according to the manufacturer's recommendation.

24. Discard blocking solution and replace it with a primary antibody solution.

25. Incubate for 2 hours at RT or overnight at 4°C using a rotator or a shaker.

Note 10. The primary antibody may be often reused several times. It can be stored for prolonged time at −20°C or with sodium azide at 4°C. Sodium azide inhibits the reaction catalyzed by HRP and therefore needs to be washed away.

1st Washing:

26. Wash the membrane three times for 10 minutes with TBST at RT.

Note 11. Washing may need to be optimized but the good starting point is to wash the membrane three times for 10 minutes with TBS buffer supplemented with Tween-20 (0.1–0.2%).

Secondary antibody incubation:

27. Prepare dilution of HRP-conjugated secondary antibody in TBST buffer accordingly to the manufacturer's recommendation. Make sure to use the antibody against the species, which was used to produce primary antibody (e.g., antimouse secondary for mouse primary).

28. Incubate membrane for 1 hour at RT using rotator or shaker.

Note 12. Secondary antibody should not be reused.

2nd Washing:

29. Wash the membrane three times for 10 minutes with TBST at RT.

ECL detection:

30. Prepare an enhanced chemiluminescence (ECL) substrate solution by mixing two components in the ratio recommended by the manufacturer.

31. Transfer the membrane on a flat surface and drain the excess liquid with absorbent paper.

32. Cover the membrane with the ECL substrate making sure that it is evenly distributed.

33. Incubate for 1–5 minutes.

34. Blot the ECL substrate with a paper tissue.

35. When using an X-ray film transfer the membrane into an exposure cassette and proceed to expose the film in the dark room followed by its processing with a developer. Otherwise use an imager to detect the chemiluminescence signal.

36. Make sure to mark the position of the protein ladder bands or acquire its image when using an imager.

Note 13. Different ECL substrates characterized by their ability to detect different minimum amount of protein are commercially available. Use the appropriate one for the expected amount of the target protein or test several.

Buffers:

NP-40 lysis buffer:

20 mM Tris, 150 mM NaCl, 0.5% NP-40, 5 mM EGTA, 5 mM EDTA, pH= 7.5.

Sample buffer:
62.5 mM Tris, 2% SDS. 25% glycerol, 0.01% bromophenol blue.

TGS:
25 mM Tris, 192 mM glycine, 0.1% SDS, pH= 8.3.

Transfer buffer:
1. **Wet transfer [1]:**
   25 mM Tris, 192 mM glycine, pH 8.3 supplemented with 20% methanol and 0.025–0.1% SDS
2. **Semidry transfer [8]:**
   48 mM Tris, 39 mM glycine, pH 9.2 with 20% methanol for semidry transfer.
   TBST:
   20 mM Tris, 500 mM NaCl, 0.1% Tween-20, pH= 7.4.
   Gel recipe:

1. 4% stacking gel (10 mL):

40% w/v acrylamide	1.25 mL
2% w/v bis-acrylamide	0.65 mL
1.0 M Tris-Cl pH 6.8	1.25 mL
MilliQ water	6.85 mL

   Supplement just before casting the gel:
   25 µL of 40% APS
   10 µL of Temed
   Note 14. Stacking gel can be prepared in larger quantity, filter sterilized and stored at 4°C. In this case, do not add APS and Temed as this will cause polymerization of the gel.
2. 10% resolving gel (10 mL):

40% Bis/Acrylamide mix (29:1)	2.5 mL
1.5 M Tris pH 8.8	5 mL
MilliQ water	2.5 mL

   Supplement just before casting the gel:
   12.5 µL 40% APS
   5 µL Temed

APPLICATIONS

The main research application of the WB assay is to detect the presence of a protein of interest in a variety of systems. It can be used to determine expression of a given protein in organelles, cells, tissues, or embryos/whole organisms.

Both endogenous and heterologously expressed proteins can be detected. When samples are collected at different time points it may allow the determination of changes in protein expression levels. Another application of WB is the assessment of posttranslational modification of the target protein such as phosphorylation (with phospho-specific antibodies), ubiquitination, or sumoylation detected with a combination of WB with immunoprecipitation (IP). The target protein is first immunoprecipitated and then probed with antibodies against ubiquitin or small ubiquitin-like modifier (SUMO) or vice versa. Similarly, one can identify protein–protein interactions by testing the presence of a potential interactor using co-IP. In this case, one protein (the bait) is immunoprecipitated using a specific antibody against the bait and the immunocomplex is run onto a SDS-PAGE. The membrane is then subjected to WB using antibodies specific to potential interactors (targets).

Determination of the presence of a protein of interest may also be useful in diagnostics [9]. The expression of a specific protein or its presence in bodily fluids may be used for diagnosis of a disease. For example, WB is used to detect the prion protein causing bovine spongiform encephalopathy, bovine spongiform encephalopathy (BSE) in cows [10]. In medicine, it is often used as a confirmatory test [usually in combination with enzyme-linked immunosorbent assay (ELISA)] in diagnosis of several diseases such as lyme disease [11] or HIV [9,12].

SCENARIO

Understanding how signaling pathways are precisely regulated to control cell fate and coordinate cell movements is one of the major challenges in developmental biology. Zhang and colleagues [13] started to address this complex problem by monitoring the activity of a variety of signaling pathways during early Xenopus development using phospho-specific antibodies. To this end, two challenges had to be overcome: Xenopus embryos are rich in yolk proteins that can disrupt protein migration in the gel and phosphatases had to be inhibited. A buffer low in salts and detergent but with high concentration of a cocktail of phosphatase inhibitors was therefore used (described in [14]). Embryos were then collected at different time points from egg to tadpole stage and processed as described [13]. Fig. 6.4 shows the temporal dynamics of the activation state of six different signaling pathways: Wnt (pLRP6), STAT3, TGFβ (pSmad2), BMP (pSmad1), Akt and Erk. Most signaling pathways are silent or have very low activity at blastula stages (stage 8 and 9). However, as gastrulation starts (stage 10), a rise of signaling activity is detected for all pathways. This corresponds to important developmental events such as mesendoderm specification and the start of cellular movements. After gastrulation (stage 12 onward), the various signaling pathways are differentially regulated. This time course analysis indicate that different signaling pathways have very precise kinetics of activation and deactivation and allow the researcher to formulate hypotheses for their potential roles during development, which can be rationally tested.

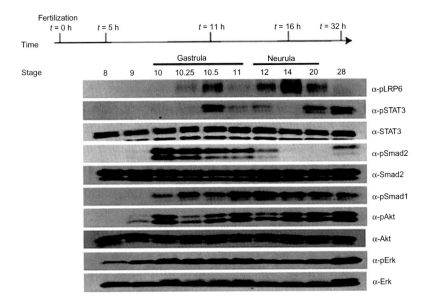

FIGURE 6.4 Time course of phosphorylation of six different signaling pathways during Xenopus development. *Xenopus laevis* embryos were collected at the time indicated and subjected to Western blot analysis. Membranes were probed with anti-phospho-LRP6 (pLRP6) to monitor Wnt signaling, anti-phospho-STAT3 (pSTAT3), anti-phospho-Smad1/5/8 (pSmad1) antibody for monitoring BMP activity, anti-phospho-Smad2 (pSmad2) antibody for TGFβ/nodal signaling, anti-phospho-Erk (pErk) for MAPK/Erk signaling, and anti-phospho-Akt (pAkt) for PI3K/Akt signaling. Anti-STAT3, anti-Smad2, anti-Akt, and anti-Erk were used as loading controls to ensure that all lanes have been loaded equally.
Adapted from Zhang S, Li J, Lea R, Amaya E, Dorey K.A functional genome-wide in vivo screen identifies new regulators of signaling pathways during early Xenopus embryogenesis. PLoS ONE. 2013;8(11):e79469.

KEY LIMITATIONS

WB has become an indispensable technique in the research toolbox and has proved to be useful in many aspects of modern science focused on proteins. It has however, as any other technique, its limitations and drawbacks.

Quantification

WB is not a quantitative technique as it will not tell how much of the protein is expressed. Only relative differences in expression levels can be assessed and even this needs to be approached with caution. A researcher has to make sure that the conditions of the assay are within the linear range of the detection method. For instance, an X-ray film has a very poor linear range and gets easily saturated therefore should be avoided in any quantification. Imaging with charge coupled device (CCD) cameras sensitive to chemiluminescence is much better for quantification purposes but is less sensitive. Additionally, as a multistep technique WB is prone to small imprecisions, coming for instance from gel loading or uneven transfer, which accumulate and results in false differences. It can be corrected to some extend by using internal controls like probing the membrane for a housekeeping protein that is constant between samples and not present in saturating amounts. Therefore to increase the accuracy of the quantification the conditions need to be established empirically by testing multiple dilutions making the assay more laborious and time-consuming.

Sensitivity

WB in general is quite sensitive, down to the low femtogram range in the best cases. In reality, however, due to accumulation of intrinsic inefficiencies of the technique (such as transfer, recognition by antibodies, etc.) the detection limit is usually higher. It is clear that not all expressed proteins can be detected, as some of them are present in a very low number of copies per single cell. For this reason, a negative result cannot be interpreted as complete lack of the target protein in the sample.

Specificity

WB is as specific as the antibody used for the membrane probing and therefore in some cases when the specificity of the antibody is poor the result may be not easy to interpret. Therefore a new antibody should be validated for a specific use by knocking out the target protein or by using recombinant or overexpressed target protein whenever possible.

TROUBLESHOOTING

As in any multistep technique, there are many steps that can go wrong resulting in poor or no interpretable results. It is difficult to predict all problems that can be encountered performing WB. The most common problems may be identified and corrected by using appropriate controls, like a previously validated sample and staining the membrane or the gel after the transfer. Below we discuss the most prevalent problems grouped in three categories, (1) no signal, (2) high background, and (3) misshaped or missing bands.

Problem	Cause	Solution
No Signal	Insufficient amount of a sample	Measure the concentration of lysates to make sure that the proper amount of total proteins is loaded on a gel (20–50 µg of cell lysate per lane or 25–100 ng purified proteins). If the target protein is expected to be at low level try to increase the amount of lysate loaded to the gel.
	Incomplete transfer	Make sure that the transfer is uniform and complete. A prestained protein ladder is a quick way to assess protein transfer but staining the membrane with Ponceau S or the gel with Coomassie Blue is more thorough. If the transfer is not complete, extend the time and/or increase electromagnetic field (voltage or current) or consider performing a wet transfer. Adjust buffer composition according to the rules provided in the "In Principle" paragraph.
	Poor detection	There are two main reasons for poor detection: (1) weak recognition and/or binding of a target protein by the antibody or (2) problems with detection system. If possible, validate the antibody by taking along a purified target protein or a sample with heterologously expressed target protein. – Increase the concentration of the antibody or extend the incubation time (overnight at 4°C). – Use less stringent washing conditions. – In case of a chemiluminescent detection system use a more sensitive substrate (e.g., ECL femto) and increase the exposure time.
High Background	The high background usually comes from the binding of the antibody to nonspecific targets, overloaded samples, insufficient washings, or poor selectivity of the antibody used.	Decrease the amount of proteins loaded per lane. Increase number and/or extend time of washings after antibody incubation. Increase stringency of an antibody incubation by using a blocking solution or increase detergent concentration. Decrease the concentration of the antibody.
	Degradation or posttranslational modifications of the target protein may result in the detection of multiple bands.	Minimize degradation by using appropriate inhibitors and handle samples at 4°C.

Problem	Cause	Solution
Misshaped Or Missing Bands	Usually this indicates problems with the electrophoresis or the transfer due to excessive heat or uneven transfer.	Run the gel and transfer on ice or decrease the voltage/current. Make sure that all the air is "rolled out" from the blot sandwich.
	Bulky bands and migration artifacts may indicate gel overloading or buffer problems ("smiling gel").	Decrease loading.

CONCLUSION

WB has been around for almost four decades and during that time it has proved to be an extremely useful and powerful tool in research. Its relatively low cost, simple principles, and accessibility made it very popular and widely used. Taking advantage of the ability of antibodies to recognize and bind a variety of antigens in a highly specific manner makes WB an extremely versatile and reliable assay if performed correctly. Although not a truly quantitative technique WB can be finetuned to provide semiquantitative results allowing to address questions about relative proteins expression and posttranslational modification and their changes in time. It is also routinely used to help in diagnosis of some diseases where high degree of confidence is necessary. It requires, however, a good understanding and initial training because it is prone to false results and incorrect interpretation.

REFERENCES

[1] Towbin H, Staehelin T, Gordon J. Electrophoretic transfer of proteins from polyacrylamide gels to nitrocellulose sheets: procedure and some applications. Proc Natl Acad Sci USA 1979;76(9):4350.

[2] Burnette WN. "Western blotting": electrophoretic transfer of proteins from sodium dodecyl sulfate – polyacrylamide gels to unmodified nitrocellulose and radiographic detection with antibody and radioiodinated protein A. Anal Biochem 1981;112(2):195–203.

[3] Southern EM. Detection of specific sequences among DNA fragments separated by gel electrophoresis. J Mol Biol 1975;98(3):503–17.

[4] Alwine JC, Kemp DJ, Stark GR. Method for detection of specific RNAs in agarose gels by transfer to diazobenzyloxymethyl-paper and hybridization with DNA probes. Proc Natl Acad Sci USA 1977;74(12):5350–4.

[5] Bradford MM. A rapid and sensitive method for the quantitation of microgram quantities of protein utilizing the principle of protein-dye binding. Anal Biochem 1976;72:248–54.

[6] Lowry OH, Rosebrough NJ, Farr AL, Randall RJ. Protein measurement with the Folin phenol reagent. J Biol Chem 1951;193(1):265–75.

[7] Smith PK, Krohn RI, Hermanson GT, Mallia AK, Gartner FH, Provenzano MD, et al. Measurement of protein using bicinchoninic acid. Anal Biochem 1985;150(1):76–85.

[8] Bjerrum OJ, Schafer-Nielsen C. Buffer systems and transfer parameters for semidry electroblotting with a horizontal apparatus. Electrophoresis. VCH Weinheim: Germany; 1986;86:315–27.

[9] Soundy P, Harvey B. Western blotting as a diagnostic method Medical biomethods handbook. Springer; 2005:43–62.

[10] Schaller O, Fatzer R, Stack M, Clark J, Cooley W, Biffiger K, et al. Validation of a Western immunoblotting procedure for bovine PrPSc detection and its use as a rapid surveillance method for the diagnosis of bovine spongiform encephalopathy (BSE). Acta Neuropathol Springer 1999;98(5):437–43.

[11] Wilske B. Epidemiology and diagnosis of Lyme borreliosis.. Ann. Med Informa UK Ltd UK 2005;37(8):568–79.

[12] Dewar R, Goldstein D, Maldarelli F. Diagnosis of human immunodeficiency virus infection Principles and practice of infectious diseases, 7th ed. New York (NY): Churchill Livingstone; 2009:1663–86.

[13] Zhang S, Li J, Lea R, Amaya E, Dorey K. A functional genome-wide in vivo screen identifies new regulators of signaling pathways during early Xenopus embryogenesis. PLoS ONE 2013;8(11):e79469.

[14] Dorey K, Hill CS. A novel Cripto-related protein reveals an essential role for EGF-CFCs in Nodal signaling in Xenopus embryos. Dev Biol 2006;292(2):303–16.

SUGGESTED FURTHER READING

[1] Bio-Rad Laboratories, "Protein Blotting Guide", Bull. No. 2895.

[2] Bio-Rad Laboratories, "A Guide to Polyacrylamide Gel Electrophoresis and Detection", Bull. No. 6040.

[3] ThermoScientific, "Western Blotting Handbook and Troubleshooting Guide".

[4] Taylor SC, Berkelman T, Yadav G, Hammond M. A defined methodology for reliable quantification of Western blot data. Mol Biotechnol 2013;55(3):217–26.

[5] Yeung Y-G, Stanley ER. A solution for stripping antibodies from PVDF immunoblots for multiple reprobing. Anal Biochem. 2009;389(1):89–91.

[6] AAlegria-Schaffer, ALodge and Kvattem in Guide to Protein Purification, 2nd Edition, "Chapter 33 Performing and Optimizing Western Blots with an Emphasis on Chemiluminescent Detection", Methods in Enzymology, vol. 463, 2009, pp. 573–99.

[7] Kurien BT, Scofield RH. Western blotting. Methods 2006;38:283–93.

GLOSSARY

Electrophoresis Motion of charged molecules (such as proteins or DNA) within a matrix (e.g., gel) driven by electromagnetic field.

Femtogram 10^{-15} gram

Fluorophore Chemical compound able to re-emit light when excited by light at a different wavelength.

Immunoprecipitation   It is a technique enabling the purification of a particular antigen through the binding of an antibody. The antigen/antibody complex is then bound to protein A/G agarose beads, allowing its isolation from the rest of the sample.

Posttranslational modifications   Modifications of proteins following their synthesis within a cell. The most common include phosphorylation, ubiquitination, and glycosylation.

Ubiquitination/sumoylation   These are examples of posttranslational modification. Ubiquitination usually occurs on a Lysine residue and affect cellular processes such as trafficking, endocytosis, and is often a tag for protein degradation. Sumoylation also occurs on Lysine residues but is not associated with protein degradation but plays a role in nuclear-cytoplasmic transport, regulation of transcription, apoptosis, and progression through the cell cycle.

Chapter 7

The Enzyme-Linked Immunosorbent Assay: The Application of ELISA in Clinical Research

Jefte M. Drijvers, Imad M. Awan, Cory A. Perugino, Ian M. Rosenberg and Shiv Pillai

Ragon Institute of MGH, MIT and Harvard, Cambridge, MA, United States

Chapter Outline

Objectives

- Familiarize the reader with the basic concepts of the enzyme-linked immunosorbent assay (ELISA).
- Discuss the available forms of ELISA and their applications and limitations.

Basic Science Methods for Clinical Researchers. DOI: http://dx.doi.org/10.1016/B978-0-12-803077-6.00007-2

119

- Describe scenarios where ELISA can be used to answer scientific questions.
- Provide clinical and translational scientists with the necessary tools and considerations to apply ELISA in their own research setting.

INTRODUCTION

Immunoassays, the currently most commonly used form of which is the enzyme-linked immunosorbent assay (ELISA), make use of antibody specificity and affinity to identify and quantify a diverse array of antigens. ELISA is based on the radioimmunoassay procedure originally described by Rosalyn Yalow and Solomon Berson in 1960 [1], that made use of a radioactive isotope to label an antigen or antibody. This technique was initially described for the measurement of plasma insulin levels. In 1966, Nakane and Pierce [2] and Avrameas and colleagues [3] described the use of enzyme-labeled antibodies in immunohistochemistry and for the visualization of precipitation reactions in immunodiffusion and immune-electrophoretic gels. The first quantitative applications of enzyme-based versions of the immunoassay were independently described in 1971 by the Swedish group of Engvall and Perlmann [4] and the Dutch group of Van Weemen and Schuurs [5,6]. The technique has since undergone many modifications that have greatly expanded its range of applications.

There are many established applications of ELISA in clinical medicine, including diagnostic assays measuring human chorionic gonadotropin (HCG), estrogen, hepatitis B antigen, and various antibodies, to name but a few. However, in this chapter we will focus on the diverse ways in which ELISA can be applied in a clinical research or translational science context, aiming to provide researchers with the necessary tools and considerations to apply ELISA in their own research setting.

IN PRINCIPLE

The initial radioimmunoassays were conducted in solution, and an antibody-precipitating reagent was used to separate the radio-labeled antigen bound to a specific antibody. There are two characteristic features that differentiate an ELISA. Not only does it involve an enzyme-labeled antigen or antibody, but it also incorporates an "immunosorbent" step, which refers to the use of a solid-phase component in the assay. In ELISA, as in a radioimmunoassay, the specificity and affinity of antibodies are utilized in order to measure the presence of a certain analyte in a sample (see Fig. 7.1). An antibody or antigen is attached to a solid surface and subsequently binds a complementary antigen or antibody, respectively. Depending on the type of assay being used, quantitation requires binding of an enzyme-labeled reagent to the solid-phase. After the addition of a substrate, enzyme activity yields a proportionate amount of a catalytic product, the magnitude of which correlates with the level of the analyte in the original sample. Actual quantification can be achieved by comparison to standard

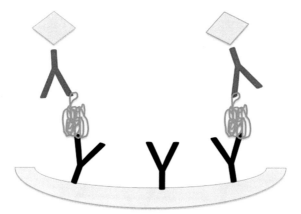

FIGURE 7.1 An antibody molecule consists of four polypeptide chains: two light chains and two heavy chains that are held together by disulfide bonds. The C-terminal parts of the heavy chains form the tail of the antibody and determine its functional characteristics. The N-terminal parts of both light and heavy chains cooperate to form the antigen-binding sites, which determine antibody specificity. ELISA exploits this antigen–antibody specificity to detect and quantify proteins in unknown samples.

curves constructed using fixed amounts of the analyte in the assay performed in parallel.

Various forms of ELISA have been developed. Based on the manner in which an analyte is immobilized, one can distinguish assays in which an antigen is directly immobilized to the solid surface from those where this occurs indirectly through a capture antibody ("sandwich ELISA"). Similarly, the detection of an analyte can be direct or indirect, depending on whether the antibody that is coupled to an enzyme binds the analyte directly or through other molecules. Lastly, assays can be qualified as competitive or not. These concepts will be explained further below.

Sandwich ELISA

In a sandwich ELISA, typically two distinct monoclonal antibodies recognizing different epitopes on a given antigen are used, one for capture and the other for detection (see Fig. 7.2). Alternatively, a polyclonal antibody mixture may be used for both capture and detection. The antigen must either be large enough to have at least two distinct physically separated and therefore available epitopes or be a structure with the same epitope repeated multiple times in a large enough analyte. The capture antibody is typically noncovalently attached to a solid-phase—usually the wells of a 96-well polystyrene plate. The plate is washed to remove antibody that did not adhere, and the mixture containing the analyte to be measured is added to the plate enabling the specific antigen of interest to

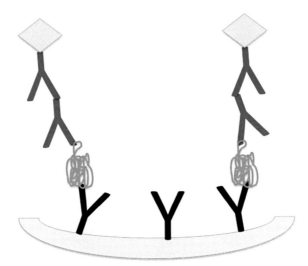

FIGURE 7.2   In sandwich ELISAs, capture antibody (black) is coated to the well and incubated with antigen of interest (blue). For detection, enzyme-conjugated developing antibody (red) is added, which binds the analyte at another epitope.

be captured. If the antigen being quantitated is large enough, another available epitope on the captured antigen is then detected by a "developing" antibody that is covalently conjugated to an enzyme. Alternatively, just like in indirect ELISAs (see below), a primary unconjugated detection antibody may be added first, followed by a secondary enzyme-conjugated antibody that binds the primary detection antibody. A higher degree of specificity can be obtained in a sandwich immunoassay because of the need for antibodies to typically recognize two different epitopes on a protein antigen. A limitation of the "sandwich" technique is that small antigens, such as peptides or drugs, cannot be detected using this approach.

Indirect ELISA

The terms "direct" and "indirect" ELISA refer to the way in which the immobilized analyte is detected. In the case of the direct technique, a detection antibody that is directly coupled to an enzyme is added. In the case of the indirect detection system, a primary antibody is used to bind the analyte and a secondary enzyme-coupled antibody is then added, which binds the primary antibody (see Fig. 7.3). While the direct method has the advantage of a slightly shorter and simpler assay protocol, it requires new specific enzyme-coupled antibodies to be designed for every single analyte that the assay is used for. An indirect assay has the advantage that the signal can be amplified if the secondary antibody recognizes several epitopes of the primary antibody's constant region.

FIGURE 7.3 In indirect assays, detection of the analyte occurs indirectly through a primary detection antibody (pink) that binds the analyte, followed by a secondary enzyme-linked antibody (purple) that binds the primary antibody.

In a sandwich approach using a labeled secondary antibody, it is essential that this labeled antibody specifically detects the primary detection antibody and not the capture antibody. This is often achieved by using capture and detection antibodies from different species. An example of this is provided later in this chapter (see "Scenario").

Competitive ELISA

A commonly used strategy to quantify small antigens that do not have the required separate epitopes for sandwich assays is competitive ELISA (see Fig. 7.4). Herein, a known amount of the target antigen (identical to the analyte being measured) is added to the analyte-containing sample and a specific antibody. In such an assay, sample analyte competes with the purified added antigen for binding to the detection antibody. There are two broad approaches that can be used, one using an enzyme-labeled antibody and a solid-phase analyte, and the other using an enzyme-labeled antigen and a solid-phase antibody.

In the first approach one might incubate an analyte-containing sample with primary detection antibodies conjugated to an enzyme, and add the mixture to a well containing immobilized antigen. The more analyte present in the sample, the fewer antibody binding sites are left available for binding to the antigen on the solid surface. Assay readouts are thus low for high analyte concentrations in the sample and vice versa. It is crucial that the amount of available enzyme-labeled detection antibody is limiting for this approach to be quantitative.

In the second competitive approach, an enzyme-labeled antigen competes with the free analyte in the mixture to be analyzed for binding to an immobilized but unlabeled specific detection antibody. The amount of enzyme-labeled antigen that binds is inversely related to the concentration of the free analyte.

FIGURE 7.4 In competitive ELISA, competition for detection antibody binding occurs between soluble antigen in the sample (the analyte) and an immobilized, known amount of the same antigen. The more analyte is present in the sample, the less antibody is available for binding the antigen coated to the well, leading to an inverse correlation between analyte concentration in the sample and assay readout.

IN PRACTICE

Below, a protocol for a common ELISA assay is provided. Since countless variations of ELISA have been developed to satisfy a researcher's specific needs, many different protocols are in use. The protocol below was adapted from Current Protocols in Immunology [7] and modified to describe a sandwich assay measuring immunoglobulin G4 (IgG4) levels in culture supernatants, as can be used in the scenario that is provided later in the chapter. Basically, wells are coated with a capture antibody that is adsorbed onto the ELISA plate. The sample containing IgG4 is then "captured" onto the ELISA wells. The detected IgG4 is then quantitated by the amount of the enzyme-detection antibody conjugate that then binds to the analyte.

Materials

- Purified goat antihuman IgG polyclonal antibody.
- Carbonate-coating buffer (0.05 M carbonate–bicarbonate, pH 9.6).
- Human IgG4 standard.

- Sample/Block buffer (0.05 M Tris, 0.14 M NaCl, pH 8.0, 1% BSA (bovine serum albumin), 0.05% Tween-20).
- ELISA wash buffer (0.05 M Tris, 0.14 M NaCl, pH 8.0, 0.05% Tween-20).
- Horseradish peroxidase-conjugated monoclonal mouse antihuman IgG4 Fc antibody.
- Horseradish peroxidase substrate solution (TMB (3,3',5,5'-tetramethylbenzidine) solution).
- Stop solution (0.18 M H_2SO_4).
- Multichannel pipettor.
- 96-well flat-bottom ELISA plates with adhesive covers.
- 96-well V-bottom microtiter plates.
- Multiwell scanning spectrophotometer with 450 nm filter.

Coat Wells with Capture Antibody

1. Make a 3.5 μg/mL solution of goat antihuman IgG polyclonal antibody (the capture antibody) in carbonate-coating buffer.
2. Add 100 μL of antibody solution to each well of a flat-bottom 96-well ELISA plate (using a multichannel pipettor). Apply adhesive cover and incubate overnight at 4°C or for 2 hours at 37°C.

   Note: Depending on the purity of the antibody being used to coat the plate, the concentration of the solution may need to be increased to adequately capture the protein of interest. This concentration may range from 0.5 to 5 μg/mL.

   Note: Warmer temperatures will maximize antibody binding to the plastic wells while precoated plates should be stored at 4°C to prevent denaturation of the antibody.

Prepare Standards and Sample Dilutions

3. Add 125 μL of sample buffer to the first three columns of rows B through G in a 96-well V-bottom plate. Add 250 μL of IgG4 standard at 1000 ng/mL to the first three wells of row A. Transfer 125 μL of the IgG4 standard from the first three wells of row A to the first three wells of row B, mix, and repeat down five more times. There are now seven standard concentrations in the first three columns of rows A through G, ranging from 1000 ng/mL to 15.6 ng/mL. Add 125 μL of sample buffer to the first three wells of row H. These are the negative controls.
4. Dilute the first sample containing an unknown amount of IgG4 by adding 175 μL of sample to the fourth, fifth and sixth well of row A. Add 140 μL of sample buffer to wells 4 through 6 of rows B through D. Transfer 35 μL from the wells containing sample to the wells below and mix. Repeat this two more times. With each step, the sample is diluted fivefold. The lowest of four rows now contains a 125-fold diluted sample concentration. Repeat these steps for each sample containing an unknown amount of IgG4.

Capture IgG4

5. Empty the now-coated ELISA plate by vigorously inverting and flicking out the solution into the sink. Wash three times using a wash squeeze bottle with ELISA wash buffer to remove any unbound antibody. Wells should be filled fully with each wash. Blot dry on paper towels after the third wash.

    Note: Quickly inverting and flicking the plate is done to prevent the contents of each well from mixing with the neighboring wells.

6. Block the residual binding capacity by filling each well with block buffer. Incubate for 1 hour at room temperature.

    Note: Optimal blocking is essential to prevent nonspecific binding of the conjugated detection antibody with the plastic wells. Alternative blocking buffers include skim milk, normal mouse serum (or any species identical to the secondary conjugated detection antibody used).

7. Transfer 100 μL of the standard and sample dilutions from the dilution plate to the ELISA plate. Cover the plate and incubate for at least 2 hours at room temperature.

8. Wash plate three times with ELISA wash buffer as in Step 5 to remove any unbound IgG4.

Detect Bound IgG4

9. Dilute the horseradish peroxidase-conjugated monoclonal mouse antihuman IgG4 Fc antibody 1,000- to 10,000-fold with sample buffer and add 100 μL to each well of the ELISA plate. Cover plate and incubate for 1 hour at room temperature.

    Note: The optimal dilution should be determined by trial and error, attempting to achieve the maximal linear absorbance curve for the standard dilutions, while keeping the background signal at a minimum.

10. Wash plate three times as in Step 5.

11. Add 100 μL of horseradish peroxidase substrate (TMB) solution to each well and incubate at room temperature for 5–30 minutes.

12. To stop the reaction, add 100 μL of stop solution to each well.

13. Read absorbance at 450 nm.

14. Generate a standard curve by plotting absorbance versus IgG4 standard concentrations. Determine which sample dilutions produce absorbance readouts within the linear portion of the standard curve (on a semi-log scale) and use those to calculate the IgG4 concentration in the undiluted sample. Since multiple values are obtained for each sample to improve reliability, the average of those values can be used as the final result.

APPLICATIONS

While many clinical applications of ELISA are used routinely, as mentioned previously, we will focus on how ELISA can be applied in clinical and translational research projects. Several applications are described more extensively elsewhere [8]. We do not intend to provide a complete list of potential applications, as the possibilities are virtually endless.

Measurement of Antigen Concentrations

The presence of antigens in a solution can be detected, and their concentration quantitated, by various ELISA techniques [9]. The assay used depends on the characteristics of the antigen and the availability of purified antigen.

Antigens with Multiple Noncompeting Epitopes

When the target antigen has multiple noncompeting (either different or repeated) available epitopes, the most sensitive choice would be the sandwich assay, discussed above. In this assay, wells are coated with a capture antibody, followed by serial incubations, interrupted by washing steps, first with samples containing (variable amounts of) antigen, then with enzyme-labeled detection antibodies and finally with a colorimetric or fluorimetric substrate. This approach is often used not only for medium- to large-sized (glyco)proteins, such as in pregnancy testing (detection of HCG) and influenza tests (detection of hemagglutinin), but also other antigens such as viruses. Many cytokine assays in clinical and translational research settings also use sandwich ELISAs.

Antigens without Multiple Noncompeting Epitopes

When the antigen in question does not have multiple noncompeting epitopes, a competitive assay can be used. This does require the availability of pure antigen either to coat the wells with antigen or to direct label antigen with enzyme as described in the section on competitive ELISAs above. Antigens that are detected and measured by competitive assays tend to be smaller. For example, many drug screens depend on this technique.

Antibody Detection

ELISA can be used to determine and quantitate the presence of antibodies in samples such as patient sera, sera from immunized animals or hybridoma supernatants. The presence of antibodies of a certain specificity is often of interest. In this case, one can either employ an indirect or a sandwich assay, depending on the availability of purified antigen [9]. Alternatively, one might be interested in the presence of antibodies recognizing a group of antigens rather than just one specific antigen, for example in order to determine the presence of autoreactive antibodies. These concepts are discussed further below.

Specific Antibody Detection

If enough (partially) pure antigen is available, one can employ an indirect assay by coating wells with antigen and incubating with antibody-containing samples, followed by the addition of enzyme-labeled antibodies specific for the Fc portions of the antibodies of interest, and finally substrate for enzymatic conversion and product detection. This approach can, for example, be used to determine the presence of HIV-specific antibodies, if plates are coated with HIV antigens and incubated with a subject's serum. When the aim of the project is to quantitate antibody isotypes against a specific antigen, different detection antibody-enzyme conjugates are used, each specific for a different antibody isotype.

If no purified antigen is available, one can use a sandwich ELISA. This is only semiquantitative and has limited utility, being of value only if the antigen of interest cannot be adsorbed on an ELISA plate. This approach is similar to the sandwich assay described previously, but requires one additional step because antibodies instead of antigens are being detected in this case. In this assay, the wells are coated with a capture antibody that binds all antibodies of the species that the sample being analyzed is derived from. This is then incubated with the sample that contains the antibodies of interest (a serum or body fluid), followed by incubation with the corresponding crude antigen. Finally, enzyme-conjugated antibodies specific for (another, noncompeting epitope of) the target antigen are added. If they bind, this indirectly demonstrates the presence of antigen-specific antibodies in the sample being tested. The practical utility of such an approach is very limited, since generally most crude antigens will adsorb and can be utilized in a nonsandwich ELISA.

Detection of Autoreactive Antibodies

Instead of detecting antibodies specific for one particular antigen, one can also use ELISA for the detection of the presence of autoreactive antibodies in general. An indirect assay, similar to the one described for specific antibody detection, would be a suitable approach. In this case, instead of individual antigens, cell lysates containing many self-antigens would be adsorbed to the plate and incubated with antibody-containing sample.

Epitope Mapping of Monoclonal Antibodies

The sandwich ELISA described for detection of specific antibodies can also be used for epitope mapping of antibodies [9]. This technique makes use of the fact that multiple antibodies cannot bind the same epitope at the same time; if a detection antibody and the antibody of interest both recognize the same or competing epitopes, there will be no or a much lower readout signal. By performing the assay multiple times with different detection antibodies, each recognizing different, known epitopes of the target antigen, one can deduce the epitope used by the antibody of interest by observing which detection antibodies yield a decreased readout signal.

Cell Surface Antigen Detection

Not only soluble, but also cell surface antigens can be analyzed by ELISA. In cellular ELISA, cells are incubated either directly with enzyme-conjugated antibodies against a surface antigen, or indirectly with a primary antibody recognizing a cell surface antigen followed by the addition of a secondary enzyme-linked antibody for detection [9]. Even though this technique can sensitively measure cell surface antigen levels, this would only be the approach of choice in the absence of access to the equipment to perform flow cytometry, since cellular ELISA is not suitable for use with a combination of different cell types or for assessing multiple surface antigens at one time. Another application of cellular ELISA is to detect the presence of antibodies against cell surface antigens. To achieve this, cells are incubated with a sample hypothetically containing antibodies against a surface antigen, followed by incubation with a secondary, enzyme-conjugated detection antibody. This approach is not very specific, since cells express many surface antigens and this technique does not determine which one of these antigens is the target of the antibodies present in the sample.

SCENARIO

The following scenario provides an example of how ELISA can provide valuable information in clinical and translational research settings. The project involves the study of the pathogenesis of IgG4-related disease (IgG4-RD), a rare systemic immune-mediated disease. As the name implies, the levels of circulating IgG4, which is generally found in very low amounts in the serum of healthy individuals, are often markedly elevated in this condition. Since the production of antibody isotypes other than IgM is the result of a process called class-switching, and since class-switching in B cells is known to be induced by collaboration of follicular helper T (T_{FH}) cells and B cells, one might decide to study this cell–cell collaboration event in order to find an answer to the question of why IgG4 levels are increased in IgG4-RD.

One approach one could take would be to coculture B cells and autologous T_{FH} cells, from either IgG4-RD patients or healthy controls. After activating the T cells and allowing several days for T_{FH}-B collaboration and the ensuing antibody production and class-switching to occur, the supernatants are collected and ELISA is employed to measure the amount of IgG4 that is present. If more IgG4 is found in the supernatants of cocultures of IgG4-RD patient lymphocytes, this would be the first step in identifying a subset of T_{FH} cells that may specifically drive the IgG4 class switch.

In order to detect IgG4 molecules with optimal sensitivity, a sandwich assay would be a logical choice, as the binding of the analyte by multiple epitopes allows for good sensitivity and specificity. In this method, the wells of a plate would first be coated with a capture antibody. The plate would then be incubated with the supernatant sample that presumably contains an unknown amount of IgG4, followed by the addition of an enzyme-conjugated anti-IgG4 antibody. In this particular case, one could use a polyclonal antihuman IgG antibody to coat the wells, which would capture all IgG molecules in the sample. As a detection antibody, one would need to select a monoclonal antibody that is specific to the Fc portion of IgG4 subclass antibodies, coupled to an enzyme such as horseradish peroxidase or alkaline phosphatase. Finally, a substrate for the chosen enzyme is added, the catalysis of which will yield a detectable product. The detected signal correlates with the amount of IgG4 present in the sample. In order to quantify the antibody amount, the signal readout is compared to a standard curve obtained from performing the same assay with known concentrations of IgG4 (see the "In Practice" section above).

As mentioned previously, it is important to validate that the capture and detection antibodies bind the analyte, in this case IgG4, through noncompeting epitopes. Also, it is crucial that the detection antibody only binds to the human IgG4 molecules and does not cross-react with the capture antibody. Therefore, the capture antibody needs to be derived from a nonhuman species that the detection antibody does not cross-react with. In the described assay, one could choose a goat-derived polyclonal antihuman IgG antibody as a capture antibody and a mouse-derived horseradish peroxidase-conjugated antihuman IgG4 antibody as a detection antibody.

In addition to measuring the IgG4 concentrations, one might also be interested in the amounts of other IgG subclasses or total IgG. For example, it might be of interest to know how much IgG4 is produced as a percentage of total IgG rather than absolute IgG4 amounts. In order to measure total IgG concentrations in the described sandwich assay, one would need to change the IgG4 detection antibody to a general antihuman IgG enzyme-conjugated antibody.

KEY LIMITATIONS

Several limitations of individual types of ELISA have been mentioned previously in the respective sections. Generally, it is important to realize that while ELISA can reliably detect and quantify the presence of antigens and antibodies, it does not provide any other information, as opposed to, say, Western blots, from which the size of a protein can be learned as well. Also, the possibilities of ELISA depend entirely on the availability of antibodies that bind the native confirmation antigen of interest specifically and with high enough affinity. Another limitation is the fact that ELISA is not suitable for analyzing multiple antigens at a time, as opposed to, for example, flow cytometry. This limitation can be circumvented to some extent by doing multiple assays, although that can easily become a labor-intensive endeavor.

TROUBLESHOOTING

Problem	Possible Cause	Solution
Absent or weak signal	Insufficient concentration of assay components	Use varying antigen, primary, or detection antibody concentrations/dilutions.

Problem	Possible Cause	Solution
	Washes are too stringent	Reduce detergent concentration in washing buffer. Alternatively, use an automated plate washer if available.
	Enzyme inhibitor present	Confirm that reagents do not interfere with enzyme reaction i.e., sodium azide inhibits horseradish peroxidase activity.
	Incorrect/Inactive substrate or conjugate	Use an appropriate substrate for the enzyme conjugate (i.e., OPD or TMB with peroxidase). Obtain fresh substrate and conjugates.
	Use of a poor antibody	Test antibodies by immunoprecipitation (not by Western blot) to determine if they effectively bring down specific protein.
High background	Insufficient washing	Ensure that wells are filled with wash buffer and fully aspirated. Use an automated plate washer if available.
	Nonspecific binding	Wash with a low concentration of Tween-20 (up to 0.2 %). Increasing salt concentration (up to 150 mM NaCl) will further reduce nonspecific interactions.
		An additional blocking step performed overnight at 4°C using either 1% BSA or 0.5 % nonfat dry milk in a Tris-buffered saline containing 0.2 % Tween-20 may help reduce noise.
	Cross-reactivity	Antibody cross-reactivity in buffer ingredients can be minimized using alternative blocking reagents such as human serum albumin or ChonBlock [8].
	Concentration of detection antibody is too high	Dilute detection antibody and determine optimal working concentration.
	Incorrect assay temperature	Confirm that assay temperature is below 37°C.
Poor standard curve	Incorrect standard solution	Confirm dilutions are made properly and double check calculations.
	Low adsorption of reagents to plate	Use an appropriate coating buffer, such as PBS pH 7.4 or carbonated coating buffer pH 9.6.

CONCLUSION

Made possible by innovative contributions by various groups in the 1960s and early 1970s, ELISA is now a very useful and established laboratory technique in clinical medicine as well as basic, translational, and clinical research. Many variations of ELISA exist, including the sandwich assay, in which an analyte is immobilized via capture antibodies, the indirect assay, in which the detection of an analyte occurs through a primary unconjugated and a secondary enzyme-linked antibody rather than just one enzyme-linked antibody, and the competitive assay, in which competition between an analyte and a known amount of added antigen takes place. The choice of assay depends on the particular purpose that ELISA is employed for as well as other factors, such as the size of the analyte, the presence of noncompeting epitopes and the availability of purified antigen. In fact, every single ELISA has to be designed individually based on the exact goals of the experiment and the characteristics of the antigens and/or antibodies that one is working with. Some of the most common applications of ELISA in clinical and translational science include detection of specific antibodies and measurement of antigen concentrations, but many other possibilities exist. If one has a good understanding of the basic concepts of ELISA and its applications, one is likely to find ways in which the technique can be of use in many clinical or translational research settings.

REFERENCES

[1] Yalow RS, Berson SA. Immunoassay of endogenous plasma insulin in man. J Clin Invest 1960;39:1157–75.

[2] Nakane PK, Pierce GB. Enzyme-labeled antibodies: preparation and application for the localization of antigens. The journal of histochemistry and cytochemistry: official journal of the Histochemistry Society 1966:929–31.

[3] Avrameas S, Uriel J. Method of antigen and antibody labelling with enzymes and its immunodiffusion application. C R Acad Sci Hebd Seances Acad Sci D 1966;262(24):2543–5.

[4] Engvall E, Perlmann P. Enzyme-linked immunosorbent assay (ELISA). Quantitative assay of immunoglobulin G. Immunochemistry 1971;8(9):871–4.

[5] Van Weemen BK. Schuurs AHWM. Immunoassay using antigen—enzyme conjugates. FEBS Letters 1971:232–6.

[6] Lequin RM. Enzyme immunoassay (EIA)/enzyme-linked immunosorbent assay (ELISA). Clin Chem 2005;51(12):2415–8.

[7] Schreiber RD. Measurement of mouse and human interferon gamma. Curr Protoc Immunol 2001 Chapter 6:Unit 6.8.

[8] Terato K. Preventing intense false positive and negative reactions attributed to the principle of ELISA to re-investigate antibody studies in autoimmune diseases. Journal of Immunological Methods 2014;407:15–25.

[9] Hornbeck P. Enzyme-linked immunosorbent assays. Curr Protoc Immunol 2001 Chapter 2:Unit 2.1.

SUGGESTED FURTHER READING

Protocols for conjugating enzymes to antibodies:
[1] Winston SE, Fuller SA, Evelegh MJ, Hurrell JG. Conjugation of enzymes to antibodies. Curr Protoc Mol Biol 2001 Chapter 11(3):Unit11.1.
More protocols for diverse ELISA applications:
[2] Hornbeck P. Enzyme-linked immunosorbent assays. Curr Protoc Immunol 2001 Chapter 2:Unit 2.1.
General handbook on ELISA:
[3] ELISA technical guide by KPL at http://www.kpl.com/docs/techdocs/KPL%20ELISA%20 Technical%20Guide.pdf.

GLOSSARY

Immunosorbent   Using a specific antibody or antigen chemically bound to a surface to selectively remove the corresponding antigen or antibody from a solution.
Monoclonal antibody   Antibody derived from the identical clones of the same parent plasma cell.
Polyclonal antibody Antibody derived from clones from different parent plasma cells.
Fc portion   Fragment crystallizable region or the tail of an antibody that binds to Fc receptors on cell surfaces and certain complement proteins (C1q).

LIST OF ACRONYMS AND ABBREVIATIONS

ELISA	Enzyme-linked immunosorbent assay
HCG	Human chorionic gonadotropin
IgG4-RD	IgG4-related disease
T$_{FH}$ cell	Follicular helper T cell

Chapter 8

Immunofluorescence

Sonali Joshi and Dihua Yu

University of Texas MD Anderson Cancer Center, Houston, TX, United States

Chapter Outline

Objectives

- Understand the basic principles of immunofluorescence.
- Develop the knowledge base to troubleshoot and optimize an immunofluorescence protocol.
- Describe the practical applications of immunofluorescence for clinical as well as research applications.
- Understand the limitations of immunofluorescence.

INTRODUCTION

Various techniques such as Western blotting, enzyme-linked immunosorbent assay (ELISA), reverse-phase protein array (RPPA), and others have been

Basic Science Methods for Clinical Researchers. DOI: http://dx.doi.org/10.1016/B978-0-12-803077-6.00008-4

developed to examine protein expression. While these techniques are effective at determining protein expression, they do not provide any information regarding the cellular or subcellular localization of the protein being studied. Immunohistochemical techniques enable protein visualization based on antibody binding to the protein of interest followed by visualization of the antibody by either conjugating the antibody to an enzyme catalyzing a color-producing reaction or by conjugating the antibody to a fluorophore. Visualization of proteins by fluorophore-conjugated antibodies is a defining feature of immunofluorescence (IF). This technique can be used to visualize proteins in cells (both in suspension and adherent cells), tissues as well as 3D culture-derived spheroids.

Immunohistochemistry (IHC) studies are routinely used for pathological clinical diagnosis; however, when an experiment requires co-localization of proteins, IF technique is the method of choice. IF imaging depends on light emission from fluorophore-conjugated antibodies. A fluorescent chemical compound that can absorb light at a specific wavelength resulting in light emission at a longer or lower energy wavelength is referred to as a fluorophore [1]. For example, a commonly used fluorophore Alexa 488 has an excitation peak of 495 nm and emits light with an emission peak at 519 nm (green spectrum), while another fluorophore Alexa 594 has an excitation peak of 590 nm and an emission peak at 617 nm in the red spectrum. As distinct fluorophores have different excitation and emission wavelengths, multiple antigens can be visualized on the same biological sample by conjugating multiple antigen-recognizing antibodies to different fluorophores with distinct excitation and emission spectrums. Another important advantage of IF is the ability to capture images on a confocal microscope to determine the cellular localization of the protein of interest. Confocal microscopy enhances optical resolution by filtering out the light emitted from the out-of-focus planes [2]. To identify whether the antigen of interest is expressed in a particular intracellular compartment, confocal IF imaging of the antigen of interest along with a well-characterized protein known to be localized to the cellular compartment of interest is performed. Confocal imaging software also allows for image capture at multiple focal planes enabling 3D imaging of the specimen of interest. This application is particularly useful in studying the structures of various organs such as the mammary gland and the vasculature [3–5]. It is important to note that while traditional IHC staining can be preserved for a long time, IF staining is highly sensitive to photobleaching and therefore stained slides can be maintained in a −20°C freezer for a limited time.

IF was first described by Albert H Coons and colleagues who reported that antigen in mammalian tissues could be detected optically under ultraviolet (UV) light with an antibody chemically linked to the fluorophore fluorescein isocyanate [6,7]. By the mid-1960s, multiple studies described the application of IF techniques to study bacterial and viral proteins [8–10]. Recent advances have led to the development of more photostable fluorophores and currently IF is widely used for multiple applications in clinical medicine as well as in cell biology and pathology.

IN PRINCIPLE

Animals are equipped with an immune system whose primary purpose is to protect the organism from attacks by pathogenic bacteria and viruses. As a part of the immune response B-cells produce protein complexes called antibodies that can detect a target antigen and elicit an immune response [11]. An antibody contains the conserved Fc (fragment that crystalizes) domain and a variable Fab (fragment having the antigen-binding site) domain that contains the antigen-binding site (antibody structure reviewed by Edelman et al. [12]). The Fc region can be recognized by effector cells, immune proteins as well as by other antibodies. The Fc domain is conserved within a species and therefore a fluorophore-conjugated antibody against the Fc fragment of one species can be used to detect all primary antibodies generated in that species. IF protocols are classified into two groups depending on whether a single antibody or two antibodies (primary and secondary) are used for fluorophore labeling of the antigen of interest. Both methods have advantages and disadvantages, which must be considered prior to initiating an IF protocol.

DIRECT IMMUNOFLUORESCENCE

An IF protocol involving a fluorophore-conjugated antibody to the target antigen of interest is referred to as direct IF. Advantages of this method include a simpler protocol for labeling of multiple antigen as well as shorter incubation times. Additionally, staining with multiple antibodies generated in the same species does not pose a problem. On the other hand, there is no signal amplification and as a result the staining intensity may be low if the antigen of interest is expressed in low abundance. Also the experiments may be limited, based on the availability and cost of the fluorophore-conjugated antibodies. This method of immune-labeling is commonly used for flow-cytometry applications.

INDIRECT IMMUNOFLUORESCENCE

An IF protocol involving detection of an antigen by a fluorophore-conjugated secondary antibody that recognizes an unlabeled primary antibody bound to the antigen of interest is classified as indirect IF. As a consequence of multiple secondary antibodies bound to the primary antibody, indirect IF protocols results in signal amplification, which is extremely useful for detecting low-abundance targets. Commercially produced secondary antibodies are quality-controlled and are available conjugated to a multitude of fluorophores. As secondary antibodies can recognize all primary antibodies derived from the host species, indirect IF protocols are more flexible and cost-effective. However, while performing labeling of multiple antigen, the primary antibodies need to be raised in different species to prevent cross reactivity. Also samples with high expression of endogenous immunoglobulin may produce a high background interfering with fluorescence imaging.

The principles of direct and indirect IF are summarized in Fig. 8.1.

Direct Immunofluorescence   Indirect Immunofluorescence

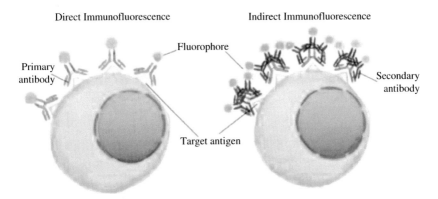

FIGURE 8.1   Direct and Indirect Immunofluorescence. Direct immunofluorescence utilizes a fluorophore-conjugated antibody to label the target antigen. An IF protocol involving detection of an antigen by a fluorophore-conjugated secondary antibody that recognizes an unlabeled primary antibody bound to the antigen of interest is classified as indirect IF.

IN PRACTICE

An IF protocol is a multistep procedure. Initially, the sample to be analyzed (cell lines, tissue samples, 3D cultures) are attached to a solid support such as a slide to facilitate visualization by a microscope. Next the sample is fixed to halt the biochemical cellular reactions to preserve cells and tissues and to enable the preparation of thin, stained sections. The samples may then be permeabilized to facilitate antibody binding to intracellular antigens. This is followed by incubating the sample with the primary antibody and multiple washes to remove the excess unbound antibody. If the primary antibody is bound to a fluorophore, the sample can then be mounted for imaging (direct IF). Alternatively, if the primary antibody is not bound to a fluorophore, the sample is incubated with a fluorophore-conjugated secondary antibody recognizing the primary antibody (indirect IF); and the excess unbound antibody is removed by multiple washes. The sample is then mounted for imaging.

It is important to note that there is no universal protocol for IF imaging. Testing and optimization is necessary to figure out the optimal conditions for every antibody, protein, and specimen combination. The sample preparation protocol may also need to be optimized for each experiment. The important steps and considerations for performing an IF experiment are discussed below.

Preparation of the Biological Sample for Immunofluorescence

IF can be performed on cell lines and primary cells grown in 2D culture, spheroids grown in 3D cultures as well as tissue samples. The main goal of this initial step is to facilitate the adherence of the specimen of interest to a solid optically suitable surface to facilitate imaging.

Primary Cells and Cell Lines Grown on a Petri Dish

Cells can be cultured on coverslip or on multiwelled chamber slides. If using coverslips, the thickness of the coverslip may have a significant impact on the image intensity and quality; it is important to note that #1.5 coverslip with a thickness of approximately 170 µm is generally compatible with most microscopes. The coverslips may be sterilized by exposure to UV light for 30 minutes prior to cell culture. For cells that do not easily attach to coverslips, the coverslip may be incubated in 50 µg/mL L-polylysine to facilitate cell attachment. Cells can also be directly cultured in multiwelled chambered slides. The advantage of using coverslips is the lower volume of antibody dilutions required for incubation with the primary or the secondary antibody, but, on the other hand, coverslips are fragile and therefore need to be handled with care during the multiple washes prior to imaging. The multiwelled chamber slides are sturdier and can accommodate multiple samples on a single slide but require a larger volume for incubation with the primary and secondary antibody. The cultured cells are then washed with 1X phosphate-buffered saline (PBS) to rinse off the cell culture media before proceeding to fixation. When using alcohol as a fixative, samples need to be air-dried prior to fixation.

Cells Grown in Suspension Culture

Cells cultured in suspension can be coated as a monolayer on a slide by the cytospin technique. About 150–200 µL of cells suspended in (1X) PBS are coated onto the microscope slide by spinning at low speeds to attach cells to the slide. The slides may be treated with L-polylysine to facilitate cell attachment. The slide is then air-dried before proceeding to fixation.

Cells Grown in a Semisolid Matrix

For IF imaging of cells grown in matrigel, the bottom of a multiwelled chamber slide is covered with a thin layer of matrigel to prevent cell attachment to the bottom of the plate. The cells are then cultured in matrigel-supplemented media to form polarized 3D structures. The liquid media is removed prior to fixation.

Tissue Samples from Humans or Animals

Prior to IF staining, tissue samples need to be cut into thin sections and attached to an imaging slide. The tissue is dissected immediately after euthanasia and is either flash-frozen or fixed prior to being embedded in paraffin. For frozen sections, the tissue is snap-frozen to avoid formation of water crystals. The freshly dissected tissue is placed on the prelabeled tissue mold and is covered with cyro-embedding media (i.e., OTC) and placed on dry ice. The samples can be stored either in liquid nitrogen or at −80°C. The blocks are then cut into approximately 6–8 µm sections and attached to a slide. For paraffin embedding, the freshly dissected tissue is fixed (typically in neutral buffered formalin overnight) and then embedded in paraffin. The paraffin blocks are then cut into 4–8 µm sections and attached to a glass slide. The section is deparaffinized by multiple xylene washes followed by hydration through graded alcohol washes. As formalin fixation and paraffin embedding may mask antigen, the samples may be subjected to antigen unmasking by heat treatment (recommended) or pepsin or saponin treatment.

Fixation of the Sample

Fixation is an important step required for optimal IF imaging. Sample fixation is required to stop the degenerative processes resulting from loss of blood supply or nutrient media; and it helps to maintain cellular architecture as close to the native state as possible. A fixative may damage antigen sites and therefore the fixation process may need to be optimized for each antibody–antigen combination. Based on the sample to be analyzed there are different options for fixatives. The advantages and disadvantages of each are listed below:

Methanol

For methanol fixation, cells are covered with a thin layer of ice cold 100% methanol and incubated at −20°C for about 15 minutes. The methanol is then aspirated and the residual methanol is removed by three washes with (1X) PBS. This is the method of choice for staining cytoskeletal proteins. Additionally, methanol fixation disturbs hydrophobic bonding and is more effective for staining with monoclonal antibodies that recognize an epitope normally buried within internal protein structures. On the other hand, methanol fixation results in reduced protein solubility and therefore this method of fixation is not useful in staining lipid-associated proteins. Reduced protein solubility also results in cell flattening making it harder to stain proteins localized to the nucleus or the mitochondria.

Acetone

For acetone fixation, the sample is covered with a thin layer of ice-cold acetone and incubated at −20°C for 3–20 minutes. Acetone fixation is more effective at maintaining antigen integrity as compared to methanol and may be used as fixative if methanol fixation is ineffective. This is commonly used as a fixative for staining cytoskeletal proteins. Acetone fixation is very effective at cell permeabilization but has the same limitations as methanol fixation. Acetone fixation is generally the method of choice for frozen sections. Note: If neither methanol nor acetone is effective, the sample may be fixed with a 1:1 acetone/methanol solution by incubating the sample for about 10 minutes at −20°C. If using alcohols as fixatives, the permeabilization step may be skipped.

Paraformaldehyde

Paraformaldehyde (polymerized formaldehyde) fixation is commonly used for staining membrane-associated proteins. Paraformaldehyde fixation results in the chemical crosslinking of free amino groups, establishing a vast network of interactions that better preserves cellular architecture. The sample is fixed in a freshly prepared 2–4% paraformaldehyde (in 1X PBS) solution for 15 minutes at room temperature. This is followed by three washes for 5 minutes each with (1X) PBS containing 100 nm glycine. Note: Paraformaldehyde is generally the fixative of choice while staining cells in 3D culture.

Formalin

Formalin is saturated 37% formaldehyde solution dissolved in water. It is important to note that a 10% formalin solution is equivalent to a 4% paraformaldehyde solution. The sample is fixed in 1% formalin in (1X) PBS for 10 minutes at room temperature followed by three washes for 5 minutes each with (1X) PBS.

Note: Fixation with paraformaldehyde or formalin may cause auto-fluorescence-mediated artifacts and therefore it is important to have a control sample that is not incubated with the primary antibody to determine nonspecific background signal while testing each antibody–antigen combination. For indirect IF, the control sample should be incubated with the secondary antibody only. Aldehyde-based fixatives do not effectively permeabilize the cell membrane and the samples need to be permeabilized prior to staining for intracellular biological molecules.

Cell Membrane Permeabilization

Permeabilization disturbs the cell membrane allowing the antibodies to bind to the intracellular antigen in the fixed cells. Fixatives such as acetone can fix as well as permeabilize cells while other fixatives are not very effective at permeabilization. IF staining for markers on the cell surface, permeabilization, will damage the cell membrane and is therefore not recommended. Detergents such as sodium dodecyl sulfate (SDS), Triton X-100, Tween-20, and saponin are commonly used for permeabilization. Different reagents, concentrations, incubation times may give different results and therefore the protocol must be optimized such that fixation and permeabilization results in minimal distortion of cellular morphology. Permeabilization with 1% triton X-100 for 1–5 minutes at room temperature followed by (1X) PBS washes is a commonly used methodology to permeabilize the sample. Note: The permeabilization step can be skipped if the protein to be stained is present on the cell membrane or if acetone is used for fixation. To achieve a specific signal it is recommended to use the mildest detergent that will allow antibody penetration. In order of increasing permeabilizing efficiency the commonly used detergents are saponin, Tween-20, triton X-100, and SDS.

Blocking to Limit Nonspecific Antibody Interactions

IF staining is enhanced by blocking nonspecific interaction of the primary and secondary antibodies with the biological sample. Nonspecific binding may result from inappropriate binding of the antibody to nonantigen molecules by excess unreacted aldehyde, trapping of the antibody in hydrophobic structures or by low-affinity polyclonal antibody binding to nonspecific molecules. Incubating the sample in a protein solution prior to incubation with the primary antibody prevents these nonspecific interactions. The sample is incubated with blocking agents such as bovine serum albumin, milk, and serum. Note: Serum is generally the reagent of choice for IF staining. It is important to note that the blocking serum should be obtained from a species distinct from the species in which the primary antibody was raised. If performing indirect IF, the blocking serum should belong to the species in which the secondary antibody was developed. As a lot of the widely used secondary antibodies are raised in goat, goat serum is a common choice for the blocking step. For blocking, incubate the sample in a 5% serum solution (in (1X) PBS with 0.05% Tween-20 or 0.05% Triton X-100) for 30 minutes to an hour. Commercially available blocking buffers containing highly purified single proteins or proprietary protein-free compounds may also be used for blocking.

Incubation with Primary Antibody

After blocking, the sample is then ready for incubation with the primary antibody. A good primary antibody with a high affinity for the antigen and good specificity is essential to obtain high-quality trustworthy images and therefore it is worth testing multiple antibodies against the same antigen. The primary antibody is diluted in the blocking buffer and is then spread over the sample. The antibody is diluted 1:100–1:10000 (or outside this range) depending on antigen abundance, concentration of the antibody, and the affinity of the antibody to the antigen. The antibody dilution needs to be optimized for obtaining good-quality images. When staining for multiple proteins, the primary antibodies can be combined provided they are derived from different species and then added to the sample. The sample can be incubated with the primary antibody for an hour at room temperature or overnight at 4°C or a combination of both. The unbound primary antibody is removed by multiple washes with (1X) PBS containing either 0.05% Tween-20 or 0.05% Triton X-100. Care must be taken to see that the sample does not dry up between all the steps. If performing direct IF, proceed to mounting and imaging. Note: In order to avoid nonspecific binding, the primary antibodies need to be derived from a species distinct from the species being studied.

Incubation with Secondary Antibody

For this step, the fluorophore-conjugated secondary antibody is diluted (1:200–1:500) in the blocking buffer and added to the sample. The sample can be incubated with the secondary antibody for an hour at room temperature or overnight at 4°C. In our experience, Alexa Fluor secondary antibody conjugates are bright and stable and exhibit good specificity. Additionally, secondary fluorophore-conjugated antibodies from Jackson ImmunoResearch and Rockland are also commonly used. Following incubation with the secondary antibody, the sample is washed three times with (1X) PBS containing either 0.05% Tween-20 or 0.05% Triton X-100 to remove the excess unbound secondary antibody. Care must be taken to ensure that the sample does not dry out between each step. Note: When staining with multiple primary antibodies, the corresponding secondary antibodies need to be conjugated to distinct fluorophores that do not overlap in their excitation and emission spectrum.

Mounting

After incubation with secondary antibody, the sample is dipped in distilled or deionized water to remove excess salts that may produce a residue after drying. After removing the excess water by capillary action with a Kim wipe, a small amount of mounting media just sufficient to cover the sample is added prior to covering the sample with a coverslip. The sample is then allowed to dry on a flat surface and the edges are sealed with nail polish. The mounting medium preserves the sample and increases the refractive index for obtaining high-quality images with an oil immersion lens. Additionally, commercially available mounting media such as Prolong Gold from Molecular Probes and Fluoromount-G(anti-fade) from Southern Biotech minimize photobleaching due to the presence of free radical scavengers. Mounting media also contains DAPI (4′,6-diamidino-2-phenylindole), a fluorophore that can interact with DNA emitting light in the blue spectrum and is commonly used to visualize nuclei. After mounting, the sample is ready for imaging. Note: It is important to note that after mounting the samples should be stored at $-20°$ until imaging to preserve the fluorescence signal.

Imaging

After IF staining the antigen of interest can be visualized by a fluorescence microscope or confocal microscope depending on the research question being addressed. A regular fluorescence microscope is generally used to examine the expression but not the co-localization of different antigens. A study that aims to examine whether two or more proteins co-localize, a confocal microscope is the instrument of choice. Confocal microscopes utilize imaging techniques that capture images with high optical resolution in the confocal plane by eliminating out-of-focus light [2]. Images captured at multiple confocal planes can be integrated to produce a three-dimensional image.

Controls and Important Considerations

The quality of IF staining may be hindered by nonspecific binding of the primary or the secondary antibody resulting in a nonspecific signal. Use of appropriate controls is essential to validate specific binding of the primary or the secondary antibody. To test the specificity of the primary antibody, IF staining should be performed either by blocking the primary antibody with the peptide/protein against which it was raised or by verifying a lack of signal in a sample with a targeted deletion or RNA interference-mediated silencing of the target of interest. To resolve issues with nonspecific staining with the secondary antibody, lack of signal needs to be confirmed in a secondary-only sample that is incubated only with the secondary antibody and is not incubated with the primary antibody. Problems with antibody specificity may be resolved by optimizing the fixation and blocking process or using different specific high-affinity antibodies against the target of interest.

Selections of fluorophore used for IF imaging is critical to obtaining good-quality IF images. An ideal fluorophore exhibits bright fluorescence, has high photostability, and its fluorescence is not altered by external factors such as antibody conjugation or changes in pH. Additionally, the fluorophore must have an excitation peak at an excitation wavelength available on the fluorescence microscope and a narrow emission spectrum that can be specifically detected by the fluorescence microscope.

APPLICATIONS

IF techniques are commonly used for both clinical and experimental pathology. For research purposes IF is used to determine organ-specific, tissue-specific, or cell-specific gene expression of both intracellular and cell membrane-associated antigen. Besides eukaryotic cells, IF techniques can also be used to identify the presence of bacteria, viruses, or other parasites in a given sample [8–10]. IF techniques are commonly used for the diagnosis of dermatological abnormalities such as bullous and connective tissue disorders, vasculitides, and conditions such as lichen planus, and also the scaling dermatoses, notably psoriasis [13]. Diagnosis of respiratory diseases resulting from viral infection can be diagnosed with direct IF-based assays [14]. Thus IF is a routinely used technique for both clinical and research purposes.

SCENARIO

A protein X is hypothesized to translocate from the nucleus to the cytoplasm in response to mitogenic stimulation. To test this hypothesis, cells are either treated with a vehicle control or with a mitogen. The cells are then fixed, permeabilized, blocked, and incubated with a primary antibody against protein X followed by incubation with a fluorophore-conjugated secondary antibody. The slide is then mounted with mounting media containing DAPI. Confocal imaging is then performed to determine whether DAPI and the fluorescent signal from protein X are colocalized. The hypothesis is supported if DAPI and the protein X are colocalized in the sample treated with the mitogenic stimulus.

KEY LIMITATIONS

As with all biochemical techniques, IF results are hindered by multiple limitations. The major issues are listed below:

Photobleaching

IF imaging requires excitation of the fluorophore at a specific wavelength resulting in light emission that can be detected by the fluorescence microscope. This process also results in the generation of reactive oxygen species (ROS) that can chemically interact with the fluorophore preventing optimal excitation of the fluorophore over time. Photobleaching may be prevented by optimizing the exposure time to the excitation wavelength or using ROS scavengers or using fluorophores specifically developed to minimize photobleaching.

Autofluorescence

The ability to obtain high-quality IF images is highly dependent on obtaining high-affinity antibodies highly specific to the target of interest. Biological samples may exhibit autofluorescence due to the presence of reduced pyridine nucleotides (NADH (nicotinamide adenine dinucleotide (reduced)): absorption 340 nm, emission 460 nm) and flavin coenzymes (FAD (flavin adenine dinucleotide) and FMN (flavin mononucleotide): absorption 450 nm, emission: 460 nm) that play an important role in regulating cellular metabolism. Therefore, detection of fluorophores that emit light in the green spectrum may result in a low signal-to-noise ratio. Often issues with autofluorescence can be eliminated by either using a different more specific high-affinity antibody or by optimizing the fixation process. Fixation methods using aldehydes such as glutaraldehyde may result in high autofluorescence that can be attenuated by washing with 0.1% sodium borohydride (a versatile reducing agent) in PBS prior to antibody incubation. Selection of appropriate probes and optical filters can minimize problems with autofluorescence.

Alternatively, for proteins that cannot be easily detected by IF, a fluorescent protein conjugated version of the protein such as GFP (green fluorescent protein) may be generated to address research questions.

Fluorophore Overlap

Fluorophore overlap is a common problem encountered while performing imaging for multiple targets in the same sample. This becomes a problem when the two or more fluorophores used emit light at similar wavelengths. To obtain IF images, fluorophore excitation by a light source results in light emission from the fluorophore at a narrow range of wavelengths. For example, Alexa fluor 430 has an excitation and emission of 434 nm and 539 nm, respectively, while another fluorophore Alexa fluor 514 has an excitation and emission of 518 nm and 540 nm, respectively. If the two fluorophores are used to stain different antigen, the fluorescence microscope will not be able to differentiate between the light emitted from the two fluorophores due to a significant overlap between their emission spectrum. Alternatively, if Alexa fluor 434 is used in combination with another fluorophore such as Alexa fluor 594 (excitation peak: 590 and emission peak: 617 nm), the emission spectra of the two fluorophores will not overlap and therefore the proteins stained with the two dyes can be optimally distinguished.

TROUBLESHOOTING

Problem	Probable Cause	Solution
No staining	Problems with secondary antibody binding to primary	Verify that the secondary antibody was raised against the species in which the primary antibody was raised
	Not enough primary antibody is bound to the antigen	Increase the concentration of the primary antibody and/or increase incubation time
	Protein is absent or expressed at low levels	Use a positive control
	Improper storage of the primary antibody	Use positive controls
	Improper storage of secondary antibody	Use a new batch of secondary antibody and limit light exposure
	Problems with sample fixation	Fixation protocol may modify antibody epitope, use alternative fixatives
	Bacterial contamination of PBS	Use freshly prepared PBS
	Problems with membrane permealization	Use a more stringent detergent for permealization
High background	Ineffective blocking	Increase blocking time or use a different blocking agent
	Primary antibody concentration is too high	Use less primary antibody for incubation
	Nonspecific secondary antibody binding	Run a control incubated only with the secondary antibody, if secondary-only control is positive use a different secondary antibody
	Improper washing, fixative not properly removed	Increase PBS washes between each step

Problem	Probable Cause	Solution
	Fixative causes auto fluorescence	Formalin and paraformaldehyde may cause auto fluorescence in the green spectrum, use a different fluorophore
	Secondary antibody concentration is too high	Dilute the secondary antibody
	Excessive membrane permealization	Use a less stringent detergent for permeabilization
Nonspecific staining	The primary antibody was raised in the same species from which the biological sample of interest is derived	Use a different primary antibody that is raised in a species distinct from which the sample being studies was obtained
	Concentration of the primary and/or secondary antibody is too high	Decrease the concentration of the primary and or secondary antibody, may also decrease the incubation time
	The sample has dried out	Keep samples covered in liquid and prevent drying

CONCLUSIONS

In summary IF techniques involve the optical detection of the antigen(s) of interest with fluorophore-conjugated antibodies. This technique is widely used for the determination of protein expression and cellular localization of antigens. In the current age of high-throughput research, technological advances have led to the development of performing IF in a multiplex manner. The newly developed tyramide signal amplification system provides improved resolution and stability for multiplex studies [15,16]. Additionally, it is important to note that in addition to antibodies, nucleotide probes can also be conjugated to fluorescent probes to visualize DNA or RNA molecules and this technique is referred to as fluorescent in situ hybridization [17]. Thus, IF techniques will be used for research and clinical purposes for a long time in the future.

REFERENCES

[1] Liu JJ, Liu C, He W. Fluorophores and Their Applications as Molecular Probes in Living Cells. Curr Org Chem 2013;17(6):564–79.

[2] Kogata N, Howard BA. A whole-mount immunofluorescence protocol for three-dimensional imaging of the embryonic mammary primordium. J Mammary Gland Biol Neoplasia 2013;18(2):227–31.

[3] Johnson MD, Mueller SC. Three dimensional multiphoton imaging of fresh and whole mount developing mouse mammary glands. BMC cancer 2013;13:373.

[4] Saatchi S, Azuma J, Wanchoo N, Smith SJ, Yock PG, Taylor CA, et al. Three-Dimensional Microstructural Changes in Murine Abdominal Aortic Aneurysms Quantified Using Immunofluorescent Array Tomography. J Histochem Cytochem 2012;60(2):97–109.

[5] Coons AH, Creech HJ, Jones RN, Berliner E. The demonstration of pneumococcal antigen in tissues by the use of fluorescent antibody. J Immunol 1942;45(3):159–70.

[6] Coons AH, Creech HJ, Jones RN. Immunological properties of an antibody containing a fluorescent group. P Soc Exp Biol Med 1941;47(2):200–2.

[7] Cherrywbmoody MD. Fluorescent-Antibody Techniques in Diagnostic Bacteriology. Bacteriol Rev 1965;29:222–50.

[8] Hers JF. Fluorescent Antibody Technique in Respiratory Viral Diseases. Am Rev Respir Dis 1963;88(SUPPL):316–38.

[9] Kraft SC, Kirsher JB. Immunofluorescent Studies of Chronic Nonspecific Ulcerative Colitis. Gastroenterology 1964;46:329–32.

[10] Edelman GM, Gall WE. The antibody problem. Annu Rev Biochem 1969;38:415–66.

[11] Chhabra S, Minz RW, Saikia B. Immunofluorescence in dermatology. Indian J Dermatol Venereol Leprol 2012;78(6):677–91.

[12] Sadeghi CD, Aebi C, Gorgievski-Hrisoho M, Muhlemann K, Barbani MT. Twelve years' detection of respiratory viruses by immunofluorescence in hospitalised children: impact of the introduction of a new respiratory picornavirus assay. BMC Infect Dis 2011;11:41.

[13] Bobrow MN, Harris TD, Shaughnessy KJ, Litt GJ. Catalyzed reporter deposition, a novel method of signal amplification. Application to immunoassays. J Immunol Methods 1989;125(1–2):279–85.

[14] Mitchell RT, E Camacho-Moll M, Macdonald J, et al. Intratubular germ cell neoplasia of the human testis: heterogeneous protein expression and relation to invasive potential. Modern pathology : an official journal of the United States and Canadian Academy of Pathology, Inc 2014;27(9):1255–66.

[15] Hu L, Ru K, Zhang L, Huang Y, Zhu X, Liu H, et al. Fluorescence in situ hybridization (FISH): an increasingly demanded tool for biomarker research and personalized medicine. Biomarker research 2014;2(1):3.

[16] Toda Y, et al. Application of tyramide signal amplification system to immunohistochemistry: a potent method to localize antigens that are not detectable by ordinary method. Pathol Int 1999;49(5):479–83.

[17] Volpi EV, Bridger JM. FISH glossary: an overview of the fluorescence in situ hybridization technique. BioTechniques 2008;45(4):385–6. 388, 390.

SUGGESTED FURTHER READING

[1] Brelje TC, Wessendorf MW, Sorenson RL. Multicolor laser scanning confocal immunofluorescence microscopy: practical application and limitations. Methods Cell Biol 2002;70:165–244.

[2] Mysorekar VV, Sumathy TK, Shyam Prasad AL. Role of direct immunofluorescence in dermatological disorders. Indian Dermatol Online J 2015;6(3):172–80.

[3] Shakes DC, Miller III DM, Nonet ML. Immunofluorescence microscopy. Methods Cell Biol 2012;107:35–66.

[4] Willingham MC, Pastan IH. An atlas of immunofluorescence in cultured cells. Orlando: Academic Press; 1985.

[5] Caul EO, Great Britain. Public Health Laboratory Service Immunofluorescence: antigen detection techniques in diagnostic microbiology. London: Public Health Laboratory Service; 1992.

[6] Al-Mulla F. Formalin-fixed paraffin-embedded tissues: methods and protocols. New York: Humana Press; 2011.

[7] Lin F, Prichard J. Handbook of practical immunohistochemistry: frequently asked questions. New York: Springer; 2011.

[8] Beutner EH, State University of New York at Buffalo. Department of Microbiology Defined immunofluorescence and related cytochemical methods. New York, N.Y.: New York Academy of Sciences; 1983.

[9] Scheffler JM, Schiefermeier N, Huber LA. Mild fixation and permeabilization protocol for preserving structures of endosomes, focal adhesions, and actin filaments during immunofluorescence analysis. Methods Enzymol 2014;535:93–102.

[10] Stadler C, Skogs M, Brismar H, Uhlen M, Lundberg E. A single fixation protocol for proteome-wide immunofluorescence localization studies. J Proteomics 2010;73(6):1067–78.

[11] Stadler C, Rexhepaj E, Singan VR, et al. Immunofluorescence and fluorescent-protein tagging show high correlation for protein localization in mammalian cells. Nat Methods 2013;10(4):315–23.

[12] Wick G, Traill KN, Schauenstein K. Immunofluorescence technology : selected theoretical and clinical aspects. Amsterdam; New York New York, N.Y.: Elsevier Biomedical Press ;Sole distributors for the USA and Canada, Elsevier Science Pub. Co.; 1982.

[13] Barbierato M, Argentini C, Skaper SD. Indirect immunofluorescence staining of cultured neural cells. Methods Mol Biol 2012;846:235–46.

[14] Chhabra S, Minz RW, Saikia B. Immunofluorescence in dermatology. Indian J Dermatol Venereol Leprol 2012;78(6):677–91.

[15] Knapp W, Holubar K, Wick G. Immunofluorescence and related staining techniques: proceedings of the VIth International Conference on Immunofluorescence and Related Staining Techniques held in Vienna, Austria on April 6–8, 1978. Amsterdam; New York New York: Elsevier/North-Holland Biomedical Press; Elsevier North-Holland; 1978.

[16] Kawamura A, Aoyama Y. Immunofluorescence in Medical Science. University of Tokyo Press; 1983.

[17] Pawley J. Handbook of Biological Confocal Microscopy. US: Springer; 2010.

[18] Muller M. Introduction to Confocal Fluorescence Microscopy. Society of Photo Optical; 2006.

[19] Klein RA, Storch WB. Immunofluorescence in Clinical Immunology: A Primer and Atlas. Birkhäuser Basel; 2012.

[20] Allan V. Protein Localization by Fluorescence Microscopy: A Practical Approach:. Oxford: OUP; 1999.

[21] WR Sanborn, Immunofluorescence, an Annotated Bibliography 4. Technical Procedures, Defense Technical Information Center, 1968.

[22] International IG Immunofluorescence: Webster's Timeline History. ICON Group International; 1960–2007.

[23] Russell J, Cohn R. Immunofluorescence: Book on Demand; 2012.

[24] Burry RW. Immunocytochemistry: A Practical Guide for Biomedical Research. Springer; 2009.

[25] Herman B. Fluorescence Microscopy. Bios Scientific Publishers; 1998.

[26] Gardner PS, McQuillin J, Grandien M. Rapid virus diagnosis: application of immunofluorescence, 2d ed. London; Boston: Butterworths; 1980.

[27] Lyerla HC, Forrester FT, Branch CfDCVT Immunofluorescence methods in virology. Dept. of Health, Education, and Welfare, Public Health Service, Center for Disease Control, Bureau of Laboratories, Laboratory Training and Consultation Division, Virology Training Branch; 1979.

GLOSSARY

Immunofluorescence (IF)   IF is an immunohistochemistry technique used to visualize proteins with fluorophore conjugated antibodies.

Fluorophore   A fluorophore is a chemical compound when excited with light at a specific wavelength, emits light as a higher wavelength.

Photobleaching   Photobleaching is the chemical modification of a fluorophore that permanently inhibits light emission upon fluorophore excitation.

Phosphate-buffered saline (PBS)   A water-based buffered solution containing sodium chloride, sodium phosphate, potassium chloride, and potassium phosphate (137 mM NaCl, 2.7 mM KCl, 10 mM Na_2HPO_4, 2 mM KH_2PO_4).

Spheroids   A sphere-shaped collection of cells grown in a semi-solid 3D matrix.

LIST OF ACRONYMS AND ABBREVIATIONS

DAPI   4′,6-diamidino-2-phenylindole
IF   Immunofluorescence
PBS   Phosphate-buffered saline

Chapter 9

Cell Culture: Growing Cells as Model Systems In Vitro

Charis-P. Segeritz[1] and Ludovic Vallier[1,2]
[1]University of Cambridge, Cambridge, United Kingdom; [2]Wellcome Trust Sanger Institute, Hinxton, United Kingdom

Chapter Outline

Basic Science Methods for Clinical Researchers. DOI: http://dx.doi.org/10.1016/B978-0-12-803077-6.00009-6

Objectives

The goal of this chapter is to familiarize the researcher with the setup of a cell culture lab and the techniques required to successfully propagate cells in vitro.
- How to ensure safety of the researcher in a cell culture laboratory setting.
- How to carry out aseptic techniques that keep cell cultures free of contaminations.
- How to detect and treat cultured cells infected by bacteria, fungi, and viruses.
- How to choose a cell line and culture conditions for different research questions.
- How to acquire the skills and techniques to maintain cells in culture.
- How cell culture is utilized in various research applications.

INTRODUCTION

Cell culture refers to laboratory methods that enable the growth of eukaryotic or prokaryotic cells in physiological conditions. Its origin can be found in the early 20th century when it was introduced to study tissue growth and maturation, virus biology and vaccine development, the role of genes in disease and health, and the use of large-scale hybrid cell lines to generate biopharmaceuticals. The experimental applications of cultured cells are as diverse as the cell types that can be grown in vitro. In a clinical context, however, cell culture is most commonly linked to creating model systems that study basic cell biology, replicate disease mechanisms, or investigate the toxicity of novel drug compounds. One of the advantages of using cell culture for these applications is the feasibility to manipulate genes and molecular pathways. Furthermore, the homogeneity of clonal cell populations or specific cell types and well-defined culture systems removes interfering genetic or environmental variables, and therefore allows for data generation of high reproducibility and consistency that cannot be warranted when studying whole organ systems.

IN PRINCIPLE

The Cell Culture Laboratory

Cell Culture Laboratory Safety

The exciting application of cell culture techniques in biomedical research requires the management of potential hazards linked to infectious agents harbored by cultured cells (e.g., HBV or HIV), but also the control of reagents that can be of toxic, corrosive, or mutagenic nature. These potential hazards can endanger the health of laboratory workers when introduced into the body (e.g., via contact of skin and mucous membranes with solids, liquids, or aerosols) and threaten the environment when handled improperly (Fig. 9.1).

Before commencing any cell culture work, the reduced or eliminated exposure to potentially hazardous agents therefore needs to be ensured to minimize infection, pathogenicity, allergic reactions, and contact with released toxins. This can be achieved by stringent training of lab personnel and implementation of standard cell culture practices (Table 9.1), which should be reviewed

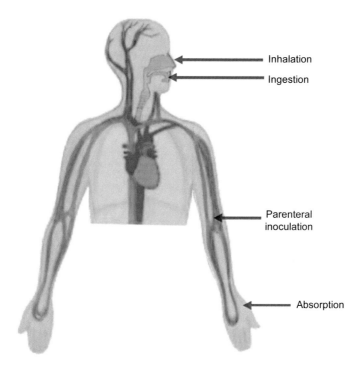

FIGURE 9.1   Routes of exposure to biohazards. Biohazards in the lab can enter the body through contaminated needles (parenteral inoculation), the consumption of food or application of make-up in the lab (ingestion), the exposure of biohazardous aerosols (inhalation), and contact of skin and mucous membrane with contaminations. Personal protective equipment and biosafety cabinets are put in place to block exposure of researchers to biohazardous agents.

and revised regularly by laboratory members and the institute's safety committee. Additionally, when working with primary cells isolated directly from human tissue, it is important to screen donors from which cells were derived for disease-causing pathogens. Up-to-date immunizations against infectious diseases such as Hepatitis B are also highly recommended for laboratory staff working with primary cells.

Safe Handling of Cell Lines

The Advisory Committee on Dangerous Pathogens (ACDP) is a national body managed by the Health and Safety Executive (HSE). It advises on hazards and risks to workers and others from exposure to pathogens and has published these recommendations [1]. Since some cell types are pathogenic or carry disease-causing agents, it is important to first determine their Hazard Group and implement appropriate safety measures. This includes a written risk assessment and review of the laboratory facilities. Microorganisms classified as Hazard Group 1

TABLE 9.1 Guidelines for Cellular Lab Safety

- Every laboratory worker is responsible for their own health and safety and that of others who may be affected by work carried out in the cell culture lab.

- Personal protective equipment must be worn upon entering the cell culture lab and removed when leaving or contaminating any personal protective equipment. Upon handling hazardous agents, potentially contaminated gloves must be removed immediately and disposed of in the biohazard waste. Wash hands.

- The exposure of skin in open-toed shoes, short trousers, and skirts is not recommended.

- The consumption of food, drink, the storage of groceries, smoking, applying cosmetics, or handling contact lenses is not permitted in the cell culture lab.

- Cell phones should not be used while working in the cell culture lab.

- Loose items of clothing (e.g., scarves, dangling necklaces) should be removed before commencing work and hair should be tied back.

- The cell culture lab must be kept tidy and cleaned routinely with a disinfectant (e.g., incubators, laminar flow hoods, and work surfaces).

- All laboratory tools in contact with potentially infectious or hazardous agents must be decontaminated before and after working with them. Potentially infectious and hazardous material must be decontaminated and disposed of via their recommended route.

- Sharp items (e.g., pipette tips) must be disposed of immediately via designated sharp's boxes.

- Hands should be washed before leaving the cell culture lab.

- The laboratory's safety officer should be notified upon exposure or spillage of infectious or hazardous agents to advise on a suitable strategy for containment and decontamination.

(e.g., *Escherichia coli* K-12) or 2 (e.g., *Staphylococcus aureus*) represent a low or moderate health risk to laboratory workers and the community, and relies on effective prophylaxis or treatment options. Cell culture work with biological agents of Hazard Group 3 (e.g., severe acute respiratory syndrome-associated coronavirus (SARS-CoV)), and 4 (e.g., Ebola viruses) involves biological agents that carry high health risks and may lack treatment options upon infection. Thus, laboratory spaces need to provide containment levels corresponding to the Hazard Group of the cultured cell types. These are referred to as biosafety levels (BSL) and carry the corresponding numbers (BSL1–4). As such, laboratories designated as BSL1 will follow standard microbiological practices, while BSL2 laboratories will need to be restricted to trained personnel, who are taught to take extreme precaution handling sharp items and to limit infectious aerosols by utilizing physical containment equipment as well as Class II biosafety cabinets.

Safe Experimental Procedures in the Cell Culture Laboratory

In order to ensure a safe working environment with cell lines and biohazardous agents, personal protective equipment (PPE) must be worn in the cell culture lab. Lab coats, gloves, and goggles create a barrier between the laboratory worker and potentially hazardous sources. Furthermore, biosafety cabinets rely on a steady, unidirectional flow of HEPA-filtered air and create an enclosed, ventilated workspace. This minimizes the exposure of researchers and the environment to hazardous material associated with the cultured cells, while simultaneously protecting the cell cultures from contaminations. While handling cell culture media and carrying out experiments in the cell culture lab, it is also recommended to review the Material Safety Data Sheet (MSDS) associated with laboratory reagents. It details the chemical and physical properties of the product, outlines suitable storage and disposal routes, informs about potential health hazards and toxicity, and advises on PPE that should be in place when handling this product.

Equipment for the cell culture laboratory

Despite the various techniques and assays carried out in different cell culture labs, the common theme of cell culture work is asepsis—the creation of a microenvironment free of unwanted pathogenic microorganisms, including bacteria, viruses, fungi, and parasites. Since asepsis is a crucial component of successful cell culture work, a separate room or designated area should be dedicated to this work and not be utilized for other purposes. Several pieces of equipment can aid in achieving such a sterile workspace and generally lead to higher efficiency, accuracy, and consistency of the cell culture performance (Table 9.2).

Aseptic Cell Culture Practices

While the previous section has explored methods aimed at decreasing the exposure of hazardous substances to the laboratory worker, this section will address the practices that should be put in place by laboratory workers to protect the cultured cells. Indeed, microbiological infections represent the main problem for the maintenance of cells in vitro. Infectious agents such as bacteria are toxic for eukaryotic cells and ultimately lead to cell death. Furthermore, even low levels of contamination can result in abnormal results and lead to wrong scientific interpretations. By adhering to several techniques that ensure asepsis in the cell culture lab, researchers can reduce the frequency and extent of contaminations and diminish loss of cells, resources, and time. This can be achieved by eliminating the entry of microorganisms into the cell culture through contaminated equipment, media, cell culture components, incubators, work surfaces, and defect or opened cell culture vessels.

TABLE 9.2   Recommended Equipment for the Cell Culture Laboratory

Equipment	Purpose
Biosafety cabinet	– To create sterile work surface; class II and III recommended
Humid CO_2 incubator	– To provide a physiological environment for cellular growth
Inverted light microscope	– To assess cell morphology and count cells
Fridge, freezers (−20°C, −80°C), liquid nitrogen storage	– To store cells, cell material, and culture components
Centrifuge	– To condense cells
pH meter	– To determine the correct pH of media components
Pipettes and pipettors	– To aliquot different volumes
Cell media and supplementary components	– To culture cells in desirable components
Hemacytometer	– To count cells, determine growth kinetics and prepare suitable plating densities
Autoclave	– To sterilize pipettes and other equipment in contact with cells
Vacuum pump	– To aspirate cell culture medium
Water bath (with adjustable temperature)	– To warm up cell culture media
Cell culture dishes	– To culture cells in different formats (e.g., flasks, Petri dishes, 96-well plates)
Containers for waste (biohazardous)	– To correctly dispose of waste

Creating an Aseptic Work Environment

Given that atmospheric air is laden with microparticles of potentially infectious nature, the biosafety cabinet is the most crucial piece of equipment to restrict nonsterile aerosols and airborne components from contaminating cultured cells. The biosafety cabinet should be located in a laboratory space that does not interrupt its airflow through external sources of wind (e.g., drafts from windows or doors). Most biosafety cabinets require a warm-up time after which the work surface should be decontaminated with an antifungal detergent (e.g., 5% Trigene) followed by 70% ethanol. All equipment entering the biosafety cabinet also needs to be sprayed and wiped with 70% ethanol. The number of items used in the biosafety cabinet, however, should be kept at a minimum to avoid any obstruction of airflow. The biosafety cabinet should only be turned off after its daily use has been completed and the ultraviolet lamp may be turned on to sterilize the exposed surface areas over night. Regular maintenance also includes cleaning of the area under the work surface onto which media may spill through the grill. Furthermore, routine servicing through biosafety cabinet engineers can ensure correct airflow and full filter capacity of this important piece of cell culture equipment.

It is critical to keep all other surfaces in contact with the cell culture vessels or media components clean. This includes the incubator, centrifuge, microscope, water bath, fridge, and freezer. Stainless steel incubators allow for easy cleaning and protect the surfaces from corrosion of the humid environment. Treatment solutions can be added to water baths to prevent the growth of microbes. On a larger scale, the equipment stored in the cell culture space should be kept free from dust and regular cleaning of cell culture floors is advisable.

Laboratory staff can contribute to a clean work surface by washing hands with soap before and after working with cell cultures. Disposable gloves sprayed with 70% ethanol and lab coats can further reduce the introduction of contaminants carried by hair, skin cells, or dust. However, gloves need to be removed when leaving the cell culture space and lab coats should also be worn only within the confinement of the cell culture laboratory. Furthermore, lab coats need to be washed at hot temperatures on a regular basis.

Using Aseptic Reagents and Media for Cell Culture

The main sources of contamination are laboratory staff, the environment, and the culture medium. Commercially sourced media and supplementary cell culture products are generally supplied in sterile condition. In addition, filter-sterilizing allows for the generation of cell culture media that are based on nonsterile culture reagents, while autoclaving is conventionally used to sterilize equipment in contact with cultured cells. The filter-sterilization of liquids can be achieved by forcing the liquid through a $0.22\,\mu M$ polyethersulfone low-binding filter system using a vacuum pump. The addition of antibiotics (e.g., Penicillin/Streptomycin) further limits the risk of bacterial growth in media bottles after opening and in cell culture vessels. However, some laboratories refrain from using antibiotics routinely since it can facilitate the emergence of resistant bacteria strains, allow for low-level background contaminations, and may lead to interference with cell metabolisms and experimental outcomes.

Contaminations

Since contaminations can generally not be avoided altogether, it is important to train cell culture laboratory staff to recognize early signs in order to prevent the spread of contaminants to other cells or cell culture products. Contaminants are most commonly of biological nature and can include bacteria, fungi, viruses, and parasites (Fig. 9.2). It is important to limit biological contaminants since

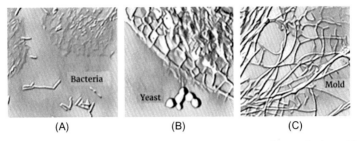

(A)   (B)   (C)

FIGURE 9.2   Microbial contaminants in cell culture. (A) Depending on the bacterial strain introduced, the morphology of bacterial contamination under the microscope can vary from rodlike shapes, cocci, flagellated to barely visible. (B) Yeast forms multicellular stringlike structures that appear ovoid in shape. (C) Mold growth is marked by the production of multicellular, highly connected, thin filaments (hyphae).

they can alter the phenotype and genotype of the cultured cell line through competition for nutrients, synthesis of alkaline, acidic or toxic by-products, and the potential interference of viral components with the cell culture genome. Other contaminants may include the introduction of undesired chemicals impurities (e.g., plasticizers in cell culture vessels) or other cell types cocultured in the lab.

Bacterial Contamination

The bacteria kingdom includes highly ubiquitous, prokaryotic microorganisms characterized by the size of a few micrometers in diameter, wide diversity in their morphologies, and fast doubling times through asexual reproduction. While the latter property allows for ready detection in cell culture supernatants shortly after infection, it also facilitates quick spread. Cell cultures affected by bacterial contamination generally appear turbid in appearance. Furthermore, the high metabolic rates of bacteria can modify the pH of the culture media and thus change the color of phenol red to yellow. While bacteria may be detected as small particles at low microscope magnification, their distinct shapes are generally detected at higher magnification. While bacterial strains such as *E. coli* can therefore be uncovered quite easily due to their size (~2 µM) and flagella-induced mobility, other strains such as *Mycoplasma* are smaller in size (<1 µM), immobile, and therefore not as easily detectable. As a result, *Mycoplasma* infections can go unnoticed for a longer time and usually only become apparent through declining quality of the cultured cells. This can manifest as reduced cell proliferation and cell death. In order to monitor cell cultures for potential infections with *Mycoplasma*, it is advisable to routinely test cultures for their presence using polymerase chain reaction (PCR), enzyme-linked immunosorbent assay (ELISA), or immunostaining [2].

Fungal Contamination

Yeasts are unicellular eukaryotes that form multicellular string-like structures during asexual reproduction. These budding cells appear ovoid in shape, can grow to approximately $4\,\mu M$ of size and are therefore easily detected at low microscope magnifications.

Molds are additional members of the fungi kingdom that can be found in cell cultures. Their growth is marked by the production of multicellular, highly connected, thin filaments (hyphae).

Cell culture supernatants contaminated with yeasts or molds appear turbid and although the pH remains stable during the initial stages of infection, it increases in high contaminant concentrations. Yeast contaminations may also be accompanied by a distinct smell. Since fungal species can spread via airborne spores, it is particularly important to identify and contain such contaminations quickly.

Viral Contamination

Viruses are infectious agents that rely on host cells for their own replication. Owing to their limited size of upto 300 nm and their intracellular lifecycle, they are not visible in generic light microscopy and very difficult to detect. While some viruses may induce morphological changes in the cultured cells (cytopathic effects), other species may integrate into the cellular genome and alter the phenotype of the investigated cell line. Viruses can enter cell cultures, for example, through the use of animal-derived cell culture products such as trypsin or fetal bovine serum and are a serious health concern for laboratory workers. The presence of viral contaminants can be challenging to confirm but generally relies on PCR, ELISA, immunocytochemistry, or electron microscopy [3].

Eliminating contaminations

Regardless of the type of contamination identified, affected cell cultures should be removed from the cell culture room and discarded to prevent the spread of infectious agents to other cultures. Furthermore, it is important to determine the source of contamination. It is advisable to dispose of culture media and other cell culture components that have been in contact with the contaminated cells and to clean the surfaces that have touched the contaminated vessel (e.g., incubator, biosafety cabinet, microscope, aspirator).

It is not recommended to treat or proceed culturing infected cells, since any handling of contaminated cultures will increase the potential spread of contaminants—especially airborne fungal spores. Furthermore, the use of antifungal compounds to contain an established infection can interfere with the metabolism of cultured cells. Similar consequences are expected if deciding to prolong the cultures using antibiotics such as 1% Ciprofloxacin to diminish the bacterial growth: the continued release of endotoxins from bacteria will impact the cellular metabolism and likely falsify the cellular readout.

The elimination of contaminants from the cell culture laboratory is a very tedious task and reinforces the importance of prophylactic, aseptic measures to prevent contaminants taking root in the first place.

The Cell Line

The choice of a cell line for cell culture depends heavily on the functional properties and specific readouts required of the cell model [4]. The selected cell lines will also need to align with the available equipments and requirements of their specific hazard group. Cells cultured in the lab can be classified into three different types: primary cells, transformed cells, and self-renewing cells. *Primary cells*, such as fibroblasts obtained from skin biopsies and hepatocytes isolated from liver explants, are directly isolated from human tissue. Biomedical and translational research oftentimes relies on using these cell types since they are good representatives of their tissue of origin. However, there are stringent biosafety restrictions associated with handling these cell types. Furthermore, primary cells are generally characterized as "finite" and therefore rely on a continuous supply of stocks since their proliferation ceases after a limited amount of cell divisions and cell expansion is oftentimes impossible. *Transformed cells* can be generated either naturally or by genetic manipulation. While the use of such immortalized cell lines leads to a cellular platform that generates fast growth rates and stable conditions for maintenance and cloning, their manipulated genotype may result in karyotypic abnormalities and nonphysiological phenotypes. On the other hand, standardized cell lines derived from human or nonhuman species or (e.g., *Chinese hamster ovary* (CHO), HeLa, *human umbilical vein endothelial cells* (HUVEC)) are oftentimes thoroughly characterized and may therefore be easier to set-up. *Self-renewing cells* include, for example, embryonic stem cells, induced pluripotent stem cells, neural and intestinal stem cells. These cells carry the capacity to differentiate into a diversity of other cells types, while their self-renewing property allows for long-term maintenance in vitro. Self-renewing cell types oftentimes act as physiologically relevant representatives of in vivo mechanisms.

Cell lines can be obtained commercially, where certain quality control measures are in place that guarantee genomic stability and absence of contaminants. Other places to source cell lines from can be cell banks or other cell culture laboratories. The introduction of new cell lines in a lab should always be accompanied by a *Mycoplasma* PCR test to ensure clean cultures.

The Cell Culture Microenvironment

Regardless of the cell line chosen, a common requirement will be the selection of suitable growth conditions. This also includes the format of cell growth. There are several advantages and disadvantages associated with culturing cells either in suspension or in plated forms. While fast-growing cells in suspension are more suitable for experiments that aim to isolate recombinant proteins, adherent cells are more appropriate for studies in which the polarity of cells is a crucial component of the cell's functionality (e.g., epithelial cells). Cells grown in suspension generally adopt spherical shapes, while adherent cells display spiked or polygonal morphologies.

The Cell Culture Medium

The goal to create an environment that allows for maximum cell propagation is achieved primarily through the incubator (i.e., temperature, humidity, O_2 and CO_2 tensions) and the basal cell culture medium and its supplements. This includes not only the supply of nutrients such as carbohydrates, vitamins, amino acids, minerals, growth factors, hormones, but also components that control physicochemical properties such as the culture's pH and cellular osmotic pressure. Additionally, the solid or semisolid growth substrate and the cell density allow for cell–matrix anchoring and cell–cell interactions respectively, which further govern the imitation of a physiologically relevant microenvironment.

A great variety of cell culture medium compositions have been created for the requirements of specific cell types and can be classified according to their level of supplemented serum. Serum in the form of fetal bovine serum (FBS) is most commonly added to basal media that already contain a standard formulation based on amino acids, vitamins, carbon sources (e.g., glucose), and inorganic salts. Serum provides cells with growth factors and hormones and acts as a carrier for lipids and enzymes, and the transportation of micronutrients and trace elements. Several labs, however, aim to reduce the supplementation of basal media with animal-derived factors such as serum since it is an undefined component that can highly vary between batches. It is also a costly cell culture product, carries the risk of causing undesired stimulatory or inhibitory effects on cellular growth and function, and may introduce contaminations if not sourced from reliable suppliers. Reduced-serum or serum-free media rely on formulations that reduce or replace serum with more defined components. This generally yields cell cultures characterized by greater consistency in growth and in downstream experimental applications. The concentration of supplements can also be adjusted according to the specific needs of the cell types.

Temperature, pH, CO_2, and O_2 Levels

The desired temperature for cell cultures depends on the body temperature of the species and the microenvironment from which the cultured cell types were isolated. While most human and mammalian cell lines are incubated at 36–37°C, cell lines originating from cold-blooded animals can be maintained at wider temperature ranges between 15°C and 26°C. The pH level for most human and mammalian cell lines cultured in the lab should be tightly controlled and kept at a physiological pH level of 7.2–7.4. In contrast, some fibroblast cell lines favor slightly more alkaline conditions between pH 7.4 and 7.7, while transformed cell lines prefer more acidic environments between pH 7.0 and 7.4 [5].

Stable temperatures for cell cultures can be achieved through incubators that tightly regulate and monitor the temperature of the cell culture environment. As the cells propagate, their growth requires energy supplied in the medium, for example in the form of glucose. When metabolized, its by-products include pyruvic acid, lactic acid, and CO_2. Since the pH level is dependent on the balance of CO_2 and HCO_3^- (bicarbonate), the addition of bicarbonate-based buffers to cell culture media can equilibrate the CO_2 concentrations. Other pH buffers can be of organic nature and include 4-(2-hydroxyethyl)-1-piperazineethane-sulfonic acid (HEPES) (10–25 mM) or 3-(N-morpholino)propanesulfonic acid (MOPS) (20 mM). Many cell culture media contain pH indicators (e.g., phenol red), which display a color range between acidic (yellow) and alkaline (pink) conditions. Furthermore, fluctuations in atmospheric CO_2 concentrations can also alter the pH level. Cells should therefore be cultured in incubators that also allow for CO_2 tensions to be adjusted to 5–7%. To delay shifts in pH, glucose may be substituted by another carbon source such as galactose or fructose. Although this will slow down the rate of cell growth, it will also reduce the accumulation of by-products such as lactic acid.

For specific cell cultures that mimic pathological or physiological conditions under low oxygen tension (hypoxia), it is recommended to use hypoxic incubators with adjustable 1–21% oxygen concentrations balanced with nitrogen.

Subculturing

When the available space in the cell culture vessel reaches ~80% confluency (coverage), cells need to be transferred to new vessels to continue their growth. This process, referred to as "passaging," generates subcultures or subclones, and requires enzymatic digestion or mechanical disruption of the adherent cell monolayer to detach cells from their tissue-culture-treated substrate (Fig. 9.3). While the growth of adherent cells is limited or enabled by the available surface area, it is the concentration of cells in the medium that creates the rate-limiting step in suspension cultures. It is therefore essential to monitor the growth rates in suspension cultures over time.

IN PRACTICE

This section explains the basic protocols required for the maintenance of cell cultures. Since some of these protocols may need to be amended to accommodate the specific requirements of various cell types, it is helpful to review the recommendations of the cell line supplier.

Dissociating Adherent Cells from Culture Vessels for Subculturing

Cells cultured in vitro over time will deplete nutrients supplied in the medium, release toxic metabolites and grow in number. In order to expand and/or

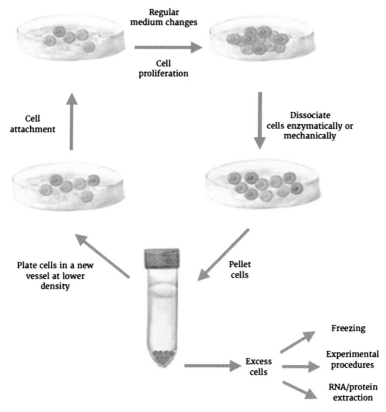

FIGURE 9.3   Cell maintenance. Adherent cell lines are maintained in flasks or plates and regular medium changes ensure healthy cell propagation. Once ~80% confluency has been reached, cells are enzymatically or mechanically dissociated from their plating substrates. The detached cells can be collected in a Falcon tube and pelleted. With a subset of these cells new cell culture vessels can be seeded, while any remaining cells can be frozen or utilized for downstream experiments.

maintain a healthy cell culture, it is therefore essential to produce a new culture with a subset of cells from the originating culture, removing toxic by-products, and replenishing nutrients with fresh medium. A suitable time for passaging is reached when the growth of adherent cells reaches ~80 % confluency. They can then be enzymatically digested or mechanically dissociated to lift off their substrate. In a biosafety cabinet, cells are washed with phosphate-buffered saline (PBS) free of Mg^{2+} and Ca^{2+} to remove dead cells and are incubated at 37°C with sufficient digestive enzymes or chelating agent to cover the monolayer (e.g., trypsin, dispase, collagenase, ethylenediaminetetraacetic acid (EDTA)). The time required to detach the anchored cells from their substrate and cell–cell interactions can take 1–60 minutes and depends on the cell type and the digestive enzymes used. The extent of dissociation can be monitored under a light microscope and once complete, tapping of the culture vessel should dislodge remaining adherent cells. The dissociated cells are collected in a sterile Falcon tube and the culture vessel should also be washed with a medium containing an inhibitor for the enzymatic digestion and dissociation of cells. Collected cells can then be concentrated and counted according to protocols 4.3 and 4.4 and seeded in new culture vessels at the desired concentrations. Lower cell concentrations ($\sim 10^4$ cells/mL) are suitable for cell lines with fast proliferation rates, while higher cell concentrations ($\sim 10^5$ cells/mL) are more adapted for cells with slower growth rates.

Note: It is good lab practice to record the number of passages that have taken place since the culture has been initiated. Some cell lines are not suitable for experimental work beyond a given passage number since chromosomal abnormalities tend to increase in mammalian lines with cell divisions over time.

Subculturing of Suspension Cultures

The subculturing of suspension cultures can be achieved by aseptically removing one-third of the cell suspension solution and replacing the volume with pre-warmed complete medium.

Pelleting Cells

In order to concentrate cells for transfer to new cell culture vessels, freezing, or other experimental assays, the cell suspension is centrifuged at $300 \times g$ for 10 minutes. After removing the supernatant, the cell pellet is resuspended in the desired medium through gently pipetting cells up and down three times.

Note: Single cells can be quite fragile and it is therefore advisable to not centrifuge at higher speeds or to pipette them vigorously.

Quantification of Cells and Determining Cell Viability

Cells can die in the process of culturing or during handling and passaging. When relying on a specific concentration of live cells to start a culture or needing a specific number of live cells for an assay it is important to distinguish between live and dead cells. Cell counting is also helpful when assessing growth rates. Since cells are commonly cultured in the millions, the number of cells are first counted in a small volume and then extrapolated to the full cell volume. To achieve this, all cells are dissociated, pelleted, and evenly resuspended in a suitable medium volume. In a 1:1 dilution with 0.4% trypan Blue, a small volume of the cell suspension is mixed in an Eppendorf tube. Trypan Blue dye permeates only nonviable cells that can therefore be excluded from the subsequent quantification [6]. This occurs by loading 10 µL of the cell mixture in Trypan Blue onto a hemacytometer (Fig. 9.4). Using an inverted microscope, phase contrast, and a magnification of at least 10X, all cells located in the four outer squares are counted. Viable cells contain a darker "halo," while nonviable

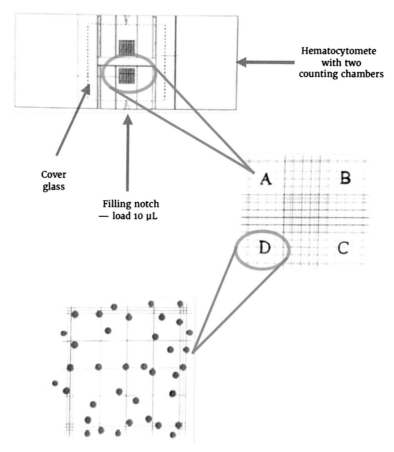

FIGURE 9.4   Cell quantification using Trypan Blue. The hemacytometer is prepared by covering both counting chambers with a cover glass. Subsequently, 10 μL of a 1:1 cell suspension with
0.4 % Trypan Blue is loaded onto the filling notch of one of the counting chambers. Through
capillary action, this volume will cover the grid that can be observed in an inverted microscope at
magnifications of at least 10X. The average of cells covering squares A–D determines the number
of cells per mm^2. Viable cells in these squares are counted by excluding nonviable cells that appear
black due to their absorption of Trypan Blue through their permeable cell membranes. Only cells
overlapping with one of the outer horizontal and vertical borders should be included.

cells stain blue/black. To determine the total number of viable cells, the number
of cells found in all four squares is divided by 4 (to determine the average cell
number in 1 mm^2), multiplied by 10^4 (to obtain the cell number per mL), multiplied by 2 (to account for the dilution factor of Trypan Blue) and multiplied
by the initial medium volume of the entire cell suspension. The percentage of
viable cells can be determined by dividing the number of unstained cells by the
total number of cells, and multiplying the ratio by 100. A healthy cell culture is
characterized by 80–95% cell viability.

Freezing Cells

When a surplus of cells becomes available during subculturing, they can be preserved at that passage through freezing with cryoprotective agents (e.g., glycerol or dimethyl sulfoxide (DMSO)) that prevent the formation of harmful extra- or intracellular crystals [7]. To that end, cells are dissociated from the culture vessel and condensed as described in protocol 4.3. The cell pellet is resuspended in 1 mL of freezing medium (e.g., knockout serum replacement medium supplemented with 10% DMSO) and ~1×10^6 cells are transferred into each cryovial. After 20–30 minutes, the cryoprotectant will have penetrated the cells. Cooled down overnight at −80°C at a controlled freezing rate of 1–2°C/min, the vials are then transferred to liquid nitrogen for long-term storage.

Note: While glycerol and DMSO are both suitable cyroprotective agents, handling of DMSO needs to be carefully monitored. In high (stock) concentrations, DMSO is toxic to personnel and cultured cells and therefore cannot be added to cells without prior dilution. This toxicity also affects cells in freezing medium containing 10% DMSO when left for several hours at room temperature, highlighting the need to transfer cells to −80°C for storage within 30 minutes. In general, chemically protective gloves should be worn to safeguard personnel from the hazards of DMSO and its solutes to easily penetrate membranes, including the skin.

Thawing Cryopreserved Cells

Most mammalian cells can be preserved in liquid nitrogen (<130°C) for numerous years since all biological processes are halted at these temperatures. To recover cells, 10 mL of complete medium is prewarmed in a water bath. After removing the frozen vial from liquid nitrogen, it is immediately placed into a 37°C water bath and gently swirled until two-thirds of the content are completely thawed. The vial is wiped with 70% ethanol and placed in a biosafety cabinet where 1 mL of the prewarmed medium is added in a drop-wise fashion to the partially thawed vial to minimize the osmotic stress imposed upon the cells when DMSO is diluted. The contents of the now completely thawed vial are transferred also in a drop-wise fashion to the remaining 9 mL of complete medium and centrifuged at $300 \times g$ for 3 minutes. After aspirating the supernatant, the cell pellet can be washed once in medium to remove residual cryopreservatives. Cells are then resuspended in complete medium and transferred to a cell culture vessel. Cell attachment should occur within 24 hours.

Note: The viability of cells after cryopreservation is impacted by their ability to cope with the stressors of freezing and thawing. It is therefore recommended to perform the thawing process as swiftly as possible. When handling cryovials that have been frozen with glycerol as the cryoprotectant, the thawing process can be simplified by diluting the cryopreserved cells ten times directly into complete, prewarmed medium, avoiding the centrifugation and washing step.

APPLICATIONS

Model Systems in Health and Disease

Cell culture is one of the most important techniques in cellular and molecular biology since it provides a platform to investigate the biology, biochemistry, physiology (e.g., aging) and metabolism of wild-type cells and diseased cells. The interaction and route of infection between wild-type cells and pathogenic agents (e.g., bacteria and viruses) can also be studied in specific cocultures. Furthermore, immortalized cancer cell lines have given researchers insight into the biology of cancer and through the selective treatment of wild-type cells with UV radiation, viruses, and toxins, causative agents of tumorigenicity have been identified. Finally, human-induced pluripotent stem cells (hIPSCs) have been derived from individuals with inherited disorders and differentiated toward the affected cell type in which the disease manifests. These hIPSC-derived somatic cells are suitable platforms for the studying molecular mechanisms of a disease in a dish.

Drug Development and Drug Testing

Cell culture tools can also be applied to screen novel chemicals, cosmetics, and drug compounds for their efficacy and assess drug cytotoxicity in specific cell types. Detoxifying cell types such as hepatocytes and kidney cells are often-times of high interest for these purposes. When using cell cocultures or diseased cells obtained from individual patients, it is also possible to screen for drugs to selectively target specific cell types (e.g., in cancer treatment), at doses that are nontoxic and with minimized side-effects for the patient. Furthermore, large-scale cells cultures can serve for the generation of genetically engineered proteins, antibodies, hormones. and biopharmaceuticals that can be isolated and used therapeutically.

Virology and Vaccine Production

Cell culture with mammalian cells offers a host for viruses to replicate, allowing researchers to study their growth rates, development, and conditions required for their infectious cycle. Furthermore, the attenuated viruses used in vaccines against polio, measles, chicken pox, rabies, and hepatitis B are raised in animal cell cultures.

Tissue Regeneration and Transplantation

hIPSCs, embryonic stem cells, and adult stem cells have the capacity to regenerate and differentiate into specialized cell types that can be used as replacement tissues or organs. These cell cultures are oftentimes performed in a 3D protein matrix that allows cells to self-organize into functional cell clusters (organoids).

Genetic Engineering and Gene Therapy

The expression of specific genes and their impact on cells can be studied by the introduction of new genetic material (e.g., DNA, RNA) into the nucleus of cultured mammalian cells. Similarly, the importance of genes in regulating specific pathways can be observed through silencing them. Oftentimes, viral vectors or specialized enzymes are used to carry out these tasks. Altering the genome of cells can also aid in restoring dysfunctional genes in patients.

The overarching benefit of using cell culture techniques to address these basic scientific and translational research questions is the homogeneity and reproducibility of data that can be generated using clonal cell lines. Studying an isolated, simplified cellular system in a well-defined and controlled environment limits the exposure of confounding effects inherent to an in vivo system and therefore allows for the generation of simplified but robust data sets.

SCENARIO

Culturing hIPSCs to study inherited liver diseases: Small pieces of skin biopsies obtained from patients with an inherited liver disease can be cultured in a Petri dish with fibroblast growth medium. Primary fibroblast cultures will emerge from the skin tissue after 2–3 days. Upon reaching ~80% confluency, the outgrown primary fibroblasts can be isolated from the skin tissue using enzymatic digestion with collagenase and subcultured in new vessels. This generates a pure population of fibroblasts, which can be scaled-up and virally transduced to express pluripotent genes (e.g., *OCT4, NANOG, TRA-1-60*), thereby "reprogramming" somatic cells to become hIPSCs [8]. hIPSCs in culture will pack together tightly in flat colonies with sharp edges and are characterized by their high nucleo-cytoplasmic ratio, ability to self-renew, and capacity to form cells of all three germ layers. As such, hIPSCs can be differentiated toward definitive endoderm using Wnt pathway activation and Activin A. Cells at this stage of differentiation will migrate from their colonies into a monolayer, downregulate their pluripotency genes and express markers of their endoderm fate (e.g., *SOX17, CXCR4, GSC*). Further changes in the media composition and exogenous growth factors added to the definitive endoderm cells will direct them toward foregut endoderm and subsequent specification to hepatic endoderm. The final stage of the hepatic differentiation medium yields hepatocyte-like cells that secrete albumin and other serum proteins, take up *low-density lipoprotein* (LDL), store glycogen, and metabolize drugs. Importantly, these cells will also display the disease phenotypes observed in the patient from which the skin biopsy was initially obtained. While this process allows researchers to study and rescue disease mechanisms ex vivo [9], it also sheds light on liver development and the emergence of disease in utero.

KEY LIMITATIONS

Discrepancies between Cellular Environments in vitro and in vivo

One of the pillars of cell culture research is the design of a defined cellular environment in which single variables can be manipulated in order to monitor cellular responses. To achieve this goal, the cellular environment in vitro is oftentimes oversimplified and relies, for example, on a single cell type cultured in a monolayer. However, data generated from such a cellular system does not truly phenocopy the intricate cellular interactions between different cell types and extracellular matrices of an in vivo environment. To address this drawback, there is currently significant research into the design of cell cocultures that allow paracrine signaling between cells that cohabitate space in vivo, as well as bioartifical matrices that facilitate cellular growth in their native 3D orientation. The goal is the design of cellular systems that mimic the complexity of the multicellular in vivo niche, yet also allow standardization for cell culture assays.

Discrepancies between Gene Expression in Primary Cells and Immortal Cell Lines

The oftentimes most relevant cell types for addressing translational research questions—primary cells—are in fact very difficult to isolate and culture in vitro due to their limited proliferation and functional capacity ex vivo. To delay senescence, viral transfection of primary cells can sequester tumor-suppressor proteins, thereby extending the number of possible passages and allowing the emergence of immortal cell lines. Although this facilitates their culture ex vivo, this technique also introduces the expression of carcinogenic genes. In addition, immortalized cell lines can acquire mutations during subculturing that can further interfere with the cellular phenotype and create a nonphysiological cell culture system.

TROUBLESHOOTING

Problem	Possible Reason	Suggested Solution
Lack of viable cells upon thawing	Incorrect storage	Obtain new stock stored in liquid nitrogen that have not been thawed
	Incorrect thawing	Thaw cells quickly but gradually, dilute frozen cells drop-wise with prewarmed medium, handle cells gently, and only centrifuge at low speeds
	Glycerol exposure to light	If the cryopreservative agent glycerol was used and exposed to light, its by-product acrolein may be toxic for the cells

Problem	Possible Reason	Suggested Solution
Lack of cell attachment to culture vessel after subculturing	Residual digestive enzyme activity	Thoroughly wash cells in prewarmed medium before replating
	Excessive digestion of cells	Reduce digestion time and block enzymatic activity using inhibitor (FBS for Trypsin)
	Mycoplasma contamination	Perform routine *Mycoplasma* PCR tests on cultures
Slow cell growth	Incorrect growth medium	Growth medium must fit the requirements of the culture cell line and (if applicable) contain serum that has been screened
	Depletion or breakdown of essential cell culture components	Ensure presence of growth-promoting factors and substitute unstable components such as glutamine with GlutaMax
	Incorrect storage of medium and supplements	Follow manufacturer's instructions closely
	Passage number is too high	The proliferation rate of cell cultures may cease with continuous subcultures—cells should be replaced with low-passage stocks
	Confluency is too low	Enhance concentration of initial plating density
	Confluency is too high	Cells should be passaged in their log-phase (at around ~80% confluency)
	Mycoplasma contamination	Perform routine *Mycoplasma* PCR tests on cultures
Rapid shift in medium pH	Incorrect CO_2 tension	Adjust CO_2 concentration of incubator based on HCO_3^- concentrations of medium: 2.0 g/L requires CO_2 levels of 5%, while 3.7 g/L rely on 10% supplementary CO_2
	Lack of gas exchange	Loosen caps of tissue culture flasks
	Insufficient bicarbonate buffering in medium	Add HEPES (10–25 mM)
	Bacterial contamination	Investigate cultures under light microscope

Problem	Possible Reason	Suggested Solution
Cell death	Lack of CO_2 or fluctuating temperature	Monitor CO_2 and temperature levels of incubator and do not leave cells outside the incubator for extended periods of time
	Accumulation of toxins	Regularly replace cell culture medium
	Incorrect osmotic pressure	Review osmolality of cell culture medium and potential effects of added drug compounds or HEPES

CONCLUSION

This chapter has described the vast possibilities to employ cell culture techniques to address basic and translational research questions and has explained the necessary considerations for setting up a cell culture lab. It has also shown essential practices and techniques for successfully working with cell lines and explained the conditions required for creating a cellular environment that mimics their in vivo niche.

REFERENCES

[1] HSE/ACDP. Biological agents: managing the risks in laboratories and healthcare premises. 2005.

[2] Drexler HG, Uphoff CC. Mycoplasma contamination of cell cultures: Incidence, sources, effects, detection, elimination, prevention. Cytotechnology 2002;39:75–90.

[3] Merten O. Virus contaminations of cell cultures – A biotechnological view. Cytotechnology 2002;39:91–116.

[4] Pan C, Kumar C, Bohl S, Klingmueller U, Mann M. Comparative Proteomic Phenotyping of Cell Lines and Primary Cells to Assess Preservation of Cell Type-specific Functions. Mol Cell Proteomics 2009;8(3):443–50.

[5] Schwartz MA, Both G, Lechene C. Effect of cell spreading on cytoplasmic pH in normal and transformed fibroblasts. Proc Natl Acad Sci USA 1989;86:4525–9.

[6] Strober W. Trypan blue exclusion test of cell viability. Current Protocols in Immunology, Appendix 3B 2001.

[7] Lovelock JE, Bishop MW. Prevention of freezing damage to living cells by diemthyl sulphoxide. Nature 1959;16(183):1394–5.

[8] Takahashi K, Tanabe K, Ohnuki M, Narita M, Ichisaka T, Tomoda K, et al. Induction of pluripotent stem cells from adult human fibroblasts by defined factors. Cell 2007;131(5):861–72.

[9] Rashid ST, Corbineau S, Hannan N, Marciniak SJ, Miranda E, Alexander G, et al. Modeling inherited metabolic disorders of the liver using human induced pluripotent stem cells. Journal Clin Invest 2010;120(9):3127–36. http://dx.doi.org/10.1172/JCI43122DS1.

GLOSSARY

Cell culture   Ex vivo maintenance of cells in a sterile environment and under favorable conditions that mimic the cellular niche in vivo.

Asepsis   Sterile cell culture technique that prevents the contact of unwanted infectious organisms with the cultured cells.

Primary cell   Cell directly isolated from a human or animal tissue.

Immortalized cell line   Cells that have been naturally or experimentally introduced to mutation(s) that overcome cellular senescence and render them capable of unlimited cell divisions.

Senescence   Arrest of cellular proliferation.

Subculturing   Technique to prolong and expand cells in culture by transferring cells from a previous culture to a new vessel with fresh medium.

Hemacytometer   Tool for manual counting of cells suspended in a liquid using a light microscope.

LIST OF ACRONYMS AND ABBREVIATIONS

HBV   Hepatitis B virus
HIV   Human immunodeficiency virus
ACDP   Advisory Committee on Dangerous Pathogens
HSE   Health and Safety Executive
BSL   Biosafety level
PPE   Personal protective equipment
MSDS   Material Safety Data Sheet
PCR   Polymerase chain reaction
ELISA   Enzyme-linked immunosorbent assay
FBS   Fetal bovine serum
DMSO   Dimethylsulfoxide
hIPSC   Human-induced pluripotent stem cells

Chapter 10

Flow Cytometry

Eleni Chantzoura and Keisuke Kaji
University of Edinburgh, Edinburgh, United Kingdom

Chapter Outline

Objectives

- Introducing the main technological advances that led to the development of flow cytometry.
- Understanding the mechanistic principal of flow cytometry and what it actually measures.
- Providing a basic protocol for using flow cytometry to measure the expression levels of cell surface proteins.
- Describing some of the most popular applications.
- Analyzing a case study, where flow cytometry was used to investigate the process of mouse fibroblasts turning into induced pluripotent stem cells (iPSCs).
- Discussing the limitations of the method followed along with some troubleshooting techniques.

Basic Science Methods for Clinical Researchers. DOI: http://dx.doi.org/10.1016/B978-0-12-803077-6.00010-2
173

INTRODUCTION

Flow cytometry is a multiparametric method for the quantification of cells and their properties. In contrast to most procedures, flow cytometry analyzes characteristics of individual cells within a heterogeneous population in a quantitative manner. The most common application is the measurement of the expression levels of cell surface proteins detected by fluorescent compound (fluorochrome)-conjugated antibodies or fluorescent proteins. Expression levels of intracellular proteins, specific mRNAs, and amount of total DNA in a cell can also be measured in a quantitative manner using various fluorescent dyes. With FACS (fluorescence-activated cell sorting), a specialized type of flow cytometry, it is also possible to isolate (sort) cells with certain features for further analysis.

The history of technologies, which contributed to the development of flow cytometry, goes back in 1590s when Antoni van Leeuwenhoek built the first simple microscope, and consists of great advances in microscopy, staining, electronics, and computers. This led to the first fluorescence-based flow cytometry device, developed in 1968 by Wolfgang Göhde from the University of Münster (Table 10.1). However, it was not until 1976 that the term "flow cytometry" was officially adopted at the 5th American Engineering Foundation Conference on Automated Cytology in Pensacola (Florida) in 1976 [1].

TABLE 10.1 Main Advances That Led to the Modern Flow Cytometry

1590	A. van Leewenhowk	1st microscope
1742	M. Lomonosov	Studies on light scatter
1833	F. Savart	Liquid drop formation
1880s	P. Erlich	Staining techniques
1880s	Rayleigh	Stability of fluidics jets
1934	A. Moldavan	"Photoelectric technique for the counting of small cells"
1947	F. Gucker	1st working flow cytometer
1950	W. and J. Coulter	Coulter principle
1965	Fulwyler	1st cell sorter
1967	L. Kamentsky	Rapid cell spectrophotometer
1968	W. Göhde	1st fluorescence-based cytometer
1972	L. Herzenberg	FACS

IN PRINCIPLE

But what does a flow cytometer actually measure? When a particle passes through a laser beam, it scatters light according to its physical properties. For instance, a larger or a more complex particle will cause higher deviation of the light. In case this particle is also fluorescent, after being excited with light of the appropriate wavelength, it will emit light at another (usually higher) wavelength. In flow cytometry an optical-to-electronic coupling system is used to record how each cell scatters incident laser light and emits fluorescence. For this reason, cells should be in single-cell suspension, yet many of them can be measured. Cells suspended in a liquid (*fluidics system*) pass through a laser beam causing light scatter that can be sensed by appropriate detectors (*optics system*). Finally, the *electronics system* process the signal from the detectors giving numbers that represent the features (size and granularity) of the cells. Similarly, if a cell expressing the green fluorescent protein (GFP), is exposed to a 488 nm laser, it will emit light with a peak wavelength of 509 nm. The intensity of the emitted light will reflect the expression levels of GFP.

More details about the three parts that compose a flow cytometer are provided below:

FLUIDICS SYSTEM: CELL TRANSPORT TO THE LASER BEAM

For accuracy reasons, one cell at a time should pass in front of the laser. Thus, cells suspended in solution are introduced into a stream of sheath flow in the flow chamber, which is the heart of the instrument and remains in the center (sample core) due to the laminar flow principle. Upto thousands of cells (or events as they are called) can be analyzed per second. Cells $1-15\,\mu M$ in size are suitable for analysis; however, with specialized systems, smaller or bigger cells can also be analyzed.

OPTICS SYSTEM: LASERS AND LENSES

When a cell passes through the laser beam it scatters incident light at all directions. Forward scatter is the light scattered in the forward direction and it reflects the size of the cell (FSC). In contrast, light scattered in higher angles (side scatter) reflects the granularity and internal complexity of the cell (SSC).

In addition to these physical features of the cells, flow cytometry can measure fluorescence emission intensity from each cell when they are labeled with fluorescent molecules, such as fluorochrome-conjugated antibodies, fluorescent proteins (GFP, YFP, etc.), *DAPI* (4′,6-diamidino-2-phenylindole), as described above.

Each fluorescent compound has its own absorption and emission spectra, which means that it has a specific range of wavelengths for excitation and emission. For instance, both PE (phycoerythrine) and PE-Cy7 can be excited by the yellow/green laser (561 nm); however, PE has maximum emission at 578, whereas for PE-Cy7 maximum emission is detected at 785 nm. This means that we can distinguish components labeled with either fluorophore by detecting light in different wavelengths (Fig. 10.1). For this reason, fluorescent light of a certain wavelength is directed to the appropriate optical detectors (which are called photomultiplying tubes or PMTs) through a system of optical mirrors and filters. Also, the intensity of the emitted light depends on the levels of the fluorophore in the cell.

Modern flow cytometers can have 3–5 lasers and can measure >15 parameters. However, many of the fluorescent compounds have overlapping absorption and emission spectra. That is why in a multiparametric analysis, the single stained control samples are necessary to apply the proper compensation. For further information on compensation, see the "In Practice" section.

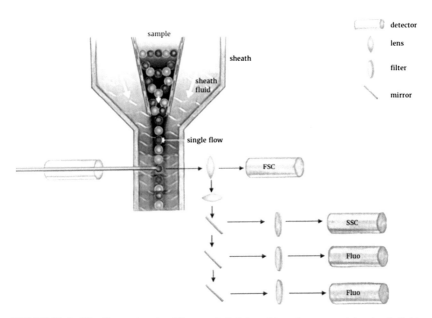

FIGURE 10.1 The flow cytometer. The sample is injected into the center of the sheath fluid, which consists of either water or buffer. The purpose of the design is to constrain the sample stream in the center of the sheath fluid at the interrogation point. As a cell passes through the laser beams, light is scattered. Forward scatter (FSC) and side scatter (SSC) give information about the size and the complexity of the cell, respectively. For instance, dead cells usually have lower FSC than live cells. Also, a neutrophil is more granular than a leukocyte, and, thus, has a higher SSC. If the cell expresses a fluorescent protein or is stained with a fluorochrome-conjugated antibody, the fluorescent molecules are excited by the laser and emit light of a specific wavelength (color). The collection optic system consists of lens and filters, which direct the emitted light to the appropriate detectors.

ELECTRONICS SYSTEM: CONVERTING THE LIGHT SIGNALS INTO NUMERICAL DATA

Finally, the detectors that collected the light signals produce an electronic pulse (the voltage pulse) that is processed by the electronics system (Fig. 10.2). Height, width, and area of the electronic pulse are determined by the size, speed, and fluorescence intensity of the cell.

The data from the flow cytometry analysis is stored in a format called Flow Cytometry Standard (FCS) which is developed by the International Society for Advancement of Cytometry (ISAC). Further details about the plots and the gating strategies can be found at the "In Practice" section.

FACS (Flow-Activated Cell Sorting)

Some cytometers can also collect (sort) the cells of interest for further analysis. The most common way for sorting is based on the deflection of the charged droplets. Those droplets break off the main stream after the interrogation point due to vibration of the flow chamber. A droplet with a cell of interest will then be electrostatically charged, and depending on the charge it will be deflected right or left into the appropriate tube. While the principal sounds complex, the

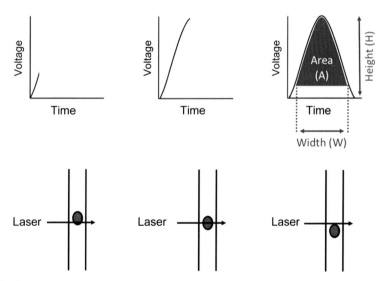

FIGURE 10.2 **The voltage pulse.** As a fluorescent-labeled cell flows through the laser, it scatters light and emits fluorescence. These signals are converted to electronic signal (voltage pulses) by the photodetectors. The highest point of the pulse reflects the point of maximum light scatter or fluorescence produced and takes place when the particle/cell is in the center of the beam (middle panel). The width of the pulse indicates the time that the cell needs to pass through the beam and is proportional to its size, whereas peak height reflects signal intensity from the cell. The voltage pulses are converted to digital values and then displayed on the data plot.

practice is simple. You indicate what type of cells you wish to isolate based on FSC, SSC, and fluorescence intensity of any channel. Viability of the cells after sorting largely depends on cell types, but many cell types are compatible with FACS and subsequent culture.

IN PRACTICE

Here we introduce a procedure to measure expression levels of cell surface proteins with fluorochrome (fluorescent dye)-conjugated antibodies using flow cytometry analysis. For this application, it is ideal to analyze >10,000 cells. Considering the loss of cells during the staining procedure, it is recommended to start with >100,000 cells per samples when possible.

Cell Harvest

1. Wash the cells with phosphate-buffered saline (PBS).

    Controls needed: (a) Unstained cells, (b) single-fluorochrome samples for each color, (c) FMO (Fluorescence Minus One) for each fluorochrome, which contains all the fluorochromes apart from one. Staining of cell populations positive or negative for the antigen of interest should be performed employing different dilutions of each antibody to find the optimal concentration.

2. Add 0.25% (v/v) trypsin 0.53 mM EDTA (ethylenediaminetetraacetic acid) in PBS to the flask and incubate at 37°C until cells detach. Note that some cell surface protein could be cleaved by trypsin and become nondetectable by antibodies. A nonenzymatic or gentler cell detachment solution (e.g., Accutase) might be more suitable for some proteins.

3. Inactivate trypsin using at least five times more serum-containing medium and transfer the cells to a universal tube.

4. Spin at 300 g for 3 minutes. Note, some cell surface proteins, such as receptors, can be internalized upon antibody binding. To prevent this, use of ice-cold buffer, and a refrigerated centrifuge is recommended.

5. Aspirate the medium and re-suspend the cells in 5 mL PBS. Count cell number with a hemocytometer.

6. Spin at 300 g for 3 minutes.

7. Aspirate the PBS and re-suspend the cells to $1-5 \times 10^6$ cells/mL in ice-cold FACS buffer (2% serum in PBS). Note that sodium azide in a final concentration of 0.05% can be added to this FACS buffer to prevent shedding or internalization of the antigen. However, do not add when cellular function is analyzed after cell sorting.

8. Transfer 100 μL/staining to a 96-well, V-bottom plate. While universal tubes can be used in this step, V-bottom plates are more convenient to handle multiple samples.

Staining

9. The primary antibody is usually diluted in 0.1–10 µg/mL in FACS buffer. 100 µL/staining is recommended. Note that because the optimal concentration of antibodies varies, serial dilutions of each antibody should be tested.

10. Spin the plate at 300 g for 3 minutes.

11. Aspirate the supernatant by placing the tip of aspirator/pipette on the edge where the V-shape starts toward the bottom. Note that it is not necessary to remove the supernatant completely and removing till the edge is sufficient.

12. Add 100 µL of primary antibody/sample.

13. Incubate the plate for 15–30 minutes at 4°C.

14. During this incubation time, prepare the fluorochrome-conjugated secondary antibody in FACS buffer in the dilution suggested by the manufacturer's instructions (100 µL/sample). Note, keep the fluorochrome-conjugated antibodies in dark. While many of recent fluorochromes are stable, some of them have lower stability under light.

15. Spin the plate at 300 g for 3 minutes, aspirate the supernatant and re-suspend the cells into 150 µL of ice-cold FACS buffer. FACS buffer, instead of PBS, is recommended for the washing steps, since it is gentler and can also minimize loss of cells by preventing them from attaching plastics. Repeat this washing process one more time. If the fluorophore is directly connected to the primary antibody go to step 21.

16. After the second wash, spin the plate at 300 g for 3 minutes, aspirate the supernatant, and re-suspend the cells in the secondary antibody-solution.

17. Incubate for 5–15 minutes at 4°C in the dark.

18. Prepare FACS tubes by adding 100–300 µL of ice-cold FACS buffer, ideally containing cell membrane impermeable DNA-binding fluorochromes (e.g., 0.1–1 µg/mL DAPI) for dead cell staining. Smaller volumes should be used if starting cell number is less. FACS tubes with cell strainer are recommended so as to exclude clusters when you add cells. Keep on ice in the dark.

19. Wash the cells with ice-cold FACS buffer two times as before.

20. Re-suspend the cells in 100 µL FACS buffer and transfer into the FACS tube prepared in step 19.

21. Store cells at 4°C in the dark till analysis (for better results it should take place as soon as possible).

22. Analysis: For best results, it is recommended to perform the analysis on the same day. However, for longer storage (>12–16 hours), cells can be fixed before staining (after step 9). Optimal fixation methods depend on the antibody, the epitope of interest, and the fluorochrome. About 0.01%–1% paraformaldehyde (PFA) for 10–15 minutes or ice-cold acetone/methanol for 5–10 minutes at −20°C could be used.

Data Acquisition and Analysis

23. FSC (Forward SCatter) axis reflects the size and SSC (Side SCatter) axis the granularity and complexity of the cells. Using the FSC and SSC parameters distinct cell populations, as well as debris, can be distinguished. Fixation usually affects the above patterns.

    In flow cytometry, grouping of cells with the same characteristics is called gating. First, make a gate in the scatter plot with the FSC and SSC parameters to exclude debris and dead cells for further analysis of the live cell population (Fig. 10.3).

24. Make scatter plot windows for all combinations of used fluorochromes and run control samples with each single color staining.

25. Adjust the voltage of the excitation laser as signals from the positive control samples are distinctive from the negative controls.

    Depending on the combination of fluorochromes, you may see positive signals in the detection channels, which are supposed to be negative. This occurs because emission spectra of some fluorochrome can be captured by filters used for other fluorochromes (Fig. 10.4). To correct these emission spectra overlaps, fluorescence compensation can be used. Setting compensation varies depending on equipment and software, but most of the modern cytometers allow to change the compensation settings after the data acquisition. In principle, after proper compensation a control sample stained with a single fluorochrome should give positive signal only in the designated detection channel. Many flow cytometry software can calculate the compensation as long as they are provided with the single stained controls for all the fluorophores.

26. After optimal acquisition parameters are set, run all samples with the same setting. Number of events (cell numbers) to record depends on proportion of cells with positive signals of interest. Make sure you record sufficient number of events for the subsequent data analysis.

    There is a variety of software to analyze the results (e.g., FlowJo or WEASEL). The distribution of the values for each parameter can be

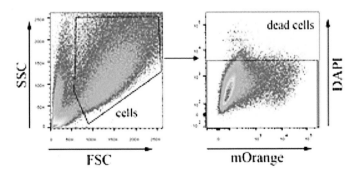

FIGURE 10.3 Gating of live cells. Using the SSC-A/FSC-A parameters, cells can be selected from debris (left plot). Then, excluding the DAPI positive cells, live cells can be used for further analysis (right plot).

presented in a histogram where the *x*-axis displays the intensity of the parameter and the *y*-axis displays the number of cells (Fig. 10.5A). Figs. 10.5B–E demonstrate the commonly used plot styles to display the two parameters (mOrange and Nanog-GFP) shown in Fig. 10.5A simultaneously. While

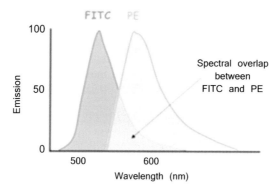

FIGURE 10.4 Spectra from fluorescein (FITC) and phycoerythrin (PE). Some of the fluorescence emission from FITC will be detected by the PE-designated detector. This is called spectral overlap or spillover and can be corrected with a process called compensation. Using single color positive controls for each fluorochrome, and through a mathematical algorithm, we can remove the spillover from one fluorochrome to the other channels. Some software can compensate automatically all the given combinations of fluorochromes, provided they are given the proper single positive controls.

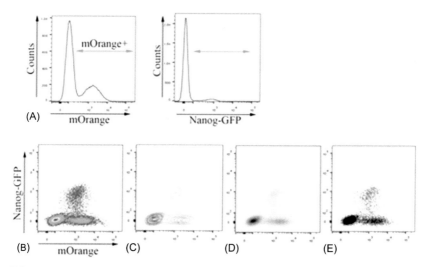

FIGURE 10.5 Flow cytometry plots. (A) Sample histograms that show the distribution of cells expressing mOrange (left) and Nanog-GFP (right) in the same sample. The *x*-axis represents signal intensity from the fluorescent proteins and *y*-axis represents the number of the cells. The same data can be presented with at least four different two-dimensional plots (B) the pseudo-color density plot, (C) the contour plot, (D) the grayscale density, and (E) the dot plot), demonstrating distinct populations that cannot be identified with the histograms.

histograms are beneficial to provide accurate numbers of cells that display certain signal intensities, two-dimensional plots (density plot, contour plot, dot plot) have the advantage to visualize the correlation of two parameters for each cell. New gates within the plot can be set and accurate proportion of cells within the gate can be displayed.

APPLICATIONS

In addition to the abovementioned cell surface protein measurement with specific antibodies, flow cytometry has evolved into a powerful tool to measure various biological parameters. Some of its applications are summarized here:

- *Intracellular antigen expression.* Transcription factors and other intracellular molecules can be stained with fluorochrome-conjugated antibodies after fixation and permeabilization of the cells. Flow cytometry, in contrast to classical microscopy techniques, can provide accurate quantification and high-throughput analysis. Expression levels of a protein in >100,000 individual cells can be measured and visualized within a few minutes. However, this internal staining tends to have higher background, whereas optimal fixation and permeabilization methods vary (such as 0.01% formaldehyde, 1–4% PFA or acetone followed by 0.1–1% NP-40 or ice-cold methanol, etc.). Proper positive and negative controls are essential.
- *Cell cycle analysis.* A fluorescent dye that binds to DNA (e.g., propidium iodide (PI) or Hoechst 33342) can be used to quantify the amount of DNA in each cell. Cells that are in S phase will be brighter than cells in G1 phase, and cells in G2 phase will be approximately twice as bright as cells in G1. Other fluorescent dyes that can stain DNA with suitable excitation and emission properties for the flow cytometry can be used, but users should pay attention whether the dye is cell membrane permeable and specific to DNA. The cells might need to be permeabilized and/or treated with RNase before the staining.
- *Apoptosis.* In the early stage of apoptosis, phosphatidyl serine (PS) residues, which normally exist only inside the plasma membrane, relocate to the outer surface. PS binding protein, annexin V, labeled with a fluorochrome can be used to detect this event. A combination of PI (permeable to dying and dead cells) and annexin V allows the detection of cells in early and late phases of apoptosis. FSC and SSC are also useful parameters to detect dead cells since they have smaller FSC and SSC values than live cells.
- *Cell proliferation assays.* Carboxyfluorescein succinimidyl ester (CFSE) is a cell permeable fluorescent dye, which binds covalently to cytoplasmic proteins and is retained within the cells. After each cell division it is diluted two fold, so its signal intensity can be used to measure cell proliferation. While a high concentration of CFSE is toxic, nontoxic levels allow detecting approximately 7–8 cell divisions before the signal becomes below a detection level.
- *Intracellular calcium flux.* Many signaling pathways stimulate the entrance of calcium into the cells. Cell permeable calcium indicator fluorochromes,

such as Indo-1, Fluo-1-4, change their fluorescence emission spectra depending on Ca^{2+} concentration. Using flow cytometry we can measure the changes of intercellular Ca^{2+} concentration, and obtain information on how cells respond to the stimuli.

Additionally, the development of smaller/cheaper instruments and the increasing number of useful antibodies have facilitated the transition of flow cytometry from a research technique to an invaluable diagnostic tool [2]. Briefly, flow cytometry is used for:

- *The diagnosis of leukemia/lymphomas*, where leukocyte surface antigens are measured in neoplastic cells.
- *Detection of Minimal Residual Disease (MDR)* in patients in remission after leukemia or cancer. CD13, CD19, CD34 are measured in blood and marrow.
- *Hematopoietic progenitor cell enumeration* for repopulating a depleted bone marrow by measuring CD34 in stem cells.
- *Histocompatibility cross-matching* measuring IgG after incubating donor's lymphocytes with donor's serum.
- *Post-transplantation monitoring* measuring the levels of circulating CD3+ T-cells. Also, a variety of antigens is measured depending on the organ being transplanted.
- *Diagnosis of immunodefficiencies*. In this case, CD4 and CD8 positive cells are measured in the blood and lymphoid tissues.
- *HIV infection diagnosis*. CD4 positive lymphocytes in the blood are measured.
- Measuring Hemoglobin F, rhesus D of maternal erythrocytes for *feto-maternal hemorrhage quantification*.
- *Reticulocyte enumeration*. RNA is quantified to distinguish reticulocytes from erythrocytes.
- *Contaminating leukocytes measurement* in blood for transfusion. Leukocytes are detecting using FSC, SSC channels, and leukocyte-antigens.
- DNA content to *detect malignancies*.
- *Auto-/allo-immune diseases*. IgG and IgM are detected for antiplatelet antibodies and IgG for antineutrophil antibodies.

SCENARIO

In 2006, Takahashi and Yamanaka demonstrated that we can reprogram differentiated somatic cells into a pluripotent state with the introduction of just four transcription factors, Oct4, Sox2, Klf4 and c-Myc [3]. While the resulting cells, called iPSCs bear a great potential for drug screening, toxicology tests, disease modeling, and regenerative medicine, the molecular mechanisms of cellular reprogramming are yet to be elucidated. The biggest problem researchers have been facing is the extremely low efficiency of reprogramming. Typically less than 1% of the cells introduced with the reprogramming factors can become iPSCs. We and others have recently developed sets of cell surface markers

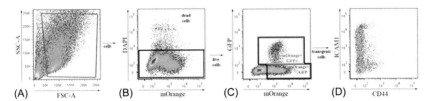

FIGURE 10.6 Gating strategy for the reprogramming analysis. Debris were excluded from the analysis using a gate based on SSC-A/FSC-A channels (A). DAPI is used to exclude the dead cells (B). Expression of mOrange and Nanog-GFP was analyzed in the live cells (C), as well as CD44 and ICAM1 expression changes in the transgenic cells (mOrange+, Nanog-GFP+ cells) (D).

whose expression dramatically changes during reprogramming to track the reprogramming process [4,5]. Here, we show that CD44, ICAM1 expression pattern along with a Nanog-GFP reporter, enables us to monitor changes of cell population undergoing reprogramming and the reprogramming kinetics.

Mouse embryonic fibroblasts (MEFs) were isolated from chimeric embryos made with embryonic stem cells that carry a pluripotency gene reporter, Nanog-GFP, as well as a doxycycline (dox)-inducible reprograming cassette. The reprograming cassette contains 2A peptide-linked c-Myc, Klf4, Oct4, and Sox2 followed by ires-mOrange fluorescent protein. mOrange expression upon administration of dox allows the identification of transgenic MEFs undergoing reprogramming. The Nanog-GFP reporter is not expressed in MEFs but becomes positive when cells regain pluripotency as a result of reprogramming. To monitor the reprogramming process, expression levels of CD44, ICAM1, and Nanog-GFP in mOrange expressing cells were analyzed every 2 days for 12 days (Figs. 10.6 and 10.7).

Table 10.2 lists the antibodies and the dilutions used, and Table 10.3 summarizes the excitation lasers and filters of BD LSR Fortessa (BD Biosciences) used for this analysis.

LIMITATIONS OF FLOW CYTOMETRY

- *Cells in suspension.* Flow cytometry is a technique that allows the analysis of chemical and physical characteristics of single cells, and demonstrates the proportion of cells with those properties in the analyzed cell population. FACS, a specialized type of flow cytometry, allows to isolate cells with features of interest for further analysis such as RNA-sequencing. However, samples need to be in single-cell suspension and this precludes the study of intact tissues, cell-to-cell interactions, and so on.
- *Overlapping between different fluorochromes.* Another practical limitation is the availability of antibodies and fluorochromes. Although some equipment has the capacity to analyze more than 10 colors simultaneously, in practice, identifying a combination of multiple antibodies with different fluorochromes and minimal spectral overlap is not trivial. It is of note that different fluorochromes conjugated with the same antibody show different

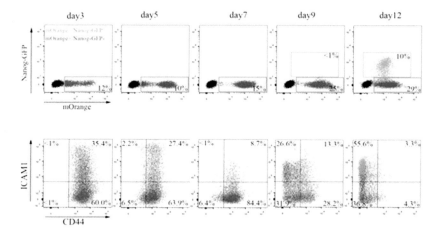

FIGURE 10.7   CD44, ICAM1, and Nanog-GFP expression on cells undergoing reprogramming. Expression of CD44 and ICAM1 in the transgenic cells, which can be identified by the expression of mOrange and Nanog-GFP (upper panels), was analyzed on days 3, 5, 7, 9, and 12 post administration of dox (lower panels). Cells initially displayed a wide range of ICAM1 expression and high CD44. Decrease of ICAM1 expression (days 5–7) was followed by downregulation of CD44 (days 7–9). CD44/ICAM1 double negative cells increased ICAM1 expression (day 9) before Nanog-GFP emergence (day12, green cells). The iPSC state was marked by ICAM1/Nanog-GFP expression and absence of CD44.

TABLE 10.2   Staining Reagents for Reprogramming Analysis

Detection	Antibody/Stain	Dilution in FACS Buffer
CD44	Antihuman/Mouse CD44 APC (eBiosciences, 17-0441)	1:300
ICAM1	Antimouse CD54 (ICAM1) Biotin (eBiosciences 13-0541)	1:100
Biotin	Streptavidin PE-Cyanine7 (eBiosciences 25-4317-82)	1:1500
DNA (for the exclusion of dead cells)	DAPI (Life Technologies 62248)	1:10,000

levels of signals due to properties of the fluorochromes and/or available lasers and filters. Expression levels of the antigens may also need to be considered before selecting fluorochromes.

- *Interpretation of the data.* To interpret the data, users also need to be aware that the proportions of cellular populations presented by flow cytometry are usually relative within the sample. For example, proportions of population A and population B were 10%, 90% in sample 1, respectively. Proportions of population A and population B were 20%, 80% in sample 2, respectively. This

TABLE 10.3 Excitation Laser and Filter Setting

LSR Fortessa		Excitation Laser			
		405 nm	488 nm	561 nm	640 nm
Filter	450 ± 40	DAPI			
	530 ± 30		GFP		
	582 ± 15			mOrange	
	780 ± 60			PE-Cy7	
	670 ± 30				APC

difference can be caused by either an increase in population A or a decrease in population B in the sample 2. Similarly, due to a common operational setting a specific cell number per sample is usually displayed, which means that flow cytometry plots often do not provide the total cell number for each sample. Thus, the data from flow cytometry need to be carefully interpreted, since any comparison based on absolute cell numbers might not be possible.

TROUBLESHOOTING

There are two main categories of problems that can arise in flow cytometry: (a) No/low signal or (b) high background/nonspecific staining.

No/Low signal	• No/low signal may be due to wrong settings on the machine. In this case fluorescent beads are useful to check and optimize the state of the equipment.
	• Occasionally, the antibody of choice is not appropriate for flow cytometry. For instance, an antibody that can be used in Western blotting, where usually a denatured protein is detected, may not work in flow cytometry, where the antigen maintains its 3D structure. Antibodies available for immunofluorescence may also need sample fixation to recognize the antigen. In the case of transmembrane proteins, you should make sure that the antibody recognizes the extracellular domain of the protein unless the cells are fixed and permeabilized. Finally, some proteins have various isoforms and the antibody might recognize only a specific isoform.
	• Trypsin might cleave the antigen. EDTA is a chelating agent and can cause conformational change to a protein (such as integrin or cadherin) that requires metal ions such as Mn^{2+}, Mg^{2+}, or Ca^{2+}. In these cases, you should use milder or nonenzymatic ways to detach the cells (such as accutase or a cell-scraper).

- Some receptors get internalized upon binding of antibody. To avoid this, keep the sample at 4°C and/or use FACS buffer containing 0.05% sodium azide that inhibits internalization.
- Choose bright fluorochromes, such as PE, APC. Titration of antibodies is recommended to identify an optimal concentration.

High background/ nonspecific staining

- Titration of the antibodies is necessary to achieve an optimal signal-to-noise ratio.
- Overcompensation also leads to high background, so it is important to use optimal voltages, which, in general, decrease the need for compensation. To this scope, FMO controls are useful for choosing the proper voltage/compensation, since they indicate the spillage of emission light in different wavelengths.
- Flavins, NADH (nicotinamide adenine dinucleotide hydride) in the media, use of PFA, some cell organelles (e.g., mitochondria) can be autofluorescent. The unstained control can give information about the level of autofluorescence in the sample. Cell debris and dead cells tend to have high autofluorescence and nonspecific staining. Exclude those from data analysis by gating. Debris usually has a low SSC and FSC, and dead cells can be identified with stains such as DAPI or 7-AAD (7-amino-actinomycin D).
- The Fc domain of antibodies binds to Fc receptors widely expressed throughout immune cells. In the case cells rich in Fc receptors are analyzed, an Fc blocking agent, monovalent F(ab) or Divalent F(ab')$_2$ antibody fragments should be used.

Clusters of cells are also not desirable, since they give misleading information. For instance, one cell in the cluster may express the marker 1, the other the marker 2, but the machine records only one cell bearing both markers. In order to avoid cell clusters, use of EDTA for the detachment of cells and FACS tubes with cell strainer is recommended. Cell height and area can also be used to gate out the cluster of cells (Fig. 10.8).

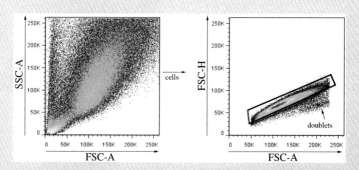

FIGURE 10.8 Sample gating of single cells. After using SSC-A versus FSC-A channels to separate the cells from debris, FSC-H, and FSC-A channels are used to gate single cells from doublets. When a doublet passes from the interrogation point, the size is twice that of a single cell. This means that both the area and the width of the pulse should be doubled, while the height remains almost the same. In BD instruments there is the option to equalize A and H for a given parameter ("Area scaling"). Thus, in the FSC-H versus FSC-A plot, the single cells should be found in the diagonal, enabling to gate only single cells. SSC-W versus SSC-H and then FSC-W versus FSW-H plots also can be used as alternatives.

CONCLUSIONS

Flow cytometry (from the greek words cyto = cell and metry = measure) is a powerful technique that can provide us with information about the properties of cells (morphology, cellular properties, cell cycle stage, etc.). Its great advantage lies on the potential to analyze individual cells in a population without averaging (in contrast to e.g., Western blot). Moreover, FACS, a version of flow cytometry, enables the isolation of the populations of interest for subsequent analysis. Recent development in image analysis software has enabled quantification of immunofluorescence data similar to flow cytometry data, which can overcome the necessity that cells need to be in suspension in flow cytometry. The image analysis is more suitable for nuclear proteins, which are easier to identify individual cells in the pictures, while flow cytometry is optimal for cell surface proteins. Combinations of research tools will allow us to address various biological questions.

REFERENCES

[1] Herzenberg LA, et al. The history and future of the fluorescence activated cell sorter and flow cytometry: a view from Stanford. Clin Chem 2002;48(10):1819–27.
[2] Davidson B, et al. The diagnostic and research applications of flow cytometry in cytopathology. Diagn Cytopathol 2012;40(6):525–35.
[3] Takahashi K, Yamanaka S. Induction of pluripotent stem cells from mouse embryonic and adult fibroblast cultures by defined factors. Cell 2006;126(4):663–76.
[4] O'Malley J, et al. High-resolution analysis with novel cell-surface markers identifies routes to iPS cells. Nature 2013;499(7456):88–91.
[5] Polo JM, et al. A molecular roadmap of reprogramming somatic cells into iPS cells. Cell 2012;151(7):1617–32.

SUGGESTED FURTHER READING

[1] Shapiro H.M. "Practical Flow Cytometry".
[2] Jahan-Tigh RR, Ryan C, Obermoser G, Schwarzenberger K. Flow Cytometry. Journal of Investigative Dermatology 2012.
[3] Darzynkiewicz, Z., Crissman H.A., and Robinson J.P., "Methods in Cell Biology: Cytometry", 3rd Edition, Part A, Vol. 63.
[4] Ormerod, M.G., "Flow Cytometry, A basic introduction".
[5] "An Introduction to Compensation for Multicolor Assays on Digital Flow Cytometers", BD Biosciences, San Jose, CA.
[6] Introduction to flow cytometry webinar: https://www.youtube.com/watch?v=o2joszUiVhM.

GLOSSARY

Autofluorescence Cell structures such as mitochondria or lysosomes can emit fluorescence if they absorb light. This naturally occurring fluorescence is called autofluorescence and can increase the background in flow cytometry and other methods that involve fluorescence detection.

Compensation   A factor to correct color spillover between different channels.

Contour plot   A display of two parameters in which the density of cells is defined by contours.

Fluorochrome   Fluorescent chemical compound that can re-emit light upon light excitation.

Density plot   Displays all cells and different frequency areas are represented with different shades (grayscale) or colors (pseudocolor).

Dot plot   A display of two parameters in which each cell is represented by a dot.

Gate   A gate is a selection of cells with specific features. A great advantage of flow cytometry is that it offers the potential to visualize a group of cells of your choice and this procedure is called gating.

Spectral overlap   The spillover between two fluorescence emission spectra. It can be corrected by applying the proper compensation.

LIST OF ACRONYMS AND ABBREVIATIONS

7-AAD 7-amino-actinomycin D
DAPI 4′,6-diamidino-2-phenylindole
FACS Fluorescence-activated cell sorting
FMO Fluorescence minus one
FSC Forward scatter
GFP Green fluorescence protein
iPSCs Induced pluripotent stem cells
PFA *Paraformaldehyde*
SSC Side scatter

Chapter 11

Transfection

Semira Sheikh, Amanda S. Coutts and Nicholas B. La Thangue
University of Oxford, Oxford, United Kingdom

Chapter Outline

Objectives

This chapter aims to

- Present the general concept of transfection and currently available transfection methods.
- Highlight the differences between transient and stable transfection.
- Discuss factors that influence the success and efficacy of transfection outcome.
- Consider techniques that can be used to optimize a given transfection protocol.
- Describe how to evaluate transfection efficiency.

INTRODUCTION

Transfection is a form of gene transfer and can be defined as the process of introducing exogenous nucleic acid material (DNA or RNA) into cells [1,2]. Genetic material that can be transfected into mammalian cells includes DNA, RNA, messenger RNA (mRNA), small interfering RNA (siRNA), microRNA (miRNA), and short hairpin RNA (shRNA) (see Glossary). Multiple methods for transfection of mammalian cells have been described; these can be subdivided into biological, chemical, and physical methods, and also into those methods that generate transient transfection or stable transfection (see Table 11.1)

Basic Science Methods for Clinical Researchers. DOI: http://dx.doi.org/10.1016/B978-0-12-803077-6.00011-4
191

TABLE 11.1 Comparison of Transient Versus Stable Transfection

Transient Transfection	Stable Transfection
Chemical or physical methods	Biological or physical
Short-term expression	Long-term expression with persistent gain-of-function or loss-of-function
No genomic integration	Risk of nonspecific integration

[2,3]. The choice of method depends on the objective of the experiment, but in general the ideal transfection method should be highly reproducible and should result in high transfection efficiency with minimal cell toxicity and no major physiological alterations to the transfected cells.

Transient Versus Stable Transfection

Transient transfection can be used to investigate the short-term impact of altered gene or protein expression. Transiently transfected genetic material is not integrated into the host genome and therefore the effect on target gene expression is time-limited and is lost during ongoing cell division. DNA introduced into cells in this way has been shown to be located in the nucleus and its impact on transcriptional gene activity is usually measurable as early as 24 hours post-transfection (sometimes earlier) with most protocols specifying between 48 and 96 hours for optimal effect; transfected RNA on the other hand is usually found to be in the cytoplasm with its optimal target effects measurable for 24–72 hours post-transfection [2].

By contrast, a stable transfection is one in which the transfected DNA is either integrated into the chromosomal DNA of the host cell, or maintained in episomal form (see Glossary). A selection marker, such as for example an antibiotic-resistance gene, can then be used to select for those cells that have integrated the genetic material of interest and to generate a cell line permanently expressing the gene of interest [2,3]. When a gene encoding resistance to a particular antibiotic is incorporated into a plasmid for transfection purposes (in certain stable transfections), this gene allows cells to grow in the presence of a particular antibiotic (e.g., kanamycin) only when the cells are expressing the plasmid containing the transfected DNA. In this way cells can be "selected" for by growing in the presence of the antibiotic at an antibiotic concentration that would normally kill the cells.

IN PRINCIPLE

The fundamental principle of transfection is to transfer negatively charged genetic material across a negatively charged cell membrane. This can be achieved by various methods (see Table 11.2), including transfer by chemical

TABLE 11.2   Overview of Advantages and Disadvantages of Different Transfection Methods

Class	Method	Pros	Cons
Chemical	• Cationic polymer • Cationic lipid • Calcium phosphate	− Relative ease of use − Relatively inexpensive	− Variable transfection efficiency − Toxicity to cells by some chemical compounds − May be difficult to transfect particular cell types e.g., small cells, nonadherent cells
Physical	• Direct injection • Biolistic particle delivery • Electroporation • Laserfection	− Can be used for hard-to-transfect cells − May allow single-cell transfection − May allow precise insertion of nucleic acid into subcellular compartment	− May cause significant toxicity to cells − May require specialized equipment and operator skill
Biological	• Virus-mediated	− High transfection efficiency − Can be used for difficult to transfect cells	− Requires biosafety-standard laboratory space − Risk of insertional mutagenesis − DNA insert size limit

methods that can neutralize the negative charge of the DNA or RNA, physical methods that may cause temporary disruption of the plasma membrane to allow passage of genetic material into the cell, or biological methods that employ viruses as agents mediating nucleic acid transfer.

CHEMICAL METHODS

Chemical methods were the first to yield successful transfection of mammalian cells and are still widely used. They include the use of cationic polymers, calcium phosphate, or cationic lipid (see Table 11.2) and have a similar underlying principle, that is, the formation of complexes with negatively charged nucleic acid, which come into close contact with the cell membrane and are then thought to be taken up by the target cell by either endocytosis or phagocytosis [2].

Diethylaminoethyl (DEAE)-dextran was one of the earliest agents used for transfection [4,5]. It is a cationic polymer that is able to form tight complexes with negatively charged nucleic acid. Its excess of positive surface charge allows the nucleic acid:polymer complex to come into close association with the negatively charged cell membrane from where it is then taken up, presumably by endocytosis. Other synthetic cationic polymers that have been used for DNA transfection include polybrene, polyethyleneimine, and activated dendrimers [2,6]. *Dendrimers* are repetitively branched, often spherical, macromolecules that carry positively charged amino acids on their surface that can interact with negatively charged nucleic acid. The resulting complex of nucleic acid and dendrimer has an overall positive charge and can therefore be bound to the surface of the target cell and subsequently endocytosed. Since the genetic material is highly condensed within the dendrimer complex, it is also protected from intracellular nucleases, and buffering of the acidic endosomes by unprotonated dendrimer amino groups also protects the nucleic acids from pH-dependent endosomal nucleases.

Calcium phosphate transfection methods are based on the formation of a DNA:calcium phosphate co-precipitate [2,7]. DNA is mixed with calcium chloride and the mixture is then added to a buffered saline/phosphate solution in a controlled fashion that generates a precipitate. The precipitate is then dispersed onto the cultured cells and taken up by endocytosis or phagocytosis. This method has been shown to be versatile and can be used in a range of different cells for both transient and stable transfection. Calcium phosphate may also be protective against intracellular and serum nucleases (see Glossary) but small changes in pH (+/−0.1 pH units) can significantly compromise its efficiency as a method.

Cationic lipids or *liposomes* are among the most popular transfection methods currently used given their high transfection efficiency and relative ease of use. Liposomal techniques can be used for both transient and stable transfections and, unlike cationic polymers or calcium phosphate, liposome-mediated nucleic acid delivery can also be used for in vivo transfer of genetic material [8,9].

The cationic portion of the lipid molecules can interact with the negatively charged nucleic acids, resulting in a complex with an overall net positive charge. Entry into cells is thought to be either by endocytosis or fusion of the liposome with the plasma membrane. Often cationic lipids are mixed with neutral lipids such as L-dioleoyl phosphatidylethanolamine (DOPE), which is thought to enhance the fusion of lipid:nucleic acid complexes with the outer cell membrane as well as facilitate the release of complexes from intracellular endosomes. More recently, the addition of biosurfactants such as mannosylerythriol lipid (MEL-A) has been shown to increase transfer efficiency while maintaining low toxicity to transfected cells [10].

PHYSICAL METHODS

Various physical methods can be used to deliver nucleic acid into cells and although these can be effective, they require specialized tools and equipment [2].

Direct microinjection delivers nucleic acid into cells by direct injection with a fine needle. Although this allows for precise delivery into the cytoplasm or nucleus of individual cells, this is a particularly labor-intensive technique that requires a lot of operator experience and skill.

Biolistic particle delivery uses gold particles coated with nucleic acid and delivers these into cells using a high velocity gene gun. The method is reliable and can be used in a wide range of cells but requires specialized equipment and causes physical damage to samples.

Electroporation is perhaps the most widely used physical method. Electrical pulses are applied to cells in order to destabilize the cell membrane temporarily to form pores. These pores allow the nucleic acid material to pass into cells. When the electric field is turned off, the pores in the membrane reseal, retaining the transfected material. Although a large number of cells can be transfected using electroporation it does require a greater starting number of cells since toxicity is relatively high compared to other methods [11,12].

In *laser-mediated* transfection, also known as laserfection or optoporation, a pulse laser is used to transiently permeabilize cells in order to transfer nucleic acid material. This method can be very efficient and a particular advantage is the fact that it allows very small cells to be transfected. However, its use can be limited by cost as it requires an expensive laser-microscope system [13,14].

BIOLOGICAL METHODS

Viruses can be used as carriers for nucleic acid material and *virus-mediated gene transfer* can also be referred to as transduction [2]. This technique is highly effective and can be used to achieve stable gene expression owing to the integration of the virus into the host genome. Often a selectable marker gene, which can either be based on antibiotic-resistance or *green fluorescent protein* (GFP) expression, is introduced alongside the nucleic acid material of interest, and allows for selection of cells that have been successfully transfected. Commonly used viral vectors include adenoviruses or adeno-associated viruses, retroviruses, and lentiviruses, all of which allow relatively large inserts of DNA (7.5–9 kb) and can also be used to transfect quiescent cells.

Drawbacks of this method, however, are the fact that it is relatively labor-intensive and requires a Biosafety Level 2 working environment for most viruses. Moreover, the exact insertion site of the viruses cannot be controlled and random insertion into the host genome can potentially disrupt important regulatory genes within the cell, and, in the case of retroviruses, lead to the activation of latent disease [15].

IN PRACTICE

Here we present a standard protocol for the transfection of adherent U2OS osteosarcoma cells (that has also been successfully applied to a wide variety of adherent cell types) with GFP plasmid DNA, using a chemical transfection method, in order to achieve transient expression of GFP in the host cell. This general

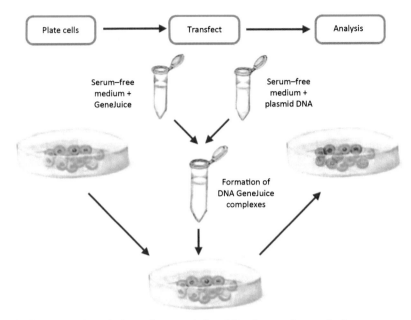

FIGURE 11.1 Schematic that outlines a typical workflow for a routine transfection.

protocol can be used in a variety of cell types, and depending on cell type, culture format, and transfection reagents used, the standard form of this protocol can be adjusted and optimized for a variety of cell lines, taking into account specific requirements of the particular method used as well as the reagent manufacturer's recommendations. For example, when using an electroporation method with nonadherent cells, the number of cells and amount of transfected material has to be scaled up significantly for a successful outcome, and if transfection efficiency is of key concern, co-transfection with a fluorescent marker that can be assessed by immunofluorescence or flow cytometry may be considered. A schematic of the transfection workflow is shown in Fig. 11.1.

I. *Materials required*

   1. Nucleic acid preparation

      a. Use high-quality DNA and RNA. For this example, we will use enhanced *green fluorescent protein* (EGFP) plasmid DNA that is commercially available (e.g., pEGFP-N1 from Clontech). Nucleic acid should be of good quality, for example we use Qiagen plasmid purification kits (Qiagen).

      b. Dilute plasmid in nuclease-free water or suitable buffer (e.g., 10 mM Tris-HCl pH 8.5 (EB provided in Qiagen kits)) to a recommended standard concentration (e.g., 200 ng/μL).

2. Cell culture and transfection reagents
   a. Cells should be grown at 37°C with 5% CO_2 in a humidified incubator unless otherwise specified. Use cells <50 passages for transfection and only use rapidly proliferating cells (unless unavoidable and thus use a transfection method that can be used for non-proliferating cells, e.g., retrovirus). For reproducibility and standardization purposes, maintain consistent conditions for cell growth and density.
   b. Cell culture medium supplemented with fetal calf serum (FCS) and antibiotic (if used), for example, Dulbecco's Modified Eagle Medium (DMEM) with 5% FCS and 1% Pen/Strep.
   c. On the day of the transfection or the day before, seed cells in 6 cm dishes (volume of complete growth medium 2 mL).
   Note: Most manufacturers will provide instructions on recommended number of cells, amount of nucleic acid to be used for transfection, and ratio of nucleic acid to transfection reagent depending on cell culture format.
   d. Serum-free medium (e.g., Opti-MEM, Invitrogen).
   e. Transfection reagent (e.g., GeneJuice, Novagen).
3. Western blotting (see also Chapter 6)
   a. Cell lysate: Determine the protein concentration using for example, a standard Bradford assay. Protein loading buffer (e.g., 4x Laemmli sodium dodecyl sulfate (SDS) sample buffer containing 2-beta-mercaptoethanol).
   b. Stacking and resolving SDS-PAGE (polyacrylamide gel electrophoresis) gels (for this example, (see Fig. 11.2) a 12% resolving gel was used as the expected molecular weight of EGFP is 27 kDa). In general,

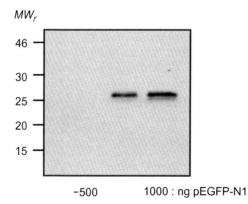

FIGURE 11.2 U2OS cells were transfected with pEGFP-N1 at the concentrations denoted, or mock-transfected (mock/(-)) using GeneJuice. Cells were left 48 hours before harvesting and processing for immunoblotting. GFP was detected using mouse-anti-GFP antibody (Covance 1/50,000).

the percentage of acrylamide depends on the molecular weight of the protein of interest.

c. SDS running buffer and transfer buffer (with 10–20% methanol or 0.1% SDS as required, note also that the pH of the transfer buffer can be adjusted depending on the pI (isoelectric point) of your protein of interest). In general, we use Towbin transfer buffer pH 8.3 with 10% methanol and 0.1% SDS as required.

d. Protein ladder/molecular weight marker (e.g., Protein Color Standard, New England Biolabs Inc).

e. Gel apparatus and transfer equipment.

f. Nitrocellose or polyvinylidene difluoride (PVDF) membrane (Immobilon, Millipore, UK).

g. Ponceau S (Sigma-Aldrich).

h. Blocking agent (e.g., 5% skimmed milk).

i. Appropriate primary and secondary antibodies.

4. Immunofluorescence (see also Chapter 8)

a. Coverslips, 13 mm diameter (VWR International), microscope glass slides.

b. Six-well plates (Costar).

c. 3.7% formaldehyde/phosphate-buffered saline (PBS) for fixing.

d. 0.5% TritionX-100 in PBS to permeabilize cells.

e. 5% FCS-PBS for blocking (if required).

f. PBS/0.025% Tween for washing coverslips.

g. Primary and secondary antibodies for the protein of interest.

h. Mounting medium, for example, vectashield with 4′,6-diamidino-2-phenylindole (DAPI) (Vector Laboratories).

i. Clear nail varnish to seal coverslips.

j. Nontransparent humidified plastic box for incubation steps.

II. *Transfection Method*

1. *Transfection* (see Fig. 11.1)

a. Cell density

i. A cell confluency of around 40–60% is generally recommended. Note: This can vary depending on individual preference, required time in culture, and success with an optimized method; confluence as low as 20% can also lead to successful transfection.

b. Ratio of transfection reagent to DNA

i. Most manufacturers will specify this in their protocol. In this example, we use GeneJuice as the transfection reagent, which in our laboratory is used at a ratio of 1.5 μL GeneJuice per 1 μg of DNA plasmid. Note: The ratio of DNA to transfection reagent can vary between 1:1 and 3:1, depending on transfection material and conditions. Similarly, siRNA transfection reagents can be used in ratios of 2:1 up to 4:1 depending on the protocol used.

c. Transfection procedure

i. Plate $10-20 \times 10^5$ U2OS cells per 6 cm dish the day before transfection and incubate overnight at 37°C (5% CO_2). Aim for cell confluency of between 20% and 40% in order to perform the experiment.

Note: Add appropriate controls to run alongside your experiment. For example, a control plasmid (e.g., empty vector) or non-targeting siRNA should be chosen to assess for nonspecific effects of the test plasmid/siRNA under the same transfection conditions. In addition a mock transfection (transfection reagent mix without nucleic acid) can be included as an additional control.

ii. For each plate to be transfected, pipette 500 μL serum-free medium into a sterile Eppendorf tube and add 1.5 μL GeneJuice per 1 μg of DNA to be used. Vortex, then incubate for 5 minutes at room temperature.

Note: We recommend making a master mix for all the individual transfections you need by adding serum-free medium and the total amount of GeneJuice needed into one tube before adding to individual tubes with DNA (see below). This can minimize variability in transfection efficiency.

iii. For each plate to be transfected, add the desired amount of DNA to a new Eppendorf tube (e.g., we have used 500 ng and 1 μg of pEGFP-N1 in Figs. 11.2 and 11.3) and then add the serum-free medium/transfection reagent mix and mix gently by

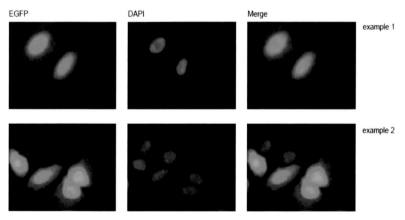

FIGURE 11.3 U2OS cells were plated onto 13 mm coverslips before transfection with pEGFP-N1 plasmid using GeneJuice. Cells were left 48 hours before fixation and processing for immunofluorescence. Note that in the case of GFP expression no primary or secondary antibody is required. In example 1, both cells shown in the picture have been transfected, whereas in example 2 both transfected and untransfected cells can be seen, giving an example of how to estimate transfection efficiency. DAPI staining was used to visualize nuclei.

pipetting up and down three times (do not vortex). Incubate at room temperature for about 15 minutes (we generally find 5–15 minutes optimal).

iv. Add the transfection mixture to the individual dishes in a dropwise fashion. Plates can be agitated gently but should not be swirled.

v. Check plates on a daily basis by fluorescence microscopy if available. EGFP should become visible in transfected cells after a few hours. Transfection efficiency can be calculated by determining the ratio of fluorescent cells to nonfluorescing cells (compare the number of cells/field using standard light microscopy versus that seen with fluorescence to visualize EGFP). Other methods to assess transfection efficiency can include f*luorescence-activated cell sorting* (FACS) and inclusion of a marker of transfection efficiency, such as luciferase or β-galactosidase, the expression levels of which can be measured.

vi. Cells can be harvested after 48–72 hours. Cell pellets are re-suspended in the appropriate lysis buffer (we routinely use TNN buffer (50 mM Tris pH = 7.4, 5 mM EDTA, 0.5% NP40, 150 mM NACl with protease and phosphatase Inhibitor)). Lysates should sit on ice for 20–30 minutes before clarification by centrifugation for 10 minutes at $15,700 \times g$ at 4°C. Lysates should be stored at −80°C until EGFP protein expression can be examined by Western blotting.

2. *Western blotting* (see Fig. 11.2)

   a. SDS-PAGE gel electrophoresis and transfer

   i. Samples stored at −80°C should be thawed and the protein concentration for each sample determined using a standard protein quantitation assay (such as Bradford reagent (Sigma-Aldrich)).

   ii. Determine the amount of protein you want to load on your gel (usually between 30 and 60 μg, depending on your protein) and add the required amount of lysate to a new tube before adding SDS sample buffer. Heat samples at 95°C in a heat block for 5 minutes, and spin down any condensation by centrifuging briefly (e.g., 10 seconds) at room temperature. Your sample is now ready for loading onto your gel.

   iii. Load samples in the preferred running order and include a protein ladder into one lane of your gel so you can compare the relative molecular mass.

   iv. Run the gels at 200 V for approximately 45 minutes, or until you can see that the loading dye has reached the bottom of the gel.

   v. Transfer gels into a transfer tank (as per manufacturer's instructions) to allow for electrophoretic transfer of proteins onto a membrane (e.g., PVDF or nitrocellulose). Transfers can be run at 400 mA for 60–90 minutes, or overnight at 150 mA, at 4°C.

    vi. Remove membranes and stain with Ponceau S for rapid detection of protein. Wash membranes with distilled water so protein bands are clearly visualized. Ponceau S is a reversible protein stain and can be easily and quickly removed before proceeding with immunodetection by a brief wash in PBS.

    vii. Block membranes in 5% skim milk (w/v) in PBS containing 0.1% Tween-20 (SM-PBST) for 20 minutes (note that the choice of blocking can also vary, for example if you were interested in detecting a phospho-protein then bovine serum albumin (BSA) as a blocking agent would be preferred, in addition the use of Tris-buffered saline (TBS) instead of PBS throughout is recommended).

  b. Incubation with primary antibody

    i. Add appropriate primary antibody to the required volume of 5% SM-PBST and incubate at 4°C overnight.

    Note: Concentrations for primary antibodies are usually 1:500–1:1000 but can be worked out during optimization of the protocol. The manufacturer's datasheet is usually a reasonable guide.

    ii. Wash membranes in PBST (e.g., a short wash of 1 minute should suffice).

  c. Incubation with secondary antibody

    i. Add appropriate secondary antibody in 5% SM-PBST. Usually this is added at a concentration of 1:10,000.

    ii. Incubate for 30–40 minutes at room temperature.

    iii. Wash membranes in PBST (e.g., three washes for 5 minutes each, but exact conditions can depend on the individual antibody and may be altered to accommodate for differences in background, strength of signals, etc.).

  d. Detection of antibody complexes

    i. We use enhanced chemiluminescence (ECL) for detection of antibody/antigen complexes. This technique requires a darkroom and film developer. In addition the reagents can be purchased commercially or homemade.

3. *Immunofluorescence* (see Fig. 11.3)

  a. Culture cells onto coverslips in six-well plates

    i. Seed cells at a density of approximately 1×10^5 cells per well. Transfect following the above protocol but using 500 ng EGFP DNA and 250 µL of serum-free medium.

    ii. After growth for the appropriate time (e.g., 24–72 hours), rinse cells with PBS x2.

  b. Fix cells by adding 1 mL 3.7% formaldehyde-PBS to each well and fix cells for 10–15 minutes at room temperature.

  c. Remove fix and wash 3x with PBS.

   d. Permeabilize cells with 0.5% TritonX-100 in PBS for 2–5 minutes at room temperature.

   e. Block cells with 5% FCS-PBS for upto 30 minutes at room temperature.

   Note: This is an optional step. Depending on the antibody used, blocking conditions can be altered or this step can be omitted completely.

   f. Add primary antibody to coverslips. Add 20 µL of primary antibody at the appropriate concentration. Incubate in a humidified dark box at 4°C overnight, or for 30–60 minutes at room temperature.

   Note: Similar to Western blotting, check manufacturer's instructions for the recommended dilution; if using an antibody for the first time without prior recommendation, try using it at 1:200 and adjust concentration in subsequent experiments depending on the results.

   g. Add the appropriate secondary antibody after washing slides 2–4x with PBS/0.025% Tween-20. Incubate in the dark for 30 minutes at room temperature. Wash coverslips 4–6x with PBS/0.025% Tween-20.

   Note: Use secondary antibodies at a concentration of 1:500 if not otherwise specified.

   h. Mount coverslips (cell-side down) onto glass slides using mounting medium with DAPI. Seal around coverslips with clear nail varnish.

   i. Visualize slides using appropriate filters on an immunofluorescence microscope.

   j. Store slides in dark boxes at 4°C.

APPLICATIONS

Transfection has become a powerful analytical tool for the study of gene or protein expression in a wide variety of cell types. The method of transfection, that is, the choice of a chemical, physical, or biological approach, depends on the cell type under investigation and the objective of the experiment. Balancing transfection efficiency with cell toxicity inherent in the transfection method is a key consideration [1,2].

Although transfection as a technique has mainly been used to study the function of a particular gene or gene product by enhancing or inhibiting its expression, constant improvement of the method has also made it possible to investigate the function of multiple genes within a cell simultaneously and to scale up transfections so as to allow the large-scale production of recombinant proteins [1,10].

Other areas of application include its use in gene therapy where a gene of interest can be delivered into particular cells in an attempt to regulate a disease phenotype, and to induce pluripotent stem cell generation by transfecting a number of different transcription factors. The advent of siRNA can be used to generate knockdown of specific genes and therefore holds great therapeutic promise for the treatment of a range of diseases [16]. In addition, newer gene editing techniques such as CRISPR-Cas9 (clustered regularly interspaced short palindromic repeats and CRISPR-associated endonuclease 9) allow gene knockouts, gene regulation and the generation of genomic mutations with greater ease, but are still reliant on the effective delivery of nucleic acids into the host cell.

Gene silencing through RNA interference and the creation of stable cell lines are two important and widespread applications of transfection and discussed in more detail below.

Gene silencing. The ability to silence genes by RNA interference (RNAi) is arguably one of the most exciting recent developments in molecular biology. RNAi-based techniques rely on the inherent cellular machinery of eukaryotic cells that function to inhibit mRNA translation (Fig. 11.4). Exogenous sequences of siRNA can be designed (e.g., using web-based tools) for a specific gene of interest and introduced into the cytoplasm of cells through transfection,

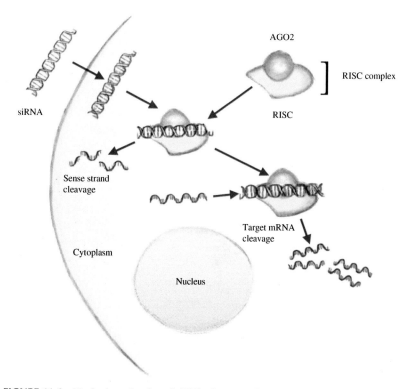

FIGURE 11.4 Mechanism of action of siRNA. Once transfected into the cell cytoplasm, siRNA is introduced into endoribonuclease-containing complexes known as RNA-induced silencing complexes (RISCs), resulting in the cleavage of the sense-strand by argonaut2 (AGO2). Activated RISCs then bind to complementary mRNA via the siRNA antisense strand, leading to degradation of target mRNA and therefore gene silencing.

leading to knockdown of gene expression. Various transfection reagents are available for this process and can be optimized depending on the cell type. For larger genome analyses, libraries of RNA molecules are available that allow analysis of many different targets at the same time. While siRNA can often have the advantage of achieving a high and reliable level of knockdown, this can be transient and lead to some unwanted off-target effects. Therefore, the introduction of shRNA sequences into cells by viral or nonviral vectors offer the opportunity of generating stable knockdown cell lines. Introduction of shRNA may be effected through nonviral vectors, using optimized chemical transfection reagents or electroporation, or through viral vectors, in particular using adeno-, adeno-associated-, or lentiviral vectors. This method offers a more stable expression, with shRNA acting through similar pathways as siRNA. However, creation of stable cell lines may be associated with lower transfection efficiency rates and can be a more time-consuming process.

Stable cell lines. The generation of stable cell lines through transfection of shRNA for gene knockdown or DNA to enhance gene expression offers advantages over transient expression techniques. It produces a system that can be used for long-term experiments and does not require repeated transfections, thereby establishing a more uniform level of gene expression in a particular cell population. In order to create a stable cell line, however, nuclear material has not only to be introduced into the cell, but also delivered to the nucleus and be integrated into the host genome. A positive selection marker, such as an antibiotic (e.g., doxycycline or puromycin) can be used to allow the selection of stable clones expressing the gene of interest. This can be a time-consuming process and also requires final verification of stable cell lines, ideally by sequencing.

SCENARIO

General considerations: A relevant scenario where both transfection with plasmid DNA and transfection with siRNA can be envisaged to be useful is one where a particular protein has been found to be implicated in a biological process. In order to test this hypothesis, cells can be transfected with plasmid DNA carrying the gene of interest, leading to protein overexpression. By using intentional gene overexpression in this way, useful mechanistic insights into the biological pathways affected may be gained [17]. However, for this experiment to yield meaningful results, it is important to run relevant controls in parallel; for example, control cells can be mock-transfected with an "empty vector" plasmid DNA or with a control plasmid DNA carrying a gene not thought to be relevant to the particular biological process under study. In addition, co-transfection with a second plasmid such as EGFP, which can be visualized by immunofluorescence also allows selection of transfected cells by flow cytometry. This can be a useful strategy to ensure that transfection has taken place and also allows estimation of transfection efficiency. Expression of a particular protein in this way can be analyzed either by immunoblotting as described above or in addition one might also choose to quantify the messenger RNA levels using, for example, quantitative polymerase chain reaction (qPCR). A complimentary experimental strategy in this context would be to subsequently knock down the protein of interest by transfecting the cell line with specific siRNA. This may result in abrogation or attenuation of the biological pathway under investigation. As before, the levels of protein or mRNA can be assessed by immunoblotting or qPCR, respectively.

If overexpression and/or knockdown of the protein in question can be shown to lead to changes in biological function, an argument may be made that this protein may be integral to the particular pathway. However, validating this outcome by reproducing this finding in other cell lines and/or an in vivo model may still be required and is generally considered good scientific practice.

Specific example: HR23B is a cytoplasmic protein which is involved in shuttling ubiquitinated proteins to the proteasome for degradation and was identified by a genome-wide loss-of-function screen to be implicated in influencing sensitivity to histone deacetylase inhibitors (HDI), an emergent class of anticancer drugs. In order to investigate its role as a candidate biomarker, HR23B was transiently depleted with siRNA against HR23B in U2OS osteosarcoma cells. Cell sensitivity to HDI subsequently was shown to be markedly reduced, suggesting that HR23B plays a role as a determinant for HDI sensitivity. A stable cell line expressing inducible HR23B was created by transfecting U2OS cells with a FLAG-HR23B construct, allowing conditional expression of HR23B. Using this system it could be shown that induced expression of HR23B led to increased sensitivity to HDI, compared to low/background only expression of HR23B. Since HDI are licensed for treatment of cutaneous T-cell lymphoma (CTCL), subsequent experiments were carried out in CTCL cell lines MYLA, HUT78, and SeAx, which are known to be more difficult to transfect with conventional transfection reagents (e.g., oligofectamine, lipofectamine) using siRNA. Therefore, electroporation was chosen as a means to transfection these cells with siRNA against HR23B. Conditions were optimized for each cell line with regards to pulse voltage and duration. Successful knockdown of HR23B cells was confirmed by Western blotting, and cell sensitivity to HDI in the context of HR23B knockdown was demonstrated by FACS and immunoblot. The above example therefore shows that different transfection methods can be used to delineate gene or protein function within different cell types [18].

KEY LIMITATIONS

Although transfection is widely used and serves as a crucial technique to most biologists, there are some important limitations to be considered. Overexpression experiments need to be interpreted with caution as the levels of protein expression achieved may far surpass what is normally found in the cell and thus could lead to off-target and/or spurious effects. In addition, ectopically expressed proteins are often detected by the addition of a tag (e.g., GFP, HA, or Myc) and so

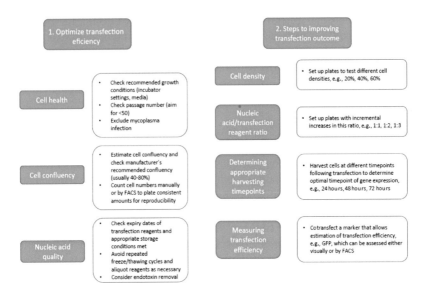

FIGURE 11.5 Troubleshooting flowchart.

one must consider that the tag may alter normal protein function and/or location within the cell. Other transfection-based techniques, such as creation of inducible stable cell lines, should also be considered as they can allow one to more finely tune and control the expression of a protein. This is also an important consideration if you are consistently unable to see protein expression as this could be due to the fact that expression of the protein is normally detrimental to the cell. For example, constitutive overexpression of a protein that causes apoptosis can result in significant cell death, making it difficult to observe expression in transient or stable constitutive expression systems. In instances such as these, it is worth considering the creation of inducible stable cell lines where the gene can be "turned on" only when required (e.g., TET ON/OFF systems (Clontech)).

Troubleshooting (see Fig. 11.5)

1. Factors affecting transfection efficiency
   A number of different parameters can influence transfection efficiency, including cell health, confluency, and nucleic acid quality.
   - *Cell health*: Cells should be grown in the required medium under the appropriate conditions, usually at 37°C in an adequately humidified incubator with 5% CO_2. Recommended optimal cell culture conditions for individual cell lines can be checked either on the ATCC or DSMZ webpages [19,20].
     Cells should be routinely checked for mycoplasma as this will not only impact on transfection efficiency but can also lead to spurious results.

Passage number may also impact on the successful outcome of transfection; thus, a passage number <50 is general recommended as cells may become less amenable to transfection after repeated passages. Sometimes it may be necessary to thaw and use new uncontaminated stocks.

- *Cell confluency*: Recommended ranges of cell confluency vary widely (40–80%) and depend on which cells are used. In general, too few cells will grow very slowly because they lack cell-to-cell contact and/or sufficient paracrine signals. By contrast, too many cells will result in contact inhibition and result in poor uptake of nucleic acid. Furthermore, actively dividing cells are better at taking up nucleic acid compared to cells that are quiescent (although, as previously described, some transfection methods may be suitable for slowly growing and/or quiescent cells).

- *Nucleic acid quality*: Ensure that nucleic acid used for transfection is free from contamination for example, with protein or microbial components such as endotoxin. We routinely prepare our DNA using Qiagen plasmid kits as these achieve a good level of purity with low endotoxin contamination. If endotoxin is a particular concern there are kits available for purification of DNA with negligible levels of endotoxin (e.g., EndoFree plasmid kits, Qiagen).

2. Optimization of transfection efficiency

There are a number of ways that you should consider when optimizing your transfection efficiency and depending on the experimental setup and/or outcome required this may be more or less of a consideration. For example, if low transfection efficiencies mean that determining a biological outcome is compromised then creating stable cell lines should be considered.

- *Cell density*: Try several cell densities to see which works best with a particular transfection reagent/method. It is also important to take into account the length of the experiment when setting up the cells for a transfection-based experiment. For example, a typical siRNA knockdown experiment would be left for 72 hours before harvesting and, in this case, if you have fast growing cells and you would ideally like your cells to be 70–80% confluent at the time of harvesting, then the starting cell density should be adjusted accordingly.

- *Ratio of nucleic acid/transfection reagent*: Each manufacturer of a particular reagent will have a range of recommended ratios. When embarking on transfections with a new reagent it is worthwhile testing several ratios (e.g., 1:1, 2:1, 3:1) of reagent:nucleic acid. Ideally, your chosen ratio should consistently give you good transfection efficiency with minimal cell toxicity while also balancing the cost benefit.

- *Cell culture conditions*: Every cell type has different culture conditions and growth parameters that may affect transfection efficiency. Moreover, some cell types are just notoriously difficult to transfect and require specific methods and/or reagents (e.g., nonadherent cells). In addition, for adherent cells the lag time between plating and transfection can also influence the efficiency of uptake of nucleic acid/reagent complexes. We normally set up our cells and transfect between 4 and 24 hours post plating, depending on cell type and experimental requirements. We have found that time points much later can decrease the efficiency of transfection substantially.

- *Measuring transfection efficiency*: There are a number of ways to measure your transfection efficiency. Direct visualization of a cell population expressing for example, pEGFP (see Fig. 11.3). In addition, a marker such as GFP expression can be used to sort cells on a FACS machine and the transfected population can then be collected. Further, cells can be transfected with constructs that encode for proteins such as β-galactosidase or luciferase, which allow for the use of colorimetric, fluorescent, chemiluminescent, or photon emission methods to determine the transfection efficiency. This information may be necessary if differences in transfection efficiency may influence the experimental outcome (e.g., if you are comparing several treatments, proteins, and/or time points).

CONCLUSION

The ability to transfect nucleic acids into cells is crucial for most aspects of cellular, molecular, and biological studies of genes and their function. Despite there being a wide range of methods and reagents for transfection into cells there still remain challenges to be overcome. Some cell types, such as nonadherent cells, are extremely difficult to transfect and better methods and reagents would help to easily include more cell types into routine studies. Additionally, cell toxicity is always a concern and care must be taken to establish the best method for each individual cell line. Lastly, the appropriate controls must always be included (e.g., vector control, mock (no nucleic acid) control) in order to determine the best treatment/protocol for any particular cell type and experimental condition.

ACKNOWLEDGMENTS

We thank the MRC and CRUK (Program Grant C300/A13058) for funding. SS was supported by an OCRC Clinical Research Training Fellowship (OCRC-CRF14-MW).

REFERENCES

[1] Geisse S, Voedisch B. Transient expression technologies: past, present, and future. Methods Mol Biol 2012;899:203–19.

[2] Kim TK, Eberwine JH. Mammalian cell transfection: the present and the future. Anal Bioanal Chem 2010;397(8):3173–8.

[3] Recillas-Targa F. Multiple strategies for gene transfer, expression, knockdown, and chromatin influence in mammalian cell lines and transgenic animals. Mol Biotechnol 2006;34(3):337–54.

[4] Pagano JS, Vaheri A. Enhancement of infectivity of poliovirus RNA with diethylaminoethyl-dextran (DEAE-D). Arch Gesamte Virusforsch 1965;17(3):456–64.

[5] Schenborn ET, Goiffon V. DEAE-dextran transfection of mammalian cultured cells. Methods Mol Biol 2000;130:147–53.

[6] Haensler J, Szoka Jr. FC. Polyamidoamine cascade polymers mediate efficient transfection of cells in culture. Bioconjug Chem 1993;4(5):372–9.

[7] Graham FL, van der Eb AJ. Transformation of rat cells by DNA of human adenovirus 5. Virology 1973;54(2):536–9.

[8] Felgner PL, et al. Lipofection: a highly efficient, lipid-mediated DNA-transfection procedure. Proc Natl Acad Sci USA 1987;84(21):7413–7.

[9] Fraley R, et al. Introduction of liposome-encapsulated SV40 DNA into cells. J Biol Chem 1980;255(21):10431–5.

[10] Nakanishi M, Inoh Y, Furuno T. New Transfection Agents Based on Liposomes Containing Biosurfactant MEL-A. Pharmaceutics 2013;5(3):411–20.

[11] Shigekawa K, Dower WJ. Electroporation of eukaryotes and prokaryotes: a general approach to the introduction of macromolecules into cells. Biotechniques 1988;6(8):742–51.

[12] Wong TK, Neumann E. Electric field mediated gene transfer. Biochem Biophys Res Commun 1982;107(2):584–7.

[13] Shirahata Y, et al. New technique for gene transfection using laser irradiation. J Investig Med 2001;49(2):184–90.

[14] Yao CP, et al. Laser-based gene transfection and gene therapy. IEEE Trans Nanobioscience 2008;7(2):111–9.

[15] Pfeifer A, Verma IM. Gene therapy: promises and problems. Annu Rev Genomics Hum Genet 2001;2:177–211.

[16] Elbashir SM, et al. Duplexes of 21-nucleotide RNAs mediate RNA interference in cultured mammalian cells. Nature 2001;411(6836):494–8.

[17] Prelich G. Gene overexpression: uses, mechanisms, and interpretation. Genetics 2012;190(3):841–54.

[18] Khan O, et al. HR23B is a biomarker for tumor sensitivity to HDAC inhibitor-based therapy. Proc Natl Acad Sci USA 2010;107(14):6532–7.

[19] http://www.lgcstandards-atcc.org/?geo_country=gb.

[20] https://www.dsmz.de/.

GLOSSARY

Episomal   Circular extrachromosomal DNA.

Nuclease   An enzyme that promotes the degradation of nucleic acid.

shRNA/siRNA   Short hairpin or small interfering RNA refers to a short stretch of double-stranded RNA that will ultimately guide the cleavage and degradation of the target RNA. siRNA sequences are usually 21–23 nucleotides long. shRNA are normally incorporated into a plasmid, and expressed and transported into the cytosol where they are cleaved into smaller siRNA molecules.

Chapter 12

In Vivo Animal Modeling: *Drosophila*

Michael F. Wangler[1,2] and Hugo J. Bellen[1,2,3]

[1]Department of Molecular and Human Genetics, Baylor College of Medicine (BCM), Houston, TX, United States; [2]Jan and Dan Duncan Neurological Research Institute, Texas Children's Hospital, Houston, TX, United States; [3]Howard Hughes Medical Institute, Houston, TX, United States

Chapter Outline

Basic Science Methods for Clinical Researchers. DOI: http://dx.doi.org/10.1016/B978-0-12-803077-6.00012-6

Objectives

This chapter should prepare the clinician training in basic science to:
- Understand the advantages of *Drosophila* genetics for answering questions underlying etiology and pathogenesis of human disease.
- Effectively evaluate experiments from the *Drosophila* literature to understand disease biology by learning to interpret *Drosophila* genetics.
- Design strategies using *Drosophila* for disease biology studies.

INTRODUCTION

Drosophila melanogaster

Drosophila melanogaster (hereafter in this chapter named *Drosophila*) a species of Dipterans are the common fruit fly or vinegar fly (Fig. 12.1). Female flies are slightly larger while male flies have more pigmentation of the lower abdomen. In the wild, *Drosophila* occupy home ranges on all the major continents, but they owe their current widespread distribution to humans as *Drosophila* are a commensal organism feeding on human fruits and other food sources to spread [1]. Prehistorically, *Drosophila* occupied a more restricted distribution in equatorial Africa and their food source before the spread of human activity is not clear. *Drosophila* can be grown on a variety of food sources that may have contributed to their initial selection as a model organism.

Use as a Research Organism

Regardless of the history leading to their initial selection, it is clear that the use of *Drosophila* in biological research has a rich history in the fields of genetics,

Wild-type Female   Wild-type Male

FIGURE 12.1 Wild-type *Drosophila melanogaster*. Images show laboratory wild-type *Drosophila melanogaster*. Female flies are slightly larger while male flies have more pigmentation of the lower abdomen.

developmental biology, and neuroscience [2]. The breadth and depth of discoveries using *Drosophila* have a substantive past and these contributions continue [3,4].

Drosophila is an excellent model system for biomedical questions and insights from studies in *Drosophila* directly impact our understanding of humans. Much of our understanding of basic genetics principles such as genetic linkage, recombination, and chromosome segregation comes from work in *Drosophila*. *Drosophila* work has aided in our understanding of a number of aspects of human health, including axonal pathfinding, cell polarity, circadian rhythms, the innate immune response, learning and memory, stem cell biology, and synaptic transmission. In addition, many signaling pathways that are important in diseases such as cancer and have emerged as drug targets were characterized first in *Drosophila* including the Hedgehog, Hippo, Notch, Wnt, and Toll pathways.

In addition to the more medical and genetic aspects of *Drosophila* research discussed here, population genetics and evolutionary biology has benefited greatly from studies on distinct species within the genus *Drosophila*. Moreover, *Drosophila* remains a key model organism for studies on genetic pathways in other insect species including vectors for tropical disease such as *Aedes aegypti* (dengue fever, yellow fever) and *Anopheles gambiae* (malaria). As a model organism, fruitflies allow genetic manipulations that are not possible in most other systems [5,6]. The clinician wishing to better understand research using *Drosophila* faces many challenges including lack of familiarity with the genetics, the husbandry, and the extensive range of options to approach disease questions. Moreover, the perceived applicability of *Drosophila* research to clinical scenarios is not always immediate or obvious. For example, seemingly vertebrate-specific phenotypes such as hypertension or schizophrenia may not seem applicable to the fly, but the applicability often exceeds expectations [3,4]. Nonetheless, the rewards for the clinician incorporating *Drosophila* into a clinical research program are many and include the possibility of achieving much greater insight into disease for less cost, the ability to use some of the most powerful genetic tools available, and the ability to study complex mechanisms in a multicellular organism. This chapter aims to provide an entry point for the physician-scientist in training to effectively understand and possibly become engaged in *Drosophila* disease-related studies, and is aimed to supplement the excellent materials and manuals already available for clinicians [7,8].

Attributes

- Short generation time (10 days from embryo to adult).
- High fecundity (females lay hundreds of eggs).
- Life cycle with a number of developmental stages allowing for studies of distinct developmental paradigms (embryos, larvae, pupae, adult flies).
- Powerful genetic tools.
- Cost-effectiveness for genetic studies.

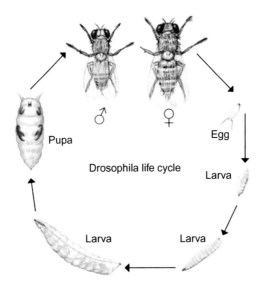

FIGURE 12.2 The life cycle of *Drosophila melanogaster*. *Drosophila* exhibit a 10-day life cycle at 25°C. The female lays numerous eggs that hatch within 24 hours to larvae. The larvae exhibit three developmental stages before undergoing pupariation. Adult flies eclose.

IN PRINCIPLE

Life Cycle

Studies on *Drosophila* often take advantage of the rapid life cycle (Fig. 12.2). Laboratory *Drosophila melanogaster* strains are grown in glass or plastic vials at room temperature on a corn-meal, yeast-based agar. A female fly lays numerous eggs and will continue to do so for weeks providing a large number of offspring for each cross. In a 24-hour period at 25°C eggs hatch and larva emerge. Over the next three days the larva progresses through three "instar" stages of development. Third instar larvae climb out of the food to undergo pupation, a process during which the whole organism undergoes metamorphosis into an adult fly, similar to caterpillars and butterflies. During this time many larval organs are absorbed to provide the energy source and substrates for adult organs. Of particular importance are the imaginal discs, tiny epithelia that will develop into adult structures such as wing, eye, and limbs. After 3–4 days the adult fly emerges in a process called eclosion. Female flies do not lay eggs in the first days of life. It is possible to separate females from one genetic strain and cross them to males from another strain. With a 10-day life cycle, adult flies can live 70–90 days.

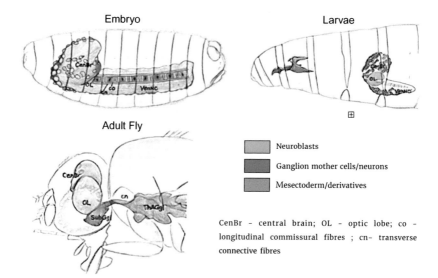

CenBr - central brain; OL - optic lobe; co - longitudinal commissural fibres ; cn- transverse connective fibres

FIGURE 12.3   The *Drosophila melanogaster* central nervous system. The *Drosophila* central nervous system begins as ectodermally derived neuroblasts in the embryo. A stage 17 embryo has a formed ventral nerve cord and central brain (left). In the larval stage the brain is larger and more complex (center), in adult flies the brain has several defined lobes for information processing and specialized functions.

Functional Anatomy of the Fly

Drosophila is a complex multicellular organism with conserved developmental processes and signaling pathways that have led to important discoveries for almost all the known signaling pathways. Many physicians may not be aware of the homologous or analogous organs present in fruitflies, which either serve equivalent functions as human organs or provide analogous functions and can provide insight into disease pathogenesis. Some selected examples of this functional anatomy are described below.

Drosophila *Nervous System and Brain*

The brain is of ectodermal origin and the development of the fly brain in the different stages of *Drosophila* development provides many paradigms to understand neurobiology [9] (Fig. 12.3). In the embryo the fly brain begins as neuroblasts that grow and produce the ventral nerve cord. Axonal projections develop for many motor neuron and peripheral nervous system populations. In larvae, the brain develops into a larger structure and peripheral synapses at the neuromuscular junction (NMJ) become tractable for microscopic and electrophysiologic studies. During pupation, the brain becomes more complex with more defined lobes with distinct functions [9]. The adult brain in the fly is accessible to genetic manipulations, which can allow one to study the neural networks and pathways that underlie adult behavior, physiology, and neurodegeneration.

Drosophila *Muscle*

Fly muscle is often studied in the body wall of larvae as well as the powerful flight muscles in adults. The molecular composition of muscle tissue is highly conserved in evolution and a number of assays for histology, live imaging, and genetics have provided important insight into muscular function [10].

Drosophila *Renal and Excretory Function*

Drosophila kidney function can be considered to be performed by three systems, the garland cells, the nephrocyte, and the malphigian tubule. *Drosophila* nephrocytes interestingly contain a number of evolutionarily conserved components of the vertebrate glomerular slit diaphragm, while the malphigian tubule is a separate system for fluid balance [11]. Nephrocytes and garland cells as endocytic clearance systems may be regarded as a primitive reticuloendothelial and renal system [12].

Drosophila *and the Gastrointestinal Tract*

The fly has a complex digestive system with a gastrointestinal (GI) tract, which is amenable to many manipulations and studies ranging from stem cell biology to immunity and recent studies have utilized the GI system in the fly to model host–pathogen interactions [13].

Drosophila *and the Cardiovascular System*

Drosophila have a primitive heart tube or dorsal vessel that is morphologically distinct from the multichambered vertebrate heart. However, the genetic programs are remarkably similar [14,15]. Detailed studies have allowed this invertebrate cardiovascular system to be used to provide insights into cardiac aging, and even diabetic cardiac disease [16,17].

Some additional examples of functional anatomical correlates are listed in Table 12.1.

Genome and Genetics

Drosophila have four pairs of chromosomes, (Fig. 12.4). One of the key advantages of using *Drosophila* is the ability to control and manipulate the genotype [5,18]. Some key features include:

- Dominant markers—Mutations that produce a visible phenotype (changes in eye color, eye shape, wing shape, bristle color, bristle size/shape, etc.).
- Balancer chromosomes—Modified chromosomes that prevent recombinants from emerging in a strain because of extensive inversions and

TABLE 12.1 Examples of Functional Analogs of Human Organs in *Drosophila*

Human Organ System	Functional Analog in *Drosophila*	Description and Function
Brain	Larval brain or adult fly brain	Optic lobes and central brain with ventral nerve cord posteriorly neuroblast lineages during larval stages. The fly brain is estimated to have one million-fold fewer neurons than the human brain but with similar complexity of neuronal subtypes [6]
Neuromuscular junction	Larval muscle and NMJ	Arrays of muscle groups in each thoracic and abdominal segment with large neuromuscular junctions [8]
Muscle	Embryonic body wall, larval body wall muscle, indirect flight muscle [10]	Mesoderm-derived muscles in developing stages and in adults [10]
Developing epithelia	Larval imaginal discs	Epithelial pouches that develop into wings, eye, limbs
Gastrointestinal tract	Digestive tract	Epithelial—lined alimentary canal
Liver and adipose tissue	Fat body and oenocytes	Oenocytes—polyploid ectodermal cells with lipid and detoxification functions
		Fat body—mesodermal adipose and secretory cells
Kidney	Malphigian tubules	Malphigian tubule—responsible for fluid balance and filtration functions
	Garland cells	Nephrocytes—responsible for filtration and excretory function
		Garland cells
Endocrine system	Ring gland	Steroid hormone secreting gland
Heart and aorta	Dorsal vessel	One-chambered pump providing circulation for the hemolymph
Respiratory system	Trachea	Epithelial tubes allowing circulation of air in internal organs
Hematologic system	Hemocytes	Head mesoderm—derived circulating phagocytic cells (*Drosophila* have an open circulation)

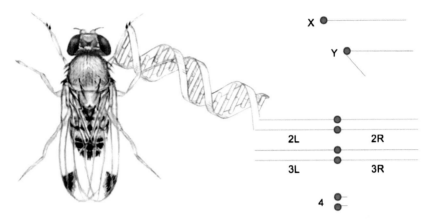

FIGURE 12.4   The *Drosophila* genome. The fly genome consists of four chromosome pairs, the sex chromosomes X and Y, second, third, and fourth chromosomes. The fourth chromosome is <4% of the genome.

rearrangements. These chromosomes also carry homozygous lethal mutations and a dominant mutation that is easily identified.

- Transposable elements—Large collections of stable strains of transposable elements are available. A range of elements are used that incorporate different features: the insertion of transposons can act as a mutagen, aid in mutagenesis through imprecise excision of the element, allow the detection of gene expression patterns via enhancer trapping, permit the tagging of proteins via protein trapping, and allow genome engineering.
- Chemical mutagenesis—Unbiased forward genetic screens can be conducted by inducing chemical mutations with ethylmethane sulfonate (EMS) that produces DNA mutations [5]. As an alkylating agent of nucleic acid, EMS produces primarily point mutations. Since EMS mutagenesis produces random mutations, mapping of the mutations can be labor-intensive, but recent advances have allowed for whole-genome sequencing to be used to aid in mapping EMS alleles [19].
- Genomic rescue—The use of transgenic genomic fragments to rescue the phenotype of mutants is a powerful tool to control for genetic background in experiments.
- Clonal analysis—*Drosophila* studies allow for the study of lethal mutations in adult tissue through the generation of mutant cells in a heterozygous animal based on the use of a yeast recombinase, FLP (flippase), that recombines between short FLP recombination target (FRT) sites [20] (Fig. 12.5).
- Transgenesis—Numerous methods for the generation of transgenic flies including transposons, site-specific integration, and recombination [18].

Collections of transposable elements integrated throughout the genome allow different manipulations

Binary systems allowing separate control of which gene is being expressed where it is being expressed

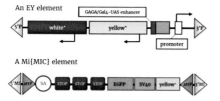

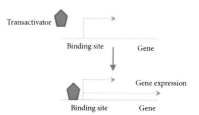

The FLP/FRT system allows for the generation of mutant mosaic clones

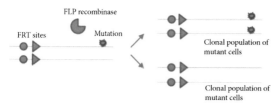

FIGURE 12.5 Tools to control genotype in *Drosophila*. Numerous methods to modify genotypes are available. These include the use of collections of transposable elements that are integrated throughout the genome. Two examples of these are shown. The EY element is a *P*-element that contains a GAL4-UAS enhancer allowing "forced" gene expression of a nearby gene. The MiMIC is a powerful transposon derived from the Minos element. It contains a strong splice acceptor and stop codons in all three reading frames along with an enhanced green fluorescent protein (EGFP) marker. The MiMIC element has attP sites allowing its cassette to be exchanged with other constructs using recombination-mediated cassette exchange (RMCE). Powerful binary systems allow the separate in vivo control of which gene is being expressed and where it is being expressed through different transactivator lines. The FLP/FRT system allows for the generation of mutant mosaic clones in specific cell populations in vivo.

- CRISPR—The most rapidly expanding tool in current use is the Cas9 nuclease from the bacterial clustered regularly interspaced short palindromic repeat (CRISPR) system. This system is now successfully employed in *Drosophila* (http://flycrispr.molbio.wisc.edu/). Injection of gRNA into *Drosophila* embryos carrying the Cas9 nuclease was initially reported to achieve mutagenesis for a target in up to 88% of the flies with germline transmission [21]. CRISPR is now widely used and can not only create deletions but can also be used to engineer elements into the fly genome.
- Binary systems—Binary systems refer to a transactivator that binds to a specific DNA sequence producing expression of the sequence downstream. Use of binary systems allows one to control (a) which gene is being expressed and (b) where the gene is being expressed. The most widely used binary system is the GAL4-UAS but other systems such as LexA and QUAS are also widely used [6] (Fig. 12.5).

- Genome-wide RNAi lines—Genome-wide libraries to produce RNA interference in *Drosophila* are available (see Table 12.2).
- MiMIC insertions—A large collection of >7000 Minos-mediated integration cassette (MiMIC) insertions has been generated [22]. This collection has allowed the tagging of over 700 genes in which green fluorescent protein (GFP) is introduced in an artificial exon. About 72% of these internally tagged proteins are functional and >90% can be imaged in unfixed preparations with live imaging in third instar larvae [22].

Phenotype

Identification of unique phenotypes in *Drosophila* has a rich history of over 100 years [2]. Subsequent sections will discuss some key organ systems and areas of particular success in understanding disease. However, some key phenotypic characteristics that can be rapidly determined and can guide more detailed characterization include:

- Lethality—Failure of flies to eclose can result from lethality during embryonic, larval, or pupal stages. Determining a lethal stage can be a key phenotype.
- Lifespan—Viable flies may not have a normal lifespan and longevity can be studied in survival assays.
- Morphological defects—Visible phenotypes in wings, bristles, eyes, and cuticle are valuable phenotypes that often connect to key developmental signaling pathways [23].
- Locomotion—Altered patterns of locomotion, climbing, flight, sensitivity to mechanical stress, or circadian behavior can be key characteristics.
- Fertility—Sterile males or females could indicate defects in spermatogenesis, oogenesis, or behavior.

Many organ systems in *Drosophila* have been studied extensively such that detailed phenotypic characterizations can be related to known pathways and mutations. For example, in the nervous system larval brain neuroblasts can be used to dissect pathways underlying neuronal proliferation [8,23]. The NMJ can be assayed ultrastructurally, electrophysiologically, and immunohistochemically to identify specific defects in synaptic growth and synaptic transmission. Likewise the *Drosophila* retina can be assayed to identify synaptic defects and neurodegenerative phenotypes, some of which are specific to retinal degeneration, but many of which relate to general mechanisms of neurodegenerative disease. Other studies examine the giant fiber a large rapidly conducting nerve fiber that can be used in studies of neuronal aging (Fig. 12.6).

TABLE 12.2 Resources and Collections for *Drosophila*

Resource	Description	URL
Flybase	The central hub for *Drosophila* research to search the genome, mutants, images, expression data, and gene-ontology	http://flybase.org/ [44]
Bloomington Drosophila Stock Center at Indiana University	Most widely used Drosophila stock center in the United States	http://flystocks.bio.indiana.edu/
Gene Disruption Project	Large collections of transposable element insertions, lines are available through Bloomington stock center	http://flypush.imgen.bcm.tmc.edu/pscreen/ [45]
Berkeley Drosophila Genome Project	The Genome Project resource for *Drosophila*	http://www.fruitfly.org/
Drosophila Genomics Resource Center	A genomics resource for the fly community providing vectors and cDNAs	https://dgrc.cgb.indiana.edu/Home
Kyoto Drosophila Genetic Resource Center	Large stock center in Japan widely used globally	http://www.dgrc.kit.ac.jp/
Vienna Drosophila Resource Center	Stock center with particular focus on RNAi lines	http://stockcenter.vdrc.at/control/main [46]
BrainTrap	An expression collection from protein trap lines	http://fruitfly.inf.ed.ac.uk/braintrap/
Transgenic RNAi Project	Stock collection with focus on RNAi lines	http://www.flyrnai.org/TRiP-HOME.html
P[acman] Resources	Resource on P[acman] clones and recombineering	http://www.pacmanfly.org/ [25]
FlAnnotator	Gene expression database	http://www.flyprot.org/
Atlas of Drosophila Development	Images of developmental stages	http://www.sdbonline.org/sites/fly/atlas/0607.htm

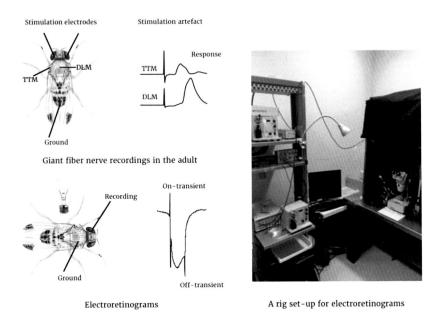

FIGURE 12.6 Drosophila electrophysiology. Two examples of paradigms for electrophysiologic recordings in adult flies. The giant fiber system (top left) records from two muscle groups, the tergotrochanter muscle (TTM) and the dorsal longitudinal muscle (DLM). These two muscle groups respond to stimulation of the giant fiber nerve that can be stimulated from implanted brain electrodes. In electroretinograms (lower left) a field potential is recorded in the eye in response to a light stimulus. The depolarization follows an "on-transient" and repolarization follows an "off-transient."

IN PRACTICE

The Use of Drosophila for Scientific Research

Laboratory *Drosophila* are grown in glass or plastic vials typically raised on a cornmeal, yeast-based agar with added sugar (Fig. 12.7). *Drosophila* can be grown at different temperatures and the temperature will affect the timing of growth. *Drosophila* stocks grown at 18°C will grow more slowly than those grown at room temperature, while stocks raised at 25°C will develop rapidly. Some experimental paradigms call for growth at 29°C to enhance transgene expression for example. Therefore, most *Drosophila* laboratories may have several incubators and storage shelves for fly stocks.

    Drosophila can be anesthetized most commonly with cold or CO_2, after which flies can be examined and sorted with a brush under a light microscope (Fig. 12.7). The ease of sorting, crossing, and screening progeny allows for rapid screening, establishment of stocks with unique genotypes, and numerous other advantages of *Drosophila* genetics.

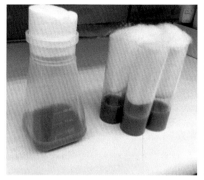

Yeast and cornmeal based food vials   Healthy stocks

Storage of strains   Scoring and sorting flies using CO_2 under a microscope

FIGURE 12.7 *Drosophila* husbandry. *Drosophila* are grown on a cornmeal- and yeast-based agar in plastic or glass vials (top left). A health stock will have larvae growing in the food and active culture of adult flies (top right). Strains can be stored on shelves or temperature-controlled incubators (bottom left). Flies can be anesthetized with cold CO_2 or ether and examined under a light microscope to set crosses or collect progeny (bottom center and bottom right).

Understanding Drosophila Genotypes

Understanding *Drosophila* genotypes can appear difficult but great care should be taken to ascertain the genotype of each stock. Failure to understand genotypes before starting a cross or experiment will lead to failure. An extensive and excellent resource on nomenclature is available (http://flybase.org/static_pages/docs/nomenclature/nomenclature3.html#4). Some key points that could help a clinician scientist understand fly genotypes are:

1. Fly gene names are generally lowercase italicized (*white;w, yellow;y*) (Fig. 12.8). Uppercase gene names indicates a dominant phenotype (*Stubble;Sb*) (Fig. 12.9). "CG" gene names indicate uncharacterized genes. Alleles are superscript (w^{1118}).

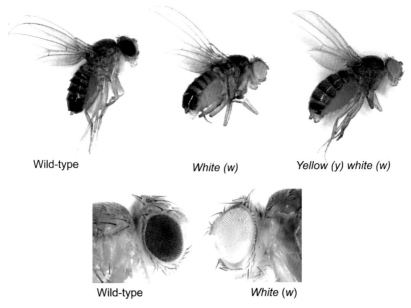

FIGURE 12.8 *Drosophila melanogaster* dominant markers white and yellow. Wild-type *Drosophila* have a pigmented eye (top left, and bottom left), *white* mutants have a loss of pigmentation due to mutations in the *white* (*w*) gene (top center and bottom right). A typical *yellow white* (*y w*) fly has loss of pigmentation in the eye and loss of body pigmentation due to mutation in the *yellow* gene (*y*).

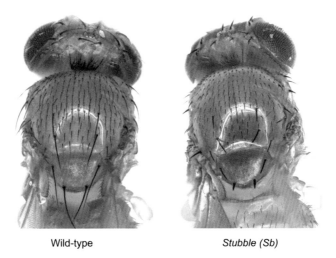

FIGURE 12.9 *Drosophila melanogaster* dominant markers Stubble. The Stubble (*Sb*) marker is an example of a visible dominant marker that can be used to score different genotypes of flies. The wild-type bristles have a characteristic length (left, *red arrows*), which is markedly shorter in *Sb* mutants.

2. Many fly genes are named for the mutant phenotype. Thus the *white* gene (*w*) is required for red-eye pigment. White mutant (such as w^{1118}) flies have white eyes.

3. When mutations involve different chromosomes the mutations are listed in order of the chromosomes with each chromosome separated by a semicolon such that first chromosome; second chromosome; third chromosome. For example, a stock with mutations in *yellow (y)* and *white* on first chromosome, *Lobe (L)* on the second, carrying the second chromosome *CyO* balancer, *Dichaete (D)* on the third, and the third chromosome *TM6B* balancer is abbreviated as y^1 w^1; *L/CyO; D/TM6B*.

4. Different types of transposon insertions are indicated by letters followed by brackets with the contents of the construct, for example *P{Epgy2}pex1* $6^{EY05323}$.

Using Existing Genetic Collections for Mapping or Other Application

Large collections of existing genetic strains derived from mutagenesis, or mobilization of transposable elements, or containing specific constructs are available from three large stock centers located in Bloomington, Kyoto, and Vienna. These and some other frequently used online resources for researchers wanting to obtain *Drosophila* strains are listed in Table 12.2. It is common practice to obtain strains from publically available stock centers and then use crossing schemes to bring different genetic tools into the same strain for experimental purposes. For example, to examine mitochondria in neurons in a mutant of interest, one could obtain stocks from the Bloomington stock center (as below) and use a crossing scheme to bring them into a mutant strain. For example,

w[*]; P{w[+ mC] = GAL4-elav.L}3 → This stock expresses GAL4 under the *elav* promoter leading to neuronal expression if a UAS transgene is present.

w[1118]; P{w[+ mC] = UAS-mito-HA-GFP.AP}3, e[1] → This stock has a mitochondrial GFP and HA marker under UAS control.

Transgenesis

The most widely used and successful method for *Drosophila* transgenesis is introduction of plasmid DNA through injection into embryos. In the past, the injected plasmid was integrated by relying on transposition of a *P*-element for which the insertion site in the genome could not be controlled, and for which the amount of DNA that could be integrated was limited. More recent technology allows for site-specific integration of transgenes. For example, many of the current methods rely on ΦC31 integration [24] (Fig. 12.10). The P[acman] transgenesis platform has allowed for recombineering in conditionally amplifiable bacterial artificial chromosomes (BACs) and has permitted ΦC31-mediated integration of large pieces of DNA in specific sites [25].

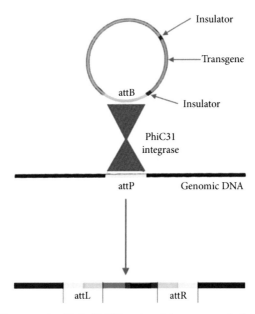

FIGURE 12.10   Transgenesis with the PhiC31 system. A transgene can be integrated into a docking site in the fly genome using PhiC31 integrase. DNA in the transgene along with a dominant maker and an attB site is integrated into the site of a genomic attP docking site by integrase.

Imaging Tools

The *Drosophila* field has numerous imaging tools to allow detailed visualization of tissues, cell structure, subcellular organelles, and pathologic aggregates. Laser-scanning confocal microscopy is the most widely used [26]. Many other imaging modalities such as live imaging, superresolution, and multicolor imaging are common [26].

Behavioral Scoring

Complex behavior is a key feature of the in vivo phenotypes that can be assayed in *Drosophila* [8]. For example, circadian rhythms can be assayed using high-throughput locomotor monitoring assays in controlled light and temperature conditions [27]. Visual learning, courtship, olfactory learning, and aggression all have well-established behavioral paradigms in *Drosophila* [8].

Signaling Pathways

Drosophila research has produced numerous key contributions to the knowledge of signaling pathways. The *Notch* signaling pathway is an excellent example as the pathway is involved in human diseases including Alagille syndrome, aortic valve disease, cerebrovascular dementia, and leukemia. Vertebrates have four paralogs of the Notch receptor (*NOTCH1-4*), whereas *Drosophila* has just a single *Notch* gene. The simpler evolutionary relationships and the numerous tools to study signaling in *Drosophila* make it a key model organism to study this and many other fundamental signaling pathways [28]. Other signaling pathways are also well studied in *Drosophila* including JAK/STAT, TGF-β, Wingless, Hedgehog, Toll, Hippo, and MAPK.

Metabolism

There has been a recent resurgence in metabolic studies and *Drosophila* has played an important role in this resurgence. Numerous studies aim at understanding mechanisms that underlie diabetes, obesity, hyperlipidemia, and inborn errors of metabolism [29]. The fly diet can be well controlled providing openings for nutrition and nutrient utilization studies. In addition, drugs can be easily incorporated into fly food. Biochemical and metabolomic analytes such as glycogen, ATP can be measured by standardized protocols, and metabolic organelles such as mitochondria and peroxisomes and other metabolic pathways are under study [29]. With the advent of metabolomic profiling, many more studies can probe into the mechanisms of disease-causing processes.

APPLICATIONS

The numerous tools and methods available allow for approaches to disease problems that can provide excellent insight into disease pathogenesis. This section is far from comprehensive but provides some key points for clinicians.

One tendency for a clinician approaching disease questions is to rely on reverse genetics of a homolog, or on overexpression of a human gene or protein in the fly. These strategies while sometimes successful, represent only a fraction of what can be done using fruitflies. Below are some examples of what has been achieved using flies in contributing to disease research.

Hippo Signaling Pathway and Regenerative Medicine

Hippo signaling was characterized in *Drosophila* [30,31] and at the same time implicated human cancer. Hippo signaling controls organ size and controls proliferation and apoptosis. Delineation of this pathway has provided key insights not only into cancer biology but also into regenerative medicine [32].

Cancer Gene Discovery

A screen was conducted in *Drosophila* for increased cell proliferation, and a gene *archipelago* was identified that leads to persistent elevations of Cyclin

E proteins. Archipelago protein was shown to bind Cyclin E, and target it for degradation. The human homolog of Archipelago was then found to be mutated in four cancer cell lines and mutations in this gene were observed in 3 of 10 ovarian carcinomas [33].

Leptin, Insulin, and Drosophila Endocrine Signaling

For years, insulin and leptin were thought to be metabolic systems that were not conserved in *Drosophila*. Subsequently, *Drosophila* insulinlike peptides (Dilps) were identified and shown to be nutrient sensors that regulate energy balance and growth. Another protein Unpaired 2 (Upd2) was identified and shown to be part of the nutrient sensor system that regulates the secretion of Dilps from the *Drosophila* brain. Human leptin was able to rescue the phenotype of *Upd2* mutants, suggesting that leptin is functionally conserved in flies, similar to insulinlike peptides [34].

Chronic Pain

A genome-wide screen using RNAi lines in *Drosophila* was conducted identifying hundreds of genes involved in heat nociception. One of these genes belonged to the alpha2delta family of calcium channels, also named *straightjacket* (*stj*). Mutant mice for the ortholog of *CACNA2D3* were then found to have heat pain sensitivity, and the mice were found to have impaired transmission of pain signals to high order pain centers in the central nervous system (CNS). In the same vein of research, specific polymorphisms in the *CACNA2D3* gene were found to be associated with altered pain sensitivity in humans allowing direct output from *Drosophila* to mouse then to human chronic pain studies [35].

Neurodegeneration

A degenerative hereditary motor neuropathy (HMN7B) due to mutations in the dynactin gene has been studied in *Drosophila*. The specific missense mutation found in patients with HMN7B alters a conserved glycine and this conserved glycine was mutated in the fly locus using homologous recombination. These flies were found to exhibit impaired neurotransmitter release, locomotor defects as well as a synaptic vesicle trafficking defect at NMJs. While axonal transport was normal there was an accumulation of endosomes and motor proteins in the terminal boutons of the NMJs. These observations suggested that the HMN7B mutation disrupts retrograde transport at synaptic terminals [36].

Parkinson's disease (PD) is often sporadic, but rare Mendelian forms of Parkinsonism have been studied because of their potential to provide insight into PD. Specifically mutations in Parkin (*PARK2)* and *PINK1* have been studied in *Drosophila* [37,38]. Characterization of these mutants in *Drosophila* led not to focus on dopaminergic neurons but rather on muscle phenotypes that suggested a dramatic and severe mitochondrial dysfunction. *PINK1* and Parkin were revealed to be involved in mitochondrial dynamics and these insights into mitochondrial from *Drosophila* muscle have helped guide PD research toward mitochondrial dysfunction [37,38].

Gene–Gene Interaction and Heart Defects

Down syndrome leads to an increase in the copy number and expression of all the genes on human chromosome 21, and a common phenotypic feature is congenital heart disease. Pairwise combinations of genes from an interval on human chromosome 21 were screened in *Drosophila* for interactions. These were expressed alone or in combination to identify pairs that interact in the *Drosophila* dorsal vessel or heart. The combination of two genes *DSCAM* and *COL6A2* were found to interact to produce heart defects in the fly. Heart defects were also found in mice overexpressing this pair of genes in the mouse heart and these genes were then tied mechanistically to adhesion and cardiac hypertrophy [39].

SCENARIO—HUMAN DISEASE GENE STUDIES

Drosophila studies can be used to understand human genetic disorders from the numerous genomic and personalized medicine efforts underway. Estimates suggest that approximately 75% of human disease genes have a homolog in the *Drosophila* genome [40]. A large forward genetic EMS mutagenesis screen was conducted on the *Drosophila* X-chromosome. Numerous phenotypes were tested to capture genes required for development, function, and maintenance of the nervous system. The mutations were mapped and rescued, and whole-genome sequencing was used to identify 165 genes [19]. A key observation of this work was that genes that are essential in flies and have multiple human homologs are the most likely to be associated with human diseases. The set of genes identified corresponded to 250 human homologs and these human genes were then studied in a genomic database that included 1,929 human exomes from families with Mendelian disease. This approach led to the identification of disease associated mutations in six families including two with Charcot-Marie Tooth disease due to *DNM2* mutations, three with Bull's eye Maculopathy due to *CRX* mutations. In addition, mutations in *ANKLE2* a gene not previously associated with any disease, and whose function was not well understood, was found to cause a severe microcephaly [23]. This provided a direct output of human disease gene discovery from genomic data starting from *Drosophila* screens.

Synaptotagmin is a crucial calcium sensor for synaptic vesicle release and *Drosophila* synaptotagmin mutants exhibit a lack of evoked release of synaptic vesicles [41]. Two multigenerational families with Lambert–Eaton myasthenic syndrome, a disorder with muscle weakness due to defects in the NMJ, were found to carry mutations in *synaptotagmin 2*. The missense mutations in the two families were noted to be conserved amino acids and they were mutated in the *Drosophila synaptotagmin* gene. A dominant disruption of neurotransmitter release was found, documenting the first human disease caused by *synaptotagmin* mutations [42].

Ongoing human disease gene studies can make use of versatile genetic tools that allow manipulation of the fly gene and expression of transgenes to study specific variants and genes from clinical genomic sequencing studies [43]. The opportunity to create collections of fly strains in which a human gene is under study is of great interest for *Drosophila* and human genetics. In the era of personalized medicine, employing these technologies to study personal variants using the wealth of genetic tools and phenotypic studies available in the fly will continue long into the future.

KEY LIMITATIONS

Understanding Phenotypes

As a model organism there is often discussion of "fly models" of human disease. We argue that flies should not be used or thought of as direct disease models but rather as excellent tools for studying pathogenic mechanisms that underlie diseases. *Drosophila* cannot model all the complex features of humans but they can be used to identify important fundamental biology, key cellular phenotypes, or molecular mechanisms that underlie many human diseases.

Determining the phenotype of strong loss of function mutants is a key step in understanding gene function in *Drosophila*. This is necessary baseline data when studying overexpression or toxic proteins, specific gain-of function mutations, or specific human mutations.

Genetic Considerations

Drosophila have a large set of defined genetic collections and many options for genetic manipulation. For the new physician researcher absorbing the vast array of genetic options can be a challenge. Luckily several excellent online resources can aid in identifying tools and reagents for almost any application (Table 12.2).

TROUBLESHOOTING

Problem	Potential Solutions
Genetic background effects or second site hits in *Drosophila* stocks	One of the key advantages of *Drosophila* is the ability to do experiments using very precise genetics. Utilizing genomic rescue constructs for any phenotypic analysis involving a genetic lesion is often a key step in controlling for second hits or genetic background. One can rescue phenotypes with genomic constructs and with the overexpression of cDNA constructs. Isolating whenever possible independent alleles for genes of interest for comparison can also provide key controls.

Problem	Potential Solutions
Difficult stock husbandry	Some *Drosophila* stocks are sick and slow growing or prone to being lost and may require growth at different temperatures, more frequent transfers to fresh food. Some stocks require the addition of yeast to the media.
Failed crosses	Crossing males and females can be more difficult with older males, males with prolonged CO_2 exposure or too few flies in the cross. More females should be placed in a cross than males. Typically 3–5 males and 5–10 females.

CONCLUSION

Drosophila melanogaster has been a key model system in genetics, developmental biology, and neuroscience. As more and more genetic tools and phenotypic and biological connections have been made, *Drosophila* has become an important system for helping unraveling the pathogenic mechanisms underlying numerous diseases. As genomic sequencing has become increasingly common as a research tool in the clinical setting, the need for model organisms that provide insight into gene function has increased. Therefore, as the clinical use of genomics increases, *Drosophila* will be a key system for understanding gene function. Hence, there will be an increasing need for clinicians who understand the advantages *Drosophila* offers. In any case, a clinician scientist should realize the unique position of *Drosophila* as a complex multicellular organism that is easily manipulated and analyzed allowing it to be used to contribute to biomedical discoveries across a range of fields.

ACKNOWLEDGMENTS

We thank Shinya Yamamoto for providing images of fly markers. We thank Hsiao-Tuan Chao for helpful advice in preparing this chapter.

REFERENCES

[1] Keller A. Drosophila melanogaster's history as a human commensal. Curr Biol 2007;17(3):R77–81.
[2] Bellen HJ, Tong C, Tsuda H. 100 years of Drosophila research and its impact on vertebrate neuroscience: a history lesson for the future. Nat Rev Neurosci 2010;11(7):514–22.
[3] Wangler MF, Yamamoto S, Bellen HJ. Fruit flies in biomedical research. Genetics 2015;199(3):639–53.
[4] Bier E. Drosophila, the golden bug, emerges as a tool for human genetics. Nat Rev Genet 2005;6(1):9–23.

[5] Venken KJ, Bellen HJ. Chemical mutagens, transposons, and transgenes to interrogate gene function in Drosophila melanogaster. Methods 2014;68(1):15–28.

[6] Venken KJ, Simpson JH, Bellen HJ. Genetic manipulation of genes and cells in the nervous system of the fruit fly. Neuron 2011;72(2):202–30.

[7] Greenspan RJ. Fly Pushing The Theory and Practice of Drosophila Genetics, 2nd ed. Cold Spring Harbor, New York: Cold Spring Harbor Laboratory Press; 2004.

[8] Zhang B, Freeman MR, Waddell S, editors. Drosophila Neurobiology A Laboratory Manual. Cold Spring Harbor, New York: Cold Spring Harbor Laboratory Press; 2010.

[9] Freeman MR. Studying neural development in Drosophila melanogaster Zhang B, Freeman MR, Waddell S, editors. Drosophila neurobiology: A laboratory manual. Cold Spring Harbor, New York: Cold Spring Harbor Laboratory Press; 2010. p. 1–3.

[10] Weitkunat M, Schnorrer F. A guide to study Drosophila muscle biology. Methods 2014;68(1):2–14.

[11] Weavers H, Prieto-Sanchez S, Grawe F, Garcia-Lopez A, Artero R, Wilsch-Brauninger M, et al. The insect nephrocyte is a podocyte-like cell with a filtration slit diaphragm. Nature 2009;457(7227):322–6.

[12] Ivy JR, Drechsler M, Catterson JH, Bodmer R, Ocorr K, Paululat A, et al. Klf15 Is Critical for the Development and Differentiation of Drosophila Nephrocytes. PLoS One 2015;10(8):e0134620.

[13] Lemaitre B, Miguel-Aliaga I. The digestive tract of Drosophila melanogaster. Annu Rev Genet 2013;47:377–404.

[14] Bodmer R. Heart development in Drosophila and its relationship to vertebrates. Trends Cardiovasc Med 1995;5(1):21–8.

[15] Bier E, Bodmer R. Drosophila, an emerging model for cardiac disease. Gene 2004;342(1): 1–11.

[16] Cannon L, Bodmer R. Genetic manipulation of cardiac ageing. J Physiol 2015 Jun 9.

[17] Diop SB, Bodmer R. Gaining Insights into Diabetic Cardiomyopathy from Drosophila. Trends Endocrinol Metab 2015;26(11):618–27.

[18] Venken KJ, Bellen HJ. Genome-wide manipulations of Drosophila melanogaster with transposons, Flp recombinase, and PhiC31 integrase. Methods Mol Biol 2012;859:203–28.

[19] Haelterman NA, Jiang L, Li Y, Bayat V, Sandoval H, Ugur B, et al. Large-scale identification of chemically induced mutations in Drosophila melanogaster. Genome Res 2014;24(10):1707–18.

[20] Golic KG, Lindquist S. The FLP recombinase of yeast catalyzes site-specific recombination in the Drosophila genome. Cell 1989;59(3):499–509.

[21] Bassett AR, Tibbit C, Ponting CP, Liu JL. Highly efficient targeted mutagenesis of Drosophila with the CRISPR/Cas9 system. Cell Rep 2013;4(1):220–8.

[22] Nagarkar-Jaiswal S, Lee PT, Campbell ME, Chen K, Anguiano-Zarate S, Cantu Gutierrez M, et al. A library of MiMICs allows tagging of genes and reversible, spatial and temporal knockdown of proteins in Drosophila. Elife 2015;4.

[23] Yamamoto S, Jaiswal M, Charng WL, Gambin T, Karaca E, Mirzaa G, et al. A drosophila genetic resource of mutants to study mechanisms underlying human genetic diseases. Cell 2014;159(1):200–14.

[24] Hillman RT, Calos MP. Site-specific integration with bacteriophage PhiC31 integrase. Cold Spring Harb Protoc 2012;2012(5).

[25] Venken KJ, He Y, Hoskins RA, Bellen HJ. P[acman]: a BAC transgenic platform for targeted insertion of large DNA fragments in D. melanogaster. Science 2006;314(5806):1747–51.

[26] Rebollo E, Karkali K, Mangione F, Martin-Blanco E. Live imaging in Drosophila: The optical and genetic toolkits. Methods 2014;68(1):48–59.

[27] Tataroglu O, Emery P. Studying circadian rhythms in Drosophila melanogaster. Methods 2014;68(1):140–50.

[28] Zacharioudaki E, Bray SJ. Tools and methods for studying Notch signaling in Drosophila melanogaster. Methods 2014;68(1):173–82.

[29] Tennessen JM, Barry WE, Cox J, Thummel CS. Methods for studying metabolism in Drosophila. Methods 2014;68(1):105–15.

[30] Kango-Singh M, Nolo R, Tao C, Verstreken P, Hiesinger PR, Bellen HJ, et al. Shar-pei mediates cell proliferation arrest during imaginal disc growth in Drosophila. Development 2002;129(24):5719–30.

[31] Tapon N, Harvey KF, Bell DW, Wahrer DC, Schiripo TA, Haber D, et al. Salvador Promotes both cell cycle exit and apoptosis in Drosophila and is mutated in human cancer cell lines. Cell 2002;110(4):467–78.

[32] Nagashima S, Bao Y, Hata Y. The Hippo pathway as drug targets in cancer therapy and regenerative medicine. Curr Drug Targets 2016 Jan 11.

[33] Moberg KH, Bell DW, Wahrer DC, Haber DA, Hariharan IK. Archipelago regulates Cyclin E levels in Drosophila and is mutated in human cancer cell lines. Nature 2001;413(6853):311–6.

[34] Rajan A, Perrimon N. Drosophila cytokine unpaired 2 regulates physiological homeostasis by remotely controlling insulin secretion. Cell 2012;151(1):123–37.

[35] Neely GG, Hess A, Costigan M, Keene AC, Goulas S, Langeslag M, et al. A genome-wide Drosophila screen for heat nociception identifies alpha2delta3 as an evolutionarily conserved pain gene. Cell 2010;143(4):628–38.

[36] Lloyd TE, Machamer J, O'Hara K, Kim JH, Collins SE, Wong MY, et al. The p150(Glued) CAP-Gly domain regulates initiation of retrograde transport at synaptic termini. Neuron 2012;74(2):344–60.

[37] Greene JC, Whitworth AJ, Kuo I, Andrews LA, Feany MB, Pallanck LJ. Mitochondrial pathology and apoptotic muscle degeneration in Drosophila parkin mutants. Proc Natl Acad Sci USA 2003;100(7):4078–83.

[38] Clark IE, Dodson MW, Jiang C, Cao JH, Huh JR, Seol JH, et al. Drosophila pink1 is required for mitochondrial function and interacts genetically with parkin. Nature 2006;441(7097):1162–6.

[39] Grossman TR, Gamliel A, Wessells RJ, Taghli-Lamallem O, Jepsen K, Ocorr K, et al. Over-expression of DSCAM and COL6A2 cooperatively generates congenital heart defects. PLoS Genet 2011;7(11):e1002344.

[40] Reiter LT, Potocki L, Chien S, Gribskov M, Bier E. A systematic analysis of human disease-associated gene sequences in Drosophila melanogaster. Genome Res 2001;11(6):1114–25.

[41] Littleton JT, Stern M, Schulze K, Perin M, Bellen HJ. Mutational analysis of Drosophila synaptotagmin demonstrates its essential role in Ca(2+)-activated neurotransmitter release. Cell 1993;74(6):1125–34.

[42] Herrmann DN, Horvath R, Sowden JE, Gonzalez M, Sanchez-Mejias A, Guan Z, et al. Synaptotagmin 2 mutations cause an autosomal-dominant form of lambert-eaton myasthenic syndrome and nonprogressive motor neuropathy. Am J Hum Genet 2014;95(3):332–9.

[43] Bellen HJ, Yamamoto S. Morgan's legacy: fruit flies and the functional annotation of conserved genes. Cell 2015;163(1):12–14.

[44] dos Santos G, Schroeder AJ, Goodman JL, Strelets VB, Crosby MA, Thurmond J, et al. FlyBase: introduction of the Drosophila melanogaster Release 6 reference genome assembly and large-scale migration of genome annotations. Nucleic Acids Res 2015;43(Database issue):D690–7.

[45] Bellen HJ, Levis RW, He Y, Carlson JW, Evans-Holm M, Bae E, et al. The Drosophila gene disruption project: progress using transposons with distinctive site specificities. Genetics 2011;188(3):731–43.

[46] Mummery-Widmer JL, Yamazaki M, Stoeger T, Novatchkova M, Bhalerao S, Chen D, et al. Genome-wide analysis of Notch signalling in Drosophila by transgenic RNAi. Nature 2009;458(7241):987–92.

SUGGESTED FURTHER READING

[1] Venken KJ, Bellen HJ. Chemical mutagens, transposons, and transgenes to interrogate gene function in Drosophila melanogaster. Methods 2014;68(1):15–28.
[2] Greenspan RJ. Fly Pushing The Theory and Practice of Drosophila Genetics, 2nd ed. Cold Spring Harbor, New York: Cold Spring Harbor Laboratory Press; 2004.
[3] Zhang B, Freeman MR, Waddell S, editors. Drosophila Neurobiology A Laboratory Manual. Cold Spring Harbor, New York: Cold Spring Harbor Laboratory Press; 2010.
[4] Venken KJ, Bellen HJ. Genome-wide manipulations of Drosophila melanogaster with transposons, Flp recombinase, and PhiC31 integrase. Methods Mol Biol 2012;859:203–28.

GLOSSARY

Instar   A developmental stage during larval development, there are three larval instar stages.
P-element   A class of transposable element.

LIST OF ACRONYMS AND ABBREVIATIONS

EMS	Ethylmethane sulfonate
CRISPR	Cas9 nuclease from the bacterial clustered regularly interspaced short palindromic repeat
GAL4-UAS	A binary system derived from yeast in wide use in *Drosophila* research
RNAi	RNA interference
ΦC31 integrase	A sequence-specific recombinase derived from bacteriophage
ATP	Adenosine triphosphate
BAC	Bacterial artificial chromosomes
RMCE	Recombination-mediated cassette exchange
NMJ	Neuromuscular junction

Chapter 13

Zebrafish as a Research Organism

Danio rerio in Biomedical Research

John Collin and Paul Martin

University of Bristol Medical School, Bristol, United Kingdom

Chapter Outline

Basic Science Methods for Clinical Researchers. DOI: http://dx.doi.org/10.1016/B978-0-12-803077-6.00013-8

Objectives

● This chapter aims to provide sufficient background information to enable a novice to begin to use zebrafish as a model of development or disease.

INTRODUCTION

Zebrafish (*Danio rerio*)

Zebrafish are small teleost freshwater fish of the carp family, named after the five horizontal blue stripes extending from gills to caudal fin tip. Zebrafish are native to the Southeastern Himalayas and thought to originate from the Ganges. In the wild they inhabit clear, shallow slow-moving bodies of water with silt and well-vegetated bottoms.

Males are torpedo-shaped, with gold stripes between the blue; females have larger, whitish bellies and silver rather than gold stripes (Fig. 13.1). Adult females also have a small genital papilla anterior to the anal fin. Zebrafish grow to around 4 cm in captivity, with a lifespan of upto five years in ideal conditions.

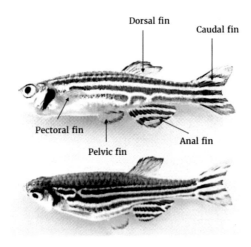

FIGURE 13.1 Photograph of adult zebrafish—male top and female bottom.

USE AS A RESEARCH ORGANISM

The popularity of zebrafish as a research organism has grown rapidly over the past two decades (Fig. 13.2).

This trend is largely due to increasing recognition of opportunities to apply invertebrate-style genetic techniques to a vertebrate with many biological similarities to humans. The zebrafish genome was fully sequenced in 2013 and contains homologues of around 70% of human genes, including 80% of those implicated in disease [1]. Many of their organs and tissues bear considerable similarity to those of humans. Rapid ex-uterine development with translucent embryonic and early larval stages makes zebrafish very amenable to live imaging, while genetic tractability initially made this organism attractive for developmental biology studies. More recently, these attributes have established zebrafish as a model organism for human disease. Furthermore, fish are relatively economical and easy to house and breed.

As an example of the value of the zebrafish model organism, Len Zon and colleagues have utilized it to elucidate the developmental basis of hematopoiesis [2]. This work led to direct clinical application with the use of prostaglandin E2 to expand hematopoietic stem cells in umbilical cord blood transplants. The same laboratory also created a transgenic zebrafish model of melanoma, identifying a novel gene (SETDB1) associated with progression to aggressive disease. Zebrafish were then used to screen for putative pharmacological agents,

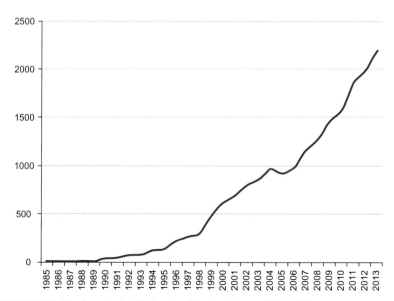

FIGURE 13.2   Number of zebrafish-based publications by year 1985–2013.

resulting in identification and clinical trials (without the need for mouse studies) of the drug leflunomide, which blocks neural crest development, as a promising therapy for metastatic melanoma [3].

Attributes

- High fecundity (several hundred eggs per female per week).
- Short generation time (three to four months).
- Rapid development and external embryogenesis.
- Translucent embryos (and potentially adults) enabling good imaging.
- Easy maintenance.

IN PRINCIPLE

Zebrafish have many anatomical and physiological similarities to higher vertebrates that makes them an excellent model organism for studies of development and disease. Their other major benefits are translucency, which makes them amenable to live imaging, and their genetic tractability.

REPRODUCTION

The generation time for zebrafish is approximately three months. A male must be present for ovulation and spawning to occur. Females are able to spawn every two to three days, laying hundreds of eggs in each clutch. Male zebrafish respond preferentially to pronounced markings on females, but in a group will mate with whichever females they can find. What attracts females is not currently understood. Embryogenesis begins immediately but stops after the first few cell divisions if sperm are absent. Fertilized eggs become transparent, which facilitates experimental manipulation by microinjection and is subsequently very useful for imaging studies.

DEVELOPMENT

Time Post Fertilization	Stage
0–72 hours	Embryo
3–13 days	Early larva
14–29 days	Late larva

Metamorphosis

30 days to 3 months	Juvenile
3 months + (sexually mature)	Adult

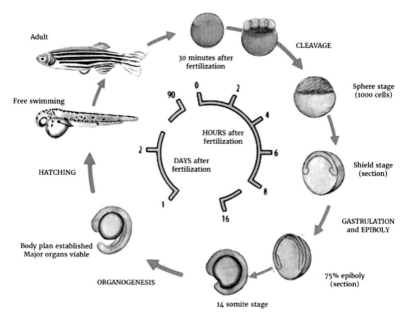

FIGURE 13.3   Zebrafish reproductive cycle.

The zebrafish embryo develops rapidly (Figs. 13.3–13.8), with precursors of all major organs present by 36 hours post fertilization. A single large fertilized egg cell attached to the yolk undergoes division to form a cap of cells that subsequently sweeps over the sides of the yolk (epiboly) during gastrulation. An elongated body axis with an obvious head and tail then develops. The yolk shrinks as it is used for nourishment until the fifth or sixth day when the larva becomes self-feeding.

Early larvae are negatively buoyant lying largely immobile on the bottom, with only occasional tail flicks. Around day 5 they swim to the surface and gulp air to inflate their gas bladders. After this they are neutrally buoyant and capable of continuous swimming as well as becoming self-feeding.

Metamorphosis occurs at around one month post fertilization and involves loss of the larval fin fold, remodeling of intestinal and nervous systems as well as acquisition of scales and secondary sexual characteristics. After around three months, growth slows as the adult fish reaches reproductive maturity and gametogenesis begins.

FUNCTIONAL ANATOMY

The following section gives a brief overview of zebrafish functional anatomy and highlights tissues and organ systems that have already shown promise for translational research.

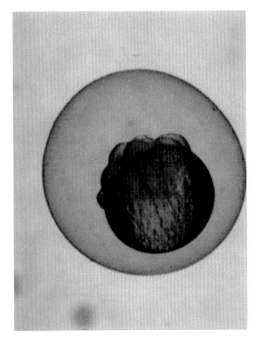

FIGURE 13.4   16-cell zebrafish embryo (1.5 hpf).

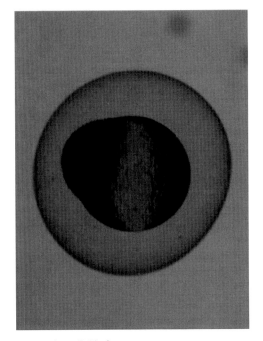

FIGURE 13.5   Cap-stage embryo (2.5 hpf).

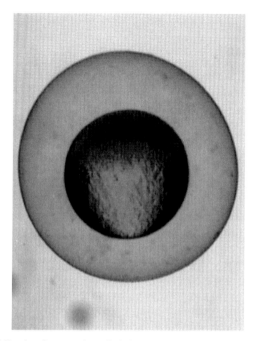

FIGURE 13.6 Cells migrating over the yolk during gastrukation (3.3 hpf).

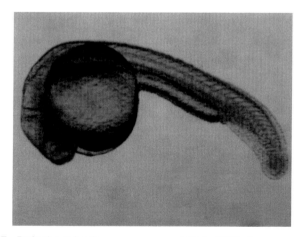

FIGURE 13.7 Body plan and major organs are now established (24 hpf).

Brain and Specialized Sensory and Endocrine Organs

Neurulation in the zebrafish is similar to other vertebrates [4]. The cerebellum processes information related to proprioception and equilibrium. The dorsal corpus cerebelli receives proprioceptive afferents, while the vestibulolateral

FIGURE 13.8 Larva at four days post fertilization.

lobe receives otolith and lateral line organ stimuli. The lateral line organs contain stereocilia that sense water movement to assist in detecting predators and prey. Studying lateral line development in the zebrafish has provided insight into cell migration, proliferation, and signaling, and is a potential model for peripheral nerve repair regeneration [5].

The mesencephalon contains the optic tectum and the tegmentum, responsible for processing visual stimuli [6]. The zebrafish eye is similar to other vertebrates, but has a relatively flat cornea compared with terrestrial animals. The lens is spherical and protrudes through the iris to afford a wide-angle view. The telencephalon is responsible for olfaction, color vision, memory, reproductive, and feeding behavior. It is connected by the olfactory bulb to the olfactory organ, through which water flows during swimming [7].

The diencephalon is subdivided into epithalamus, thalamus, and hypothalamus. The epithalamus surrounds the photosensitive pineal gland, projecting from the roof of the telencephalon. As in mammals, this secretes indolamine, melatonin, and neurotransmitters. It appears to control diurnal and seasonal rhythms, regulating reproduction, growth, and migration [8]. The pituitary gland is similar to that of mammals [9]. Isotocin and arginine vasotocin, are produced by the pars nervosa, but their exact function is unknown. Prolactin and corticotrophic hormone are synthesized in the rostral pars distalis, and thyroid-stimulating hormone in the proximal pars distalis. The pars intermedia contains melanotrophic cells that secrete α-MSH, β-endorphin, and β-hypotrophic hormones.

Cardiorespiratory System

The zebrafish heart is much used as a model of perfect myocardial regeneration following tissue damage [10]. It is situated just posterior to the gills and ventral to the esophagus. It is not four chambered as the mammalian heart. Rather, venous blood enters the atrium via the thin-walled sinus venosus and sino-atrial valve. Atrial contraction forces blood into the thicker-walled ventricle via the atrioventricular valve. Ventricular contraction pumps blood into the bulbus arteriosus via the ventricular-bulbar valve. The bulbus arteriosus has a thick fibroelastic wall that dampens the systolic pulse pressure. The ventral aorta distributes blood to the gills via the afferent branchial arteries.

The gills are essential for oxygenation, acid–base balance, osmoregulation, and excretion of waste products [11]. Water is drawn through the mouth, over the gills, and out through the opercula by alternate expansion and contraction of buccal and opercular chambers. Blood flows via the afferent filament arteries of the primary lamellae into the blood spaces of the secondary lamellae, where oxygen and carbon dioxide exchange takes place. The secondary lamellae consist of one layer of epithelial cells, supported, and separated, by pillar cells. Oxygenated blood leaves the secondary lamellae via lamellar arteries into the dorsal aorta, from where it is distributed to all other tissues.

Hematopoietic Tissue and Blood

The zebrafish is a well-established model for studying hematopoiesis, with good conservation of human lineages and genes [12]. A number of hematological diseases including malignancies have been modeled. Larval hematopoiesis begins by 48 hours post fertilization in tissue around the ventral wall of the dorsal aorta. In the adult, hematopoiesis occurs primarily in the renal interstitium and splenic stroma. Hematopoietic stem cells reside within reticuloendothelial stroma, similar to mammalian bone marrow.

Erythrocytes and Thrombocytes

Zebrafish erythrocytes are oval, nucleated, and use aerobic metabolism to generate adenosine triphosphate. Thrombocytes are nucleated cells and serve similar functions to mammalian platelets.

Innate Immune Cells

Neutrophils are the most abundant leukocyte, with morphology and function similar to mammalian neutrophils [13]. They originate at 24 hours post fertilization from axial tissue near the yolk sac and can be seen in the circulation at 34 hours post fertilization.

Macrophages originate from the rostral lateral plate mesoderm anterior to the heart at around the 13 somite-stage, entering the circulation once established. Monocytes are similar to their mammalian counterparts and develop into mature macrophages under appropriate conditions. Zebrafish eosinophils have small, nonsegmented, peripherally located nuclei. Their function is not clear, but may span those of mammalian eosinophils and mast cells.

Adaptive Immunity

Functional equivalents of antigen-presenting cells, natural killer cells, and B and T lymphocytes are found in zebrafish. Lymphopoiesis occurs in the thymus, pancreas, and kidney [12]. T-cells can be identified in the thymic region at three days postfertilization, with B-cells identified at four days post fertilization.

Musculoskeletal System

A portion of adaxial cells adjacent to the notochord migrate to the lateral surface of newly formed somites and differentiate into slow muscle fibres. The remainder differentiate into muscle pioneer cells, thought to act as targets for motor axon growth cones. Lateral, nonadaxial muscle precursor cells in the segmental plate differentiate into fast muscle cells [14]. Ribs articulate with transverse processes of the main body vertebrae and skeletal muscles insert to ribs and skin, generating a propulsive force on contraction.

Skin

Zebrafish are covered with mucous membrane except for small areas of keratinized epithelium on the jaw and fins. Larvae have a bilaminar epithelium with numerous goblet cells [15]. Overlapping scales covered by dermal fibroblasts and a multilayered epidermis develop during metamorphosis. Basement membrane separates epidermis and dermis and a thin layer of subdermal adipose tissue covers the underlying muscle.

Thymus, Thyroid, and Ultimobranchial Glands

The thymus is a paired organ in the dorsomedial branchial cavity that contains maturing lymphocytes and other immune cells, mainly macrophages. Thyroid follicles, distributed along the ventral aorta, produce hormones that regulate metabolism. A pair of ultimobranchial glands, in the transverse septum between heart and abdomen, produce calcitonin, and are analogous to parathyroid glands [16].

Kidney

The zebrafish kidney has distinct head and trunk regions. Initially, two nephrons form the embryonic kidney (pronephros). Hundreds then form from 10 days post fertilization to constitute the adult kidney (mesonephros). Nephron structure and function is similar to mammals. Hematopoietic cells are found in the parenchyma, while endocrine interrenal and chromaffin cells are distributed along larger blood vessels in the rostral kidney. Interrenal cells are analogous to the mammalian adrenal cortex and produce corticosteroids. Cortisol in particular is an important mediator of the zebrafish stress response [17]. Chromaffin cells synthesize catecholamines similar to the mammalian adrenal medulla. The caudal kidney contains corpuscles of Stannius, which secrete hypocalcin, to block calcium uptake by the gills, raise blood pressure, and affect osmoregulation.

Spleen

Teleosts do not have lymph nodes, with the kidney and spleen instead performing reticuloendothelial roles. The spleen lies adjacent to the liver, consisting mainly of erythrocytes and thrombocytes (red pulp). Periarterial sheaths of macrophages and reticular cells form ellipsoids at the termination of splenic arterioles that trap pathogens and defective blood cells.

The Gastrointestinal System

The mouth is lined by mucoid epithelium with goblet cells and taste receptors. The esophagus contains a pharyngeal pad and teeth, with enamel, dentine, and a neurovascular pulp. The intestine narrows caudally, folding twice in the abdomen. There is no distinct stomach, small or large intestine, but the epithelial morphology varies. Columnar absorptive enterocytes are the most numerous cells, followed by goblet cells [18]. The liver has three lobes that lie along the intestinal tract and has similar functions to the mammalian liver [19].

The biliary system develops from intracellular canaliculi that join to form bile ducts. The bile ducts fuse to form the gall bladder, connected to the intestine via the common bile duct. Pancreatic tissue is scattered diffusely along the intestinal tract with an acinar structure very similar to that of mammals. The cells have a very dark, basophilic cytoplasm with large numbers of bright eosinophilic, secretory granules during feeding. Brockman bodies are the equivalent of Islets of Langerhans and have been used to model regeneration and diabetes [20].

Gas Bladder

The gas bladder is derived embryologically from the upper digestive tract and regulates buoyancy. Located ventral to the kidney, it comprises two compartments of columnar, surfactant-producing epithelium. Rodlet cells, found in many teleost fish, are most prevalent within the gas bladder, gills, intestine, bile ducts, and bulbus arteriosis. Once postulated to be parasites, they are more commonly viewed as similar to mammalian mast cells or eosinophils with roles against parasitic infection. The pneumatic duct connects the bladder to the esophagus, allowing filling by gulping air.

Reproductive System

The gonads are lateral, paired organs. Testes contain tubules lined with spermatogenic epithelium [21]. Spermatogenesis occurs in cysts formed when cytoplasmic projections of one or two Sertoli cells surround a single spermatogonium. In contrast to mammalian spermatogenesis, a given Sertoli cell is in contact with only one germ cell clone. Mature spermatozoa leave the genital orifice via two ducti deferens that merge caudally. Ovaries contain eggs at different stages of development. A short oviduct conducts the eggs to the outside.

IN PRACTICE

The Use of Zebrafish for Scientific Research

In most countries where zebrafish studies are undertaken, there will be regulation and licenses required for their use as an experimental model. In the United Kingdom, their use from five days post fertilization is regulated under the Animals (Scientific Procedures) Act 1986 (https://www.gov.uk/research-and-testing-using-animals).

FISH HUSBANDRY

Fish are generally housed in dedicated aquaria with tanks arranged in racking systems (*Techniplast* systems, for example). A single rack houses around 50 tanks with each tank holding up to 20 fish. Optimal environmental conditions promote health and reproduction as indicated by clutch sizes (200+ eggs), embryo survival rates (80–95%), and steady growth (1.0–1.5 cm by 21 days post fertilization).

Water Quality and Stocking Density

Water quality is vital for maintaining healthy, disease-free fish. Oxygen (6.0 ppm), temperature (28.5°C), pH (6.8–7.5), and salinity levels need to be closely monitored and controlled. Stocking density should be limited and filtration adequate to prevent build-up of toxic excretory products (ammonia and nitrate). Crowded conditions also increase cortisol levels, reduce egg production, and retard development.

Appropriate lighting reduces stress and facilitates breeding. Light triggers spawning and periods of darkness are important for rest. Larvae reared in constant light show behavioral and visual deficits, while constant darkness retards development. A cycle of 14 hours light, 10 hours dark is commonly used. Some facilities have separate cohorts on opposing cycles to provide freshly laid eggs more than once a day. Zebrafish appear to habituate to their surroundings, including vibrations such as water pumps, but react strongly to sudden noise and vibration, which should be avoided.

Feeding

Zebrafish are omnivorous, feeding on small crustacea, insect larvae, and algae. Larvae largely hunt prey (Paramecia) visually and darkness impairs this. Juveniles onward should be fed mornings and evenings, with no more food than is consumed in 10 minutes. During early larval stages and metamorphosis, more frequent feeding may be beneficial. Adults tolerate a few days without food but require daily feeding for optimal egg production. Crushed flake food is suitable, but alternating this with brine shrimp maintains breeding efficiency (https://zebrafish.org/documents/protocols/pdf/Fish_Feeding/Schedules_and_ Suppliers/Feeding_Schedule.pdf).

Transportation

Relevant international, EU, and UK laws govern animal welfare during transportation. Responsibility usually falls on the researchers and institute from which the fish originate. http://www.fao.org/docrep/009/af000e/af000e00.HTM

In most countries authorization is required to import live fish. In the United Kingdom, licensing is administrated by the Fish Health Inspectorate https://www.gov.uk/government/groups/fish-health-inspectorate.

Newly arrived fish should be health-checked and quarantined for three to four weeks to avoid potential infection of established lines.

Note: To avoid all contact, new fish can be mated, and the embryos surface-sanitized with bleach solution (35 mg/L sodium hypochlorite for five minutes) with only these bleached embryos subsequently introduced into the main aquarium.

PROCEDURES ON FISH

Surgery and Anesthesia

Zebrafish embryos and larvae are best kept in Danieau's medium (58 mM NaCl, 0.7 mM KCl, 0.4 mM MgSO4, 0.6 mM Ca(NO3)2, 2.5 mM HEPES, pH 7.6). Larvae and adults can be anesthetized relatively easily for imaging, genotyping, or surgical procedures by addition of tricaine methanesulfonate (50 mg/L) to the water. Fish should be kept under anesthesia for as short a time as possible and placed in clean aerated water for recovery. An effective concentration of anesthesia should be maintained throughout the procedure.

Note: Careful titration of anesthetic concentration is required for longer periods of anesthesia in adult fish.

HUMANE KILLING

Zebrafish should be culled by immersion in a fatal concentration of anesthetic solution (tricaine 500 mg/L pH7) for 10 minutes following the cessation of opercular movements. Death should be confirmed by decapitation or physical destruction of the brain.

GENETIC MANIPULATION

Crossing Lines of Fish

In healthy fish that breed well, crossing lines is a simple mechanism for combining the desired genes of two different lines. A plastic mating box with a mesh or slotted bottom is placed inside a slightly larger tank (around 1 L). Breeding

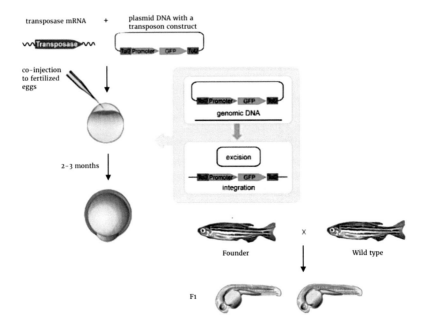

FIGURE 13.9   Schematic of transgenic construction.

Transposase mRNA is coinjected with a transposon donor plasmid containing a Tol2 construct, promoter for the target tissue (e.g., spinal cord) and gene for integration (e.g., GFP) into eggs at the single-cell stage.

The construct is excised from the donor plasmid and integrated into the genome of the founder fish. The founder is crossed with wild-type fish to produce an F1 generation of nontransgenic and heterozygous transgenic fish.

pairs or small groups of fish are added in the evening. Spawning is triggered by light in the morning and fertilized eggs fall through the mesh of the inner box, which prevents them being cannibalized.

Note: Fertilized eggs can be collected and washed by pouring the spawning tank water through a small plastic sieve.

Microinjection

Microinjection is probably the most frequently employed technique in zebrafish-based research and is required for the methods described in the following sections. Synthesized DNA, RNA, proteins, drugs, or microbes are injected at early stages of embryogenesis with fine glass needles controlled by micromanipulators (Figs. 13.9 and 13.10). With experience, a few hundred embryos can be injected in one hour.

Note: Addition of a small amount of phenol red to the injected material helps to visualize successful injection.

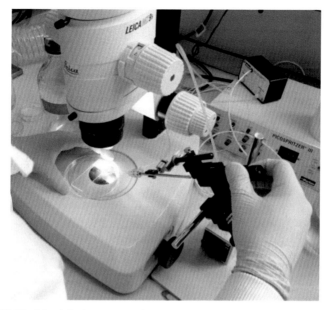

FIGURE 13.10 Microinjection apparatus in use.

Morpholinos

Morpholinos are synthetic antisense oligonucleotides (around 25 nucleotides) designed to bind and block the translation initiation complex of messenger RNA (mRNA) sequences. This technology has been used to test the role of specific genes by transient blocking, particularly during development. The morpholine ring that the nucleotides are arranged on is not recognized by nucleases or immune cells and is therefore stable and nonimmunogenic.

- ATG morpholinos block initiation of protein translation at the ribosome, thereby rendering embryos devoid of a particular protein.
- Splice blocking morpholinos interfere with RNA splicing resulting in truncated proteins and can be used to study the function of a particular protein domain.
- Target protector morpholinos block specific endogenous microRNAs (miRNA) and therefore stabilize specific mRNA transcripts.

Morpholinos are precise and efficient at interfering with genes or miRNA function, but have several potential drawbacks. They are normally injected into the yolk and affect every cell in the embryo, so may compromise normal development if the disrupted protein is required ubiquitously. The effects also only last up to five days, making them unsuitable for studying gene function at later stages of development. Morpholino effects do not always correspond to the

FIGURE 13.11 Close-up of microinjection. The glass microinjection capillary tube on the right is loaded with genetic material and phenol red tracer. The tip is inserted into one of a row of eggs lined up against a glass slide.

observed phenotype of true mutants [22] and therefore, in part because of the development of CRISPR technology (see below), morpholino use has decreased in recent years.

http://www.gene-tools.com

TRANSGENIC CONSTRUCTION

Transgenic fish can be created by injecting bacterial artificial chromosomes (*BACs*) or plasmids containing the desired DNA into single-cell embryos. Including sequences with recognition sites for DNA-modifying enzymes allows the DNA to insert randomly into the genome. Most commonly, the Tol2 *transposon* is used in zebrafish (Fig. 13.11). Successful integration of DNA results around half of the embryos dying compared with noninjected controls over 24 hours. Integration can be confirmed in the surviving embryos, often by a fluorescent marker (e.g., red heart). After three months, adult fish are screened for transgene integration. Germline transmission is confirmed by breeding and genotyping the offspring. Genotyping is best performed by polymerase chain reaction (PCR) analysis of 2–3 mm of caudal fin tip excised from an anesthetized fish using a sterile razor blade.

GENE ACTIVATION

Synthesized mRNA can be injected to cause transient (for up to three days) over-expression of the encoded protein. The function of the encoding gene can be determined by observing the resulting phenotype. To achieve expression throughout the organism, mRNA encoding the protein of interest can be injected into one-cell embryos. Alternatively, mosaic expression can be achieved by injecting single blastomeres at the eight-cell stage or later.

CRISPR-BASED GENOME EDITING

CRISPR *(clustered regularly interspaced short palindromic repeats)* are thought to be part of a prokaryotic adaptive immune response to the exogenous DNA of plasmids and bacteriophages. This system is currently the most common method used to permanently alter the zebrafish genome. There are two distinct components: a guide RNA and an *endonuclease*—the CRISPR-associated nuclease (Cas9).

Genes can be disrupted by creating double-stand breaks (DSBs), which are then repaired by *nonhomologous end joining* (NHEJ). During this process, small insertions or deletions (indels) often lead to loss-of-function mutations in the target gene.

CRISPR can also be used to introduce specific nucleotide modifications at within the target sequence. While NHEJ repair is imperfect and often results in disruption of the open reading frame of the gene, cells can utilize the less error-prone *homology-directed repair mechanism* (HDR). A DNA repair template containing the desired sequence must be present for HDR and is normally transfected into the cell along with the guide RNA and Cas9. If the template has a high degree of homology to the sequence immediately upstream and downstream of the Cas9-induced DSB, HDR can accurately introduce specific nucleotide changes at this location.

PHYSICAL MANIPULATION

Transplantation and other types of surgery are some of the more classic techniques employed in developmental biology. Cells are removed from a donor embryo and introduced into a recipient embryo by micropipetting. This technique can show whether a gene function is cell autonomous or not. Mutant cells are transplanted into a wild-type embryo and vice versa and the phenotypes are compared. Transplantation can also be used to study the effect of diffusible factors, such as growth factors on neighboring cells [23].

Transplantation may be useful when a morpholino injection is lethal. Cells from the morpholino-injected embryo are transplanted at early stages of development into a wild-type embryo to determine the effect of the protein knockdown on the cells in a wild-type environment.

GENETIC SCREENS

Forward Genetic Screens

Zebrafish are ideal vertebrates for forward genetic screens (Fig. 13.12). Exposure of male fish to mutagens such as ethylnitrosourea (ENU) generates hundreds of point mutations in their premeiotic germ cells (four-to-six weekly one-hour exposures to 3–3.5 mM ENU). These males are subsequently crossed with wild-type females to produce F_1 heterozygous progeny.

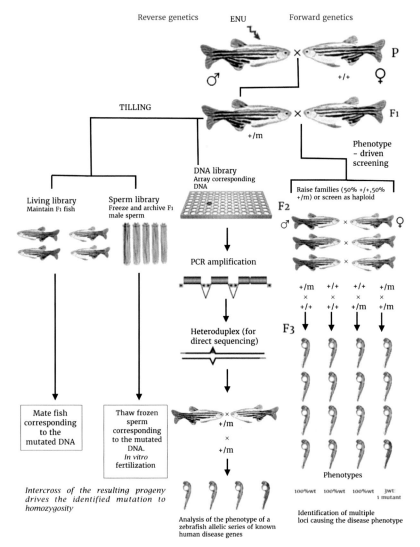

FIGURE 13.12 Schematic summarizing forward genetic screens and TILLING.

Addition of ethylnitrosourea to the tank water creates point mutations in the genome of adult male zebrafish. These mutagenized fish are outcrossed to create an F_1 founder population with a high density of mutations.

In forward-genetics the F_1 fish are incrossed through two generations to produce mutants in one-quarter of the F_3 embryos. Screening can be performed by transgenes that express fluorescent proteins in specific tissues or by phenotype analysis. Disease models are identified by similarity to human pathology.

Alternatively, males of the F_1 founder population can be used to create linked DNA and sperm libraries. These libraries can be used to target the screening of individual loci by TILLING (targeting induced local lesions in genomes), a reverse-genetic process. This method uses PCR to amplify exons of a specific disease gene of interest from individual or pooled DNA from the F_1 DNA library. The fish corresponding to the DNA library can either be a living library or stored as a library of frozen sperm. Mutations are identified by heteroduplex-detecting methods or direct sequencing. The pedigree carrying the mutation is recovered by breeding directly from the corresponding fish in a living library or by thawing the corresponding frozen sperm to use in vitro fertilization. The resulting fish are intercrossed to generate heterozygous and homozygous mutants for the disease gene.

The F_1 fish are then in-crossed with siblings to create an F_2 generation, half of which are heterozygous for a specific mutation (m) and the other half are wild-type. F_2 siblings are crossed, and the resulting F_3 progeny are 25% wild-type (+/+), 50% heterozygous (+/m) and 25% homozygous (m/m) for a recessive mutation.

Together, the large-scale Boston and Tübingen screens, began with around 300 ENU-treated males, raised more than 5,000 F_2 families, analyzed over 6,000 mutated genomes and selected more than 2,000 new developmental mutants for characterization [24].

Haploid Screens

To avoid the need for screening thousands of fish for recessive mutations in conventional screens, recessive mutations can be revealed in a single generation by taking advantage of the zebrafish's ability to survive for several days as haploid organisms. Female F_1 fish (from crossing wild-type females and ENU-exposed males as above) are squeezed gently to release their eggs, which are then fertilized with ultraviolet (UV)-treated sperm to generate haploid embryos. UV treatment destroys the parental DNA, without affecting its ability to activate the egg. A haploid clutch derived from a heterozygous female will contain 50% mutant and 50% wild-type embryos.

Targeting Induced Local Lesions in Genomes (TILLING)

Forward genetic screens in zebrafish have been very successful at identifying genes important in development and disease through examining phenotypes. TILLING is a reverse genetics approach whereby rare mutations in genes of interest can be identified irrespective of the phenotype they cause (Fig. 13.12). First a "library" of F_1 progeny is created from ENU-mutated fish (as described previously). Kept either as living fish or their cryopreserved sperm and corresponding genomic DNA, this library can be screened by PCR for specific mutations in a gene of interest. Once F_1 fish with mutations in the gene of interest have been identified in the library, F_2 heterozygotes are created by crossing or in vitro fertilization (depending whether the library is live or cryopreserved). In-crossing this generation allows the F_3 mutant phenotypes to be studied. Ongoing TILLING projects still have many genes to screen and will accept requests to screen particular genes:

http://www.sanger.ac.uk/Projects/D_rerio/mutres/
http://www.zfishtilling.org/zfish/

The series of point mutations generated during TILLING can more accurately mirror those seen in human congenital disease compared with standard homologous recombination that also often leaves a complex genomic footprint. TILLING can also create mis-sense and gain-of-function mutations that are informative when investigating gene function [25].

APPLICATIONS

Developmental Biology

The ex-utero development and optical clarity of zebrafish allows the production of high-resolution developmental fate maps by injecting cells of interest with fluorescently labeled dextrans. Alternatively, transgenic fish can be constructed that express fluorescent proteins in the cell, tissue or organ of interest. The effects of genetic or physical manipulation on development can then be observed.

Genetic Screening

Quick breeding, ease of housing large numbers, and optical clarity give zebrafish a huge advantage over other vertebrates, particularly for large-scale genetic screens. Disease phenotypes can be identified by simple microscopy, assisted by transgenic expression of fluorescent proteins in specific tissues and organs if necessary.

Disease Modeling

Once a zebrafish orthologue related to human disease has been identified through genetic screening or TILLING, a fish model of the disease can potentially be created. Disease-related developmental, cellular, and even molecular defects can be identified due to the ease of genetic and physical manipulation of embryos and ability to perform in vivo imaging at a cellular level. For example, the zebrafish mutant *sapje* suggested the myotendinous junction as a possible site of pathophysiology in muscular dystrophy (Bassett et al., 2003).

Of course, many human diseases have a complex polygenic etiology such as heart disease and cancer. In these cases, the zebrafish can be used to perform genetic suppressor and enhancer screens in the presence of already identified disease-related genes. Gene knock-out or overexpression via morpholino or mRNA injection, respectively, can be used to explore genetic interactions that affect disease severity or penetrance, and for assessing the therapeutic effects of removing or adding selected gene products [25].

Wound Healing and Regeneration

There has been great interest in exploring the mechanisms underlying healing and regeneration using the zebrafish due to their incredible regenerative potential [15]. Cardiac and neural tissue, in particular, heal without scarring in contrast to mammals. These processes can be analyzed in vivo at cellular, molecular, and genetic levels.

Immunology and Inflammation

Zebrafish have innate and adaptive immune systems similar to mammals [26] and are susceptible to infection by bacteria, mycobacteria, protozoa, and viruses. In vivo analysis of the integrated immune response has been facilitated by the use of transgenic zebrafish with green fluorescent protein (GFP)-labeled macrophages, neutrophils, and endothelia, and of transgenic pathogens expressing GFP [13]. This response can be observed in the context of an experimentally induced wound, infection, or neoplasm.

Drug Screening

The zebrafish is the ideal vertebrate for large-scale screening of molecules with therapeutic potential, particularly when zebrafish disease models are employed. Using whole-organisms assesses bioavailability and toxicity too, in contrast to in vitro or biochemical screens. Furthermore, without presumption of the molecular mechanisms involved, previously unsuspected target proteins and pathways may be identified.

SCENARIO—THE INNATE IMMUNE RESPONSE TO CANCER RESECTION

The immune system is involved in surveillance and killing of oncogene-transformed cells within tissues, but is also believed to contribute to tumor progression of established cancers. This raises questions about the effects that acute inflammation triggered by biopsy or surgery may have on residual tumor. A zebrafish model of cancer surgery has been developed to investigate this relationship.

Transgenic Construction

A zebrafish model of melanoma is ideal due to the reliability of tumor development and ease of identifying the tumors macroscopically (i.e., heavily pigmented tumors arising on the surface of the animal). The fish we used expresses the oncogene HRAS in melanocytes driven by the melanocyte specific mitfa promoter [27]. Larvae show ectopic hyperpigmentation and develop melanoma that correlates well with human melanoma in terms of histopathology, immunology, and epigenetics. Inclusion of GFP in the genetic construct also means that the tumors fluoresce, aiding detection and imaging (Fig. 13.13).

To enable identification and live-imaging of innate immune cells responding to tumor and surgery, this fish is crossed with lines that express fluorescent proteins in macrophages Tg(mpeg:FRET) and neutrophils Tg(Lysc:dsRed). The resulting tri-cross line is reared to adulthood and observed for tumor development.

Surgery

Adult fish developing tumors and with appropriately labeled cells are anesthetized in tricaine (50 mg/L) and checked for expression of fluorophores, obtaining images using a dissecting microscope and fluorescent light source. A sterile

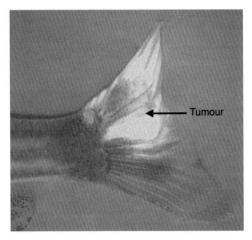

FIGURE 13.13   GFP expressing tumor (melanoma) in the caudal fin of an adult kita:RasG12VeGFP; LysC:dsRed transgenic zebrafish.

surgical punch is used to remove tissue at the tumor margin leaving a wound with edges that are part tumor and part healthy tissue (Fig. 13.14).

Live-Imaging and Immunohistochemistry

The surgical punch wound can be imaged daily (by confocal or multiphoton microscopy with the anesthetized fish mounted by the tail in 1.5% agarose) to determine the time course of innate immune cell influx and efflux. Comparison can be made between the healthy tissue and tumor wound edges. After five days, the fish are culled and immunohistochemistry is performed to compare proliferation rates in the wounded tumor and the non-wounded tumor.

These experiments can be repeated in the presence of beclamethasone to inhibit neutrophil migration and with morpholino knockdown of macrophages or neutrophils to determine the contribution of these cells to tumor proliferation [28].

KEY LIMITATIONS

A Novel Model Organism

The zebrafish is a relatively new research organism, which means that background data and resources are fewer by comparison with more established models. Genetic techniques are not yet as refined as for classic invertebrate model organisms such as *C. elegans* and *Drosophila*, while antibody availability is currently not yet as good as for the mouse.

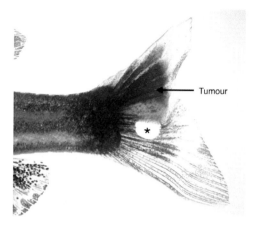

FIGURE 13.14 Photograph immediately after a 1 mm surgical punch (*) has been taken at the margin of the tumor. This will allow study of how the acute inflammatory response to tissue damage might impact on subsequent tumor growth.

Genetic Considerations

Zebrafish are excellent vertebrate research organisms, but obviously do not possess homologues of all human genes or organs. In addition, the zebrafish genome contains duplications of genes and chromosomal segments, thought to be the result of a genome-wide duplication in the ray-fin fish lineage around 400 million years ago. This means that many human genes will have two orthologues in the zebrafish. The phenotype may therefore not be as expected after knockout of one of these genes due to some degree of genetic redundancy.

Aquatic Environment

An aquatic environment demands specialized equipment, particularly when using large numbers of fish. Drug delivery can be easier, but can also have inherent problems; some small molecules require a carrier solvent that would reach toxic levels before solubility is achieved, while immersion in other agents can affect organs other than the desired target.

TROUBLESHOOTING

Problem	Potential Solutions
Slow development of fish to adulthood and sexual maturity	Careful husbandry promotes healthy fish that develop and breed well. Regular delivery of age-appropriate feed will facilitate growth, but care must be taken to avoid uneaten food leading to dirty tanks and infection.

Problem	Potential Solutions
Low numbers of eggs during matings	Fish are unlikely to lay large numbers of eggs if mated too frequently—generally no more than every three days.
Failure of fish to lay eggs	Check if you have sexed the fish to be crossed correctly. Spawning may be encouraged by tanks with shallow sloping bottoms that mimic the shore, natural light cycles, and plastic plants. Usually two to three males and one or two females are placed in the breeding tank at the end of the preceding day and then kept dark so that mating is triggered by light in the morning.
Difficulty with microinjection	At room temperature, fertilized eggs remain at the one-cell stage for around 40 minutes. Injecting is easiest when the cell rounds in the final 20 minutes prior to division. Clear dividers between males and females to be crossed can be removed in the morning to optimize timings for microinjection. Eggs can be lined up against the edge of a glass slide or in a groove scored in an agar plate with the animal pole aligned toward the injector. Care should be taken to avoid eggs drying out as they lose turgor, which makes needle penetration of the chorion more difficult. Gently tapping the microinjector can help advance the tip of the needle through the chorion and cell membrane.
Failure to recover from anesthesia	Achieving a stable level of anesthesia can be tricky in adults and will vary between individual fish. The concentration of tricaine often needs fine titration with respiratory rate. Minimizing handling reduces stress and promotes recovery. A recovery tank placed in a quiet, darkened area is also helpful. Respiratory activity can often be regained in overanesthetized fish by infusing tank water over their gills using a Pasteur pipette.

CONCLUSION

The zebrafish is a powerful tool for investigating development, wound healing, inflammation, and cancer biology. It is ideally suited to genetic screening, disease modeling, and pharmacological screening. There have already been a number of fish tank to clinic success stories and there is huge potential still to be exploited.

ACKNOWLEDGMENTS

We thank David Gurevich, Josie Morris, Rebecca Richardson, and Ilona Aylott for their advice on preparation of this chapter. The Martin Laboratory is funded by the Wellcome Trust, Cancer Research UK, the British Heart Foundation, and the Biotechnology and Biological Sciences Research Council. John Collin is supported by a research grant from the British Association of Oral and Maxillofacial Surgeons.

REFERENCES

[1] Howe K, Clark MD, Torroja CF, Torrance J, Berthelot C, Muffato M, et al. The zebrafish reference genome sequence and its relationship to the human genome. Nature 2013;496(7446):498–503.

[2] North TE, Goessling W, Walkley CR, Lengerke C, Kopani KR, Lord AM, et al. Prostaglandin E2 regulates vertebrate haematopoietic stem cell homeostasis. Nature 2007; 447(7147):1007–11.

[3] White RM, Chech J, Ratanasirintrawoot S, Lin CY, Rahl PB, Burke CJ, et al. DHODH modulates transcriptional elongation in the neural crest and melanoma. Nature 2011; 471(7339):518–22.

[4] Lowery LA, Sive H. Strategies of vertebrate neurulation and a re-evaluation of teleost neural tube formation. Mech Dev 2004;121(10):1189–97.

[5] Chitnis AB, Nogare DD, Matsuda M. Building the posterior lateral line system in zebrafish. Dev Neurobiol 2012;72(3):234–55.

[6] Sumbre G, Poo MM. Monitoring tectal neuronal activities and motor behavior in zebrafish larvae. Cold Spring Harb Protoc 2013;2013(9):873–9.

[7] Miyasaka N, Wanner AA, Li J, Mack-Bucher J, Genoud C, Yoshihara Y, et al. Functional development of the olfactory system in zebrafish. Mech Dev 2013;130(6–8):336–46.

[8] Ben-Moshe Z, Foulkes NS, Gothilf Y. Functional development of the circadian clock in the zebrafish pineal gland. Biomed Res Int 2014;2014:235781.

[9] Lohr H, Hammerschmidt M. Zebrafish in endocrine systems: recent advances and implications for human disease. Annu Rev Physiol 2011;73:183–211.

[10] Poss KD, Wilson LG, Keating MT. Heart regeneration in zebrafish. Science 2002;298(5601):2188–90.

[11] Evans DH, Piermarini PM, Choe KP. The multifunctional fish gill: dominant site of gas exchange, osmoregulation, acid-base regulation, and excretion of nitrogenous waste. Physiol Rev 2005;85(1):97–177.

[12] Davidson AJ, Zon LI. The 'definitive' (and 'primitive') guide to zebrafish hematopoiesis. Oncogene 2004;23(43):7233–46.

[13] Henry KM, Loynes CA, Whyte MK, Renshaw SA. Zebrafish as a model for the study of neutrophil biology. J Leukoc Biol 2013;94(4):633–42.

[14] Ochi H, Westerfield M. Signaling networks that regulate muscle development: lessons from zebrafish. Dev Growth Differ 2007;49(1):1–11.

[15] Richardson R, Slanchev K, Kraus C, Knyphausen P, Eming S, Hammerschmidt M. Adult zebrafish as a model system for cutaneous wound-healing research. J Invest Dermatol 2013;133(6):1655–65.

[16] Alt B, Reibe S, Feitosa NM, Elsalini OA, Wendl T, Rohr KB. Analysis of origin and growth of the thyroid gland in zebrafish. Dev Dyn 2006;235(7):1872–83.

[17] Pavlidis M, Digka N, Theodoridi A, Campo A, Barsakis K, Skouradakis G, et al. Husbandry of zebrafish, Danio rerio, and the cortisol stress response. Zebrafish 2013;10(4):524–31.

[18] Wallace KN, Akhter S, Smith EM, Lorent K, Pack M. Intestinal growth and differentiation in zebrafish. Mech Dev 2005;122(2):157–73.

[19] Cox AG, Goessling W. The lure of zebrafish in liver research: regulation of hepatic growth in development and regeneration. Curr Opin Genet Dev 2015;32:153–61.

[20] Moss LG, Caplan TV, Moss JB. Imaging beta cell regeneration and interactions with islet vasculature in transparent adult zebrafish. Zebrafish 2013;10(2):249–57.

[21] Rupik W, Huszno J, Klag J. Cellular organisation of the mature testes and stages of spermiogenesis in Danio rerio (Cyprinidae; Teleostei)–structural and ultrastructural studies. Micron 2011;42(8):833–9.

[22] Kok FO, Shin M, Ni CW, Gupta A, Grosse AS, van Impel A, et al. Reverse genetic screening reveals poor correlation between morpholino-induced and mutant phenotypes in zebrafish. Dev Cell 2015;32(1):97–108.

[23] Cavodeassi F, Houart C. Brain regionalization: of signaling centers and boundaries. Dev Neurobiol 2012;72(3):218–33.

[24] Nusslein-Volhard C. *The* zebrafish iss*ue of Development*. Development 2012;139(22): 4099–103.

[25] Jones RA, Feng Y, Worth AJ, Thrasher AJ, Burns SO, Martin P. Modelling of human Wiskott-Aldrich syndrome protein mutants in zebrafish larvae using in vivo live imaging. J Cell Sci 2013;126(Pt 18):4077–84.

[26] Keightley MC, Wang CH, Pazhakh V, Leischke GJ. Delineating the roles of neutrophils and macrophages in zebrafish regeneration models. Int J Biochem Cell Biol 2014;56:92–106.

[27] Lister JA, Capper A, Zheng Z, Mathers ME, Richardson J, Paranthaman K, et al. A conditional zebrafish MITF mutation reveals MITF levels are critical for melanoma promotion vs. regression in vivo. J Invest Dermatol 2014;134(1):133–40.

[28] Antonio N., Bønnelykke-Behrndtz M.L., Ward L.C., Collin J., Christensen I.J., Steiniche T., et al., *The wound inflammatory response exacerbates growth of pre-neoplastic cells and progression to cancer* 10.15252/embj.201490147.

SUGGESTED FURTHER READING

[1] Huang P, Zhu Z, Lin S, Zhang B. Reverse genetic approaches in zebrafish. J Genet Genomics 2012;20:421–33.

[2] Phillips JB, Westerfield M. Zebrafish models in translational research: tipping the scales toward advancements in human health. Dis Model Mech. 2014;7:739–43.

[3] Sertori R, Trengove M, Basheer F, Ward AC, Liongue C. Genome editing in zebrafish: a practical overview. Brief Funct Genomics 2015 Dec 9 pii: elv051.

[4] Essential Zebrafish Methods (Reliable Lab Solutions), 2009 by Monte Westerfield Academic Press Inc.

[5] The Zebrafish: Disease Models and Chemical Screens (Methods in Cell Biology) 2011 by H. William Detrich and Monte Westerfield Academic Press Inc.

GLOSSARY

BAC (bacterial artificial chromosomes)   A DNA vector created from bacterial F' plasmids used in nature to transfer genetic material between some bacteria.

Clustered regularly interspaced short palindromic repeats (CRISPR)   A prokaryotic defense system against foreign genetic elements such as plasmids and phages. They recognize and excise this exogenous genetic material and have therefore been adapted to use as genetic tools, particularly in the zebrafish.

Double-stranded breaks (DSB) Highly toxic genetic lesions due to breakage of both strands of the DNA. Defective DSB repair is associated with various developmental, immunological, and neurological disorders, and is a major driver in cancer.

Endonuclease An enzyme that cleaves DNA.

Epiboly Thinning and spreading of a sheet of cells during gastrulation.

Gastrulation An early event in embryogenesis when the single-cell layer blastula organizes into a trilaminar gastrula (ectoderm, mesoderm, and endoderm).

Metamorphosis In the zebrafish, the point of change from larval to adult fish. Larval features such as the fin fold are lost, remodeling (e.g., gut and nervous system) occurs, and adult features including scales are acquired.

Micro RNA (miRNA) Highly conserved 22 nucleotide long RNA sequences found in plant and animal genomes that regulate the expression of genes by binding to the 3′-untranslated regions (3′-UTR) of specific mRNAs.

Morpholino A synthetic antisense sequence of nucleic acids bound to morpholine rings rather than ribose or deoxyribose and nonionic phosphorodiamidate linkages instead of the anionic phosphates of DNA and RNA. Used to block specific RNA target sequences.

Neurulation Initial development of the spinal cord and brain by folding of neural plate ectoderm.

Nonhomologous end joining (NHEJ) The predominant and evolutionarily conserved DNA double-strand break repair pathway in eukaryotic cells.

Orthologue Genes in different species that evolved from a common ancestral gene by speciation and usually retain the same function.

Plasmid Self-replicating extra-chromosomal DNA found in many bacteria that can be used as vectors for amplification or insertion of sequences.

TILLING (targeting induced local lesions in genomes) A high-throughput reverse genetics technique to detect single basepair mutations using endonucleases that cleave heterodimers formed by PCR of the DNA in the target region. The resulting fragments are analyzed by gel electrophoresis.

Tol2 transposon A transposon or transportable element is a small sequence of DNA that is able to insert itself into another point in the genome. The Tol2 transposon is an autonomous transposon, meaning that it encodes a functioning transposase enzyme capable of identifying, excising, and reinserting the DNA element defined by its inverted terminal repeats (ITR) or other elements with the same ITRs.

Chapter 14

Xenopus as a Model Organism for Biomedical Research

Shoko Ishibashi[1], Francesca Y.L. Saldanha[2] and Enrique Amaya[1]

[1]*University of Manchester, Manchester, United Kingdom;* [2]*Massachusetts Institute of Technology, MA, United States*

Chapter Outline

Basic Science Methods for Clinical Researchers. DOI: http://dx.doi.org/10.1016/B978-0-12-803077-6.00022-9

Objectives

After reading this chapter the clinician training in basic science should:
- Understand the practical advantages of using *Xenopus* in basic science research to understand disease biology.
- Have a working knowledge of the diverse experimental approaches that are possible using *Xenopus* and how these are best applied to answer questions regarding pathogenesis of human disease.
- Be able to suggest and design appropriate assays that address etiology and mechanisms of disease processes.

INTRODUCTION

Xenopus is a genus in the order, *Anura*, which means "tailless" in Greek. The genus takes its name from the Greek for "strange foot" (ξενος and πους) owing to the webbed, five-toed, three-clawed rear feet present in the many species in the genus. Although the genus contains more than a dozen species, of varying ploidy, the two species most commonly used in biomedical research are *Xenopus laevis* (the South African clawed frog) and *Xenopus tropicalis* (the West African clawed frog) [1]. These two species of African aquatic frogs are members of the family of tongue-less frogs, known as Pipidae. *X. laevis* first arrived into the biomedical arena after the observation in the 1930s in South Africa that injection of urine from pregnant women would induce the females to lay eggs. This discovery led to their export to hospitals throughout the world and their adoption as part of a cheap and easy pregnancy test (reviewed in [2]). In the mid-20th century, embryologists began exploiting Xenopus as a model organism, given the ease by which one could obtain eggs and embryos via simple hormone injections at any time of the year, unlike more traditional amphibian models, which produced embryos only seasonally. Later on, Xenopus was adopted as a convenient model system for most biomedical research fields, including biochemistry, molecular and cellular biology, endocrinology, physiology, and neurobiology [2,3].

USE AS A RESEARCH ORGANISM

For the past half century, Xenopus has been a staple model organism in the biomedical sciences. The reasons for this are multiple, but include the facts that Xenopus can be raised and housed in large numbers and cheaply in simply fish aquaria; one can obtain eggs and embryos any time of the year, following simple hormone injections; Xenopus is highly prolific, producing many thousands of eggs, from each ovulation, thus the number of eggs/embryos is hardly ever a limiting factor when designing experiments; the relatively large size of the embryos allows researchers to perform microsurgery on them easily; their large size also make the embryos easy to inject with various substances, such as nucleic acids, proteins, or drugs at any stage of development; the great abundance of eggs and embryos that one can obtain from these animals, combined with the large size of the eggs/embryos, facilitate any large-scale –omic approach, such as genomic, proteomic, and metabolomic approaches; and finally, that it is possible to modify Xenopus genetically, either through transgenesis or by gene-editing techniques, such that gain of function and loss of function experiments can be performed on any gene of interest easily and rapidly using this model system.

Given these many advantages, many fundamental biological principles were first established in Xenopus. For example, it was using Xenopus, where Sir John Gurdon was able to show for the first time, that nuclei from differentiated cells could be reprogrammed, following transplantation into enucleated eggs, and using this method, one can produce clones of animals [4,5]. This work, which was first published in 1962, eventually led him to share the Nobel Prize in Medicine in 2012 with Shinya Yamanaka [6]. After Sir John Gurdon's pioneering experiments on nuclear reprogramming using Xenopus, a long list of major advances followed using this system, including the observation that the nucleolus is the site of rRNA synthesis and that mitochondria have DNA. In addition, this is also the system where the first eukaryotic gene was isolated and growth factors were first shown to play key roles in embryonic development and patterning, to mention but a few discoveries pioneered in Xenopus. As such, many basic scientific principles were first established using this model system, and later confirmed to be conserved in other systems, including humans (for summary of the pioneering findings, arising from studies using Xenopus, see Harland and Grainger [3]).

Although *X. laevis* is an ideal model system in many aspects, its allotetraploid genome and long generation time makes it cumbersome for genetic manipulations. Amaya and colleagues (1998) introduced the Western clawed frog, *X. tropicalis* to address these disadvantages [1]. This species retains all the advantages of its allotetraploid cousin, yet has a diploid genome and a considerably shorter generation time. In addition, the adult animals are smaller, and thus can be housed in larger numbers at lower cost. These attributes make *X. tropicalis* a more suitable model for genetic and genomic experiments [1,3]. Importantly, the complete genome sequence of *X. tropicalis* has revealed that the genome organization is very similar to that of humans, including the identification of orthologs of 1700 human disease genes [7]. For this reason, *X. tropicalis* is now used as a model of congenital human diseases [8].

IN PRINCIPLE

Strains

Two "wild" type *X. laevis* strains (a pigmented strain and an albino strain) and one inbred strain (J-strain) are commonly used in laboratories across the world. Both the pigmented and albino *X. laevis* strains were collected from the wild. The inbred J-strain was established through inbreeding over more than 40 years by Tochinai and Katagiri in Japan and this was the strain, which was chosen for genome sequencing [9].

TABLE 14.1 Basic characteristics of *X. laevis* and *X. tropicalis*

	X. laevis	*X. tropicalis*
Ploidy	Allotetraploid	Diploid
Haploid	18 chromosomes	10 chromosomes
Genome size	3.1×10^9 bp	1.7×10^9 bp
Optimal temp	16–22°C	23–28°C
Adult size	10 cm	4–5 cm
Egg size	1–1.3 mm	0.7–0.8 mm
Brood size	700–2000+	1000–3000+
Generation time	1–2 years	6–12 months

There are primarily two wild-type strains for *X. tropicalis* (one collected initially from Nigeria and the other from the Ivory Coast) [3]. The Nigerian strain has been inbred for more than 10 generations, and it was an inbred Nigerian strain, which was used for genome sequencing [7]. No albino *X. tropicalis* strains have been found in the wild, therefore we and others have generated several albino *X. tropicalis* strains using gene-editing technologies, whereby null mutations in the tyrosinase (*tyr*) gene, which is responsible for pigmentation in the melanophores, were introduced using either TALENs or CRISPR technologies [10–12]. Table 14.1 summarizes the basic characteristics of *X. laevis* versus *X. tropicalis*.

Development

A comprehensive staging system and description of embryonic development for *X. laevis* was established by Nieuwkoop and Faber in 1956 [13] (see also http://www.xenbase.org/anatomy/alldev.do). The same staging system is also used for *X. tropicalis*. A summary of the key developmental feature and stages in *X. laevis* and *X. tropicalis* are shown in Table 14.2.

Fig. 14.1 shows images and diagrams of key events that occur between the unfertilized egg and the eight-cell stage. The unfertilized egg (Fig. 14.1A) is radially symmetrical and contains only one axis, the animal-vegetal axis, observed in pigmented embryos by the animal (pigment) and vegetal (nonpigmented) halves. Within 10–15 min after fertilization, the pigmented half of the embryo undergoes a contraction event, which coincides with the extrusion of the second polar body (Fig. 14.1B). Coincident with this, the vitelline membrane lifts and the zygote reorients, such that the pigmented side is up, and nonpigmented side is down. This reorientation occurs due to the vegetal pole/nonpigmented side of the zygote containing greater density of yolk granules

TABLE 14.2 Key developmental features and stages in *X. laevis* and *X. tropicalis*

Stage	Features	Time Postfertilization	
		X. laevis (23°C)	*X. tropicalis* (25°C)
st. 0–st.7	Fertilization to early cleavage stages	0–6 h	0–5 h
Blastula st. 8–9	Animal cap explant mid-blastula transition	7 h	5 h
Gastrula st. 10–13	Mesoderm induction gastrulation	11–14 h	8–10 h
Neurula st. 13–24	Neural tube is formed Somitogenesis starts Distinct protrusion of eyes	15–30 h	11–18 h
Tailbud st. 25	Organogenesis	31 h	19–20 h
Tadpole st. 45	Feeding stage	4 days	3 days
~st. 65	Metamorphosis	2–3 months	2–3 months
Juvenile		3 months to 2 years	3–6 months
Adult		1–2 years	4–6 months

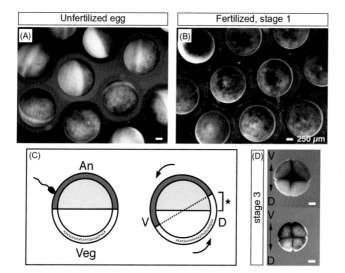

FIGURE 14.1 The sperm entry point defines the dorsal–ventral axis in Xenopus. (A) Unfertilized *X. laevis* eggs. (B) 1-cell (st. 1) *X. laevis* embryos (zygotes), 10 min after fertilization. (C) Diagram depicting the cortical rotation after fertilization, resulting in the movement of a dorsal determinant (*blue*) toward to future dorsal side of the embryo (*). (D) 4-cell (st. 3) embryos, *X. laevis* (top) and *X. tropicalis* (bottom). The darker blastomeres at the top are fated to become the ventral region of the embryo, while the paler blastomeres at the bottom are fated to give rise to the dorsal region of the embryo. D, dorsal; V, ventral; An, animal pole; Veg, vegetal pole. Scale bar: 250 μm.

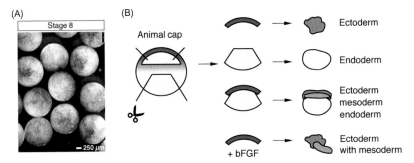

FIGURE 14.2 Blastula stage embryo and the animal cap. (A) St. 8, blastula *X. laevis* embryos. (B) Diagram of animal cap assay for mesoderm induction. Animal cap explants isolated from blastula stage embryos differentiate into atypical epidermis if they are cultured on their own. Animal caps cultured with vegetal explants or basic FGF give rise to some mesodermal tissue (*pink*). Scale bar: 250 μm.

and thus, being heavier than the pigmented/animal pole of the zygote, which is less dense. Approximately 30 min after fertilization, a cortical rotation occurs, which defines the future dorsal/anterior–ventral/posterior axis of the embryo. The cortical rotation is mediated by a subcortical microtubule network, whose orientation is influenced by the sperm entry point and the incoming centrosome from the sperm. Thus, the sperm entry point marks the future ventral/posterior axis, and the opposite side to the sperm entry point marks with the future dorsal/anterior axis (Fig. 14.1C). By the four- and eight-cell stages, the consequence of the cortical rotation becomes more apparent, as two of the blastomeres (the future dorsal/anterior side) appear less pigmented than the two prospective ventral/posterior blastomeres (Fig. 14.1D; also see time lapse movie of early cleavage stages through to the mid-blastula stage in Movie 1 (https://youtu.be/r44yZ1In7yI)). Such subtle, but clear differences in pigmentation permits the researcher to target injection of RNA into a particular region of interests in the embryo. The early embryo undergoes around 10–12 rapid and synchronous divisions, when the cell cycle has no G1 and G2 phases. Then the speed of cell cycle slows down after 12 cell divisions, when the embryo has reached the mid-blastula stage containing thousands of cells (Fig. 14.2A). At this time point, the G1 and G2 phases of the cell cycle begin and zygotic transcription starts, which is known as the mid-blastula transition (MBT). Intriguingly, the cells in the animal pole at this stage are undifferentiated and pluripotent. For this reason, these cells have been used since the mid-1980s to identify molecules, which can influence cell-fate choices in pluripotent cells, in ways very similar to what mammalian embryonic stem (ES) cells are used today. Namely, "animal cap" explants from mid-blastula stage embryos are competent to give rise to any cell type of the embryo (with the possible exception of germ cells), if given the right conditions. Thus, experiments using animal cap explants from Xenopus gave rise, for the first time, to our understanding that growth factors mediate mesoderm induction and patterning in the early vertebrate embryo [14] (Fig. 14.2B). A fate map of mid-blastula stage embryo is shown in Fig. 14.3A.

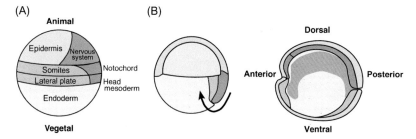

FIGURE 14.3   Formation of germ layers during gastrulation. (A) Schematic of fate map for the late blastula embryo. Derivatives of the ectoderm are shown in *blue*, derivatives of the mesoderm are shown in *red*, and derivatives of the endoderm are shown in *yellow*. (B) Gastrulation movement forms three germ layers, ectoderm, mesoderm, and endoderm, organizing the anteroposterior axis of the embryo.

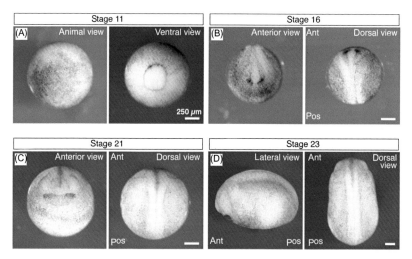

FIGURE 14.4   Xenopus embryos at gastrulation and neurulation. (A) st. 11 *X. laevis* embryo. (B) st. 16 *X. tropicalis* embryo. (C) st. 21 *X. laevis* embryo. (D) st. 23 *X. laevis* embryo. Scale bar: 250 μm.

During gastrulation the embryo undergoes its first major morphogenetic event, whereby a series of coordinated cell movements, starting with involution of cells in the dorsal blastopore lip, result in the eventual rearrangement of cells whereby the ectoderm ends on the outside, the mesoderm in middle, and the endoderm on the inside, the three germ layers of the embryo (Fig. 14.3B). Ectodermal lineages give rise to the epidermis of the skin and the nervous system; mesodermal lineages form the notochord, muscle, heart, bone, and cartilage, urogenital system, connective tissues, and blood; and endodermal lineages give rise to the digestive system, liver, pancreas, and pharynx. Photographs of embryos from the gastrula stages through to the tadpole stages are shown in Figs. 14.4 and 14.5. After gastrulation, neurulation begins whereby the neural ectoderm folds, generating

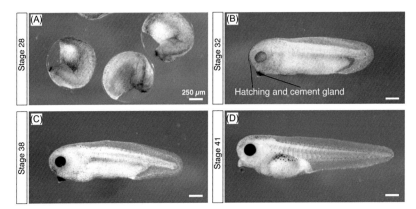

FIGURE 14.5 **Tailbud embryos.** (A) st. 28 unhatched *X. tropicalis* embryo. (B) st. 32 hatched *X. tropicalis* embryo. (C) st. 38 *X. tropicalis* embryo. (D) st. 41 *X. tropicalis* embryo. The hatching and cement glands are transient glands. The hatching gland secrete proteinases, which degrade the vitelline membrane, required for hatching. The cement gland secretes mucus, which allows the embryos to stick to surfaces. Scale bar: 250 μm.

the neural tube (Figs. 14.4B and 14.4C; see also time-lapse movie of *X. laevis* gastrulation and neurulation in Movie 2 (https://youtu.be/TmEoaKWoxTY) and *X. tropicalis* from the neurula to early tailbud stages in Movie 3 (https://youtu.be/q1SaLbfVanA)). During the tailbud stages, the somites begin to form and the embryos elongate along their anteroposterior axis (Fig. 14.4D; Fig. 14.5). During the mid-tailbud stages (Fig. 14.5A), the embryos start to twitch within the vitelline membrane and then they hatch (see time-lapse movie of neurula through hatching stage *X. laevis* embryo in Movie 4 (https://youtu.be/jVLfBCfvOv8)). All embryos will hatch by the late tailbud stage (stage 30–36) (Figs. 14.5B and 14.5C). They become active and start feeding from stage 45. Stage 42–44 tadpoles are the first stages, which can be used for tail regeneration studies (see live movie of *X. tropicalis* stage 43 and stage 47 tadpoles in Movie 5 (https://youtu.be/h-uwyJGvT2w) and time-lapse movie of *X. tropicalis* head and face morphogenesis in Movie 6 (https://youtu.be/rm3ARN6Zz6Q)). One should consider, however, that there is a refractory period where the tail cannot regenerate efficiently between stage 46 and 47 [15] (Fig. 14.6).

IN PRACTICE

Husbandry

Animals

X. laevis and *X. tropicalis* can be obtained from the European Xenopus Resource Center (EXRC) (https://xenopusresource.org/) in the United Kingdom and the National Xenopus Resource (NXR) Center (http://www.mbl.edu/xenopus/) in the United States. There are two major commercial suppliers, NASCO and

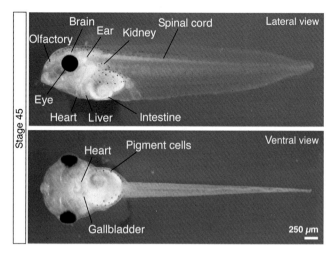

FIGURE 14.6 **Anatomy of stage 45 *X. laevis* tadpole.** Scale bar: 250 µm.

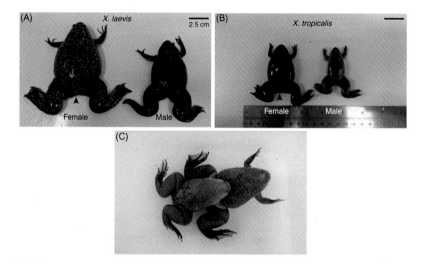

FIGURE 14.7 **Adult Xenopus frogs.** (A) *X. laevis*, female (left) and male (right). (B) *X. tropicalis*, female (left) and male (right). Arrowheads indicate cloacae. (C) Mating of *X. tropicalis*. Scale bar: 2.5 cm.

Xenopus I in the United States. Frozen sperm from transgenic animals can be shipped in dry ice from a laboratory or Resource Center. *X. laevis* embryos can be also transported internationally as long as the embryos have not yet reached the feeding tadpole stage.

Sex in adults can be identified easily by the existence of a protruding cloaca in females in both *X. laevis* and *X. tropicalis*. Fig. 14.7 shows *X. laevis* (A) and *X. tropicalis* (B). Females, shown on the left in both figures, are bigger than males on the right, and have a protruding cloaca, the posterior orifice for releasing eggs (arrowhead in Figs. 14.7A and 14.7B). Sexually mature males have black pads in their forearms, which they use to amplex with females, while mating (Fig. 14.7C).

Housing and Water

1. Tanks: For temporary housing of a small number of frogs (one to three), plastic tanks with still water can be used. More permanent housing of a larger number of frogs requires a recirculating system. Tap water should be left for more than one day to dechlorine before use, or should have been dechlorinated using other methods, such as via the use of carbon filters. Static water can be changed weekly or more depending on feeding. Recirculating systems (Fig. 14.8A) are more convenient for long-term housing, as they require less daily maintenance. However, it is important to check daily that all frogs are healthy as outbreak of disease may affect all frogs quickly. If a frog is suspected of being unhealthy, it should be isolated immediately.

2. Density of adult frogs: A 60 cm × 40 cm × 20 cm height (48 L) tank with 16–17 cm water depth accommodates approximately 10 X. laevis females or 15–20 males. A 50 cm × 30 cm × 20 cm height (30 L) tank with 15 cm water depth accommodates approximately 10–12 X. tropicalis females or 15–20 males. For both species, the minimum depth of water should be around 15 cm.

   Juvenile animals can be put at higher density tanks depending on size. However, it is important to keep similar-sized frogs together, so that all frogs can eat food equally. During raising juveniles, it may be necessary to regularly sort animals by size until they reach adulthood (maturity).

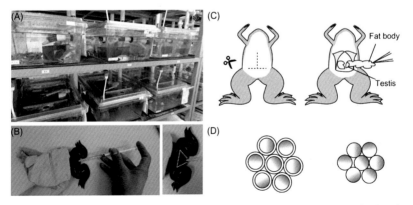

FIGURE 14.8   **Housing and basic techniques in Xenopus.** (A) A typical recirculating housing aquaria for X. laevis. (B) Hormone injection in X. laevis. Hormone is injected subcutaneously in the dorsal, posterior part of frog. The insertion points for needle are shown by dashed lines (right panel). (C) Isolation of testes. (D) Schematic diagram of embryos before and after dejellying with cysteine solution.

3. Temperature: The ideal temperature for adult *X. laevis* is 16–18°C, although tadpoles and juveniles can tolerate higher temperatures (20–23°C). However, egg quality in adults is optimal when frogs are kept in more cool conditions (16–18°C). Embryos can be kept at temperatures as low as 14°C to a maximum of 23°C. Their rate of development is dependent on temperature, such that it generally takes embryos about twice the time to develop at 14°C versus 22–23 °C, and their rate of development is 1.4 times slower at 16°C than at 22–23°C, which is useful to control timings of experiments, or to spread a single fertilization over many stages. Adult *X. tropicalis* frogs should be kept at 25–26°C. Embryos can cultured between 22°C and 28°C; however, frogs tend to become ill easily if kept below 24°C. This is probably due to their immune system being compromised at lower temperatures. On the other hand, raising temperature above 28°C increases the risk of bacterial growth in the tanks.

4. Tadpoles grow better in static water with an air pump for oxygenation. Any container with minimal depth of 10 cm can be used. One or two tadpoles can be successfully raised even in 1000 mL beaker without air pump. Water should be changed at least once a week. Especially it is important to clean the bottom of tank, as tadpoles tend to graze for food at the bottom of their containers, but this is also the area that accumulates their excrement.

Feeding

5. Adult frogs are fed a diet of Reptomin (Tetra) combined with frog pellets, such as NASCO brittle in 2–3:1 ratio three times a week. It is important to feed an amount, which the frogs can eat within 30 min to 1 h to ensure that they are not overfed. Leftover food should be removed using a net to help keep the water clean.

6. Juvenile frogs can be fed Reptomin broken into small pieces daily. In this case, some extra food can be left, so that the juvenile frogs can eat for longer and thus grow faster.

7. Tadpoles are fed spirulina (ZM Fish Food) mixed in water daily. If they are well-fed and healthy, blood vessels in the head are bright red, and the tadpoles grow well. If the head is pale, one should feed more. Increase the amount of food as they grow. One may need to separate tadpoles into another tank as they become bigger, since their density affects their growth rate. Overfeeding is the most common cause of tadpole death, especially at the early feeding tadpole stages. Keep monitoring the transparency of water after feeding. If the water does not become clear half a day after feeding, this will suggest overfeeding. They can be fed once a day, but if they start growing one may increase the feeding to twice a day. If one feeds every day, one may skip feeding at the weekend, as less food is better than overfeeding, which could lead to tadpole deaths.

Reuse of Frogs

After ovulation, females need time to regenerate the next batch of mature oocytes, prior to the next ovulation. This usually takes about 3 months, which is also the time required by UK regulations for resting between ovulations. It is also important to ovulate female frogs every 3–6 months to ensure that the oocytes do not get too old, between ovulations, which can adversely affect egg quality. Although Xenopus frogs can live more than 10 years, egg quality often declines as the females get older. Usually a female can be ovulated for 2–6 years with little loss of egg quality. Males can be reused for a natural mating after about a month; however, local regulations may require a longer period between matings.

Humane Killing

Xenopus tadpoles, juveniles, and adults can be sacrificed humanely via an overdose of anesthetic by immersion (0.1% MS-222 adjusted to pH with the same amount of sodium bicarbonate) for 15–30 min. Decapitation or physical destruction of the brain in the adults should be performed to confirm death.

FERTILIZATION AND EMBRYO MANIPULATION

Materials

Materials for Hormone Injection

1. Pregnant mare serum gonadotropin (PMSG, 100 U/mL; P.G.600, Intervet, Inc., 021825). Dissolve in water and stored at $-20°C$.
2. Human chorionic gonadotropin (hCG), lyophilized powder (SIGMA, CG5). Dissolve in 10 mL water (500 U/mL) and stored in 4°C.
3. 25 G needle and 1 mL syringe.
4. Paper towels.

Materials for in vitro Fertilization in X. laevis

1. Dissection tools, scissors, and forceps.
2. 0.1% MS-222 (ethyl 3-aminobenzoate methanesulfonate, tricaine; SIGMA A5040) with 0.1% sodium bicarbonate.
3. 90 mm petri dish.
4. Wheaton glass scintillation vials (20 mL VWR).
5. Pellet pestles (SIGMA Z359947) and 1.5 mL Eppendorf tube.
6. 0.1X Marc's Modified Ringer (MMR): 10 mM NaCl, 0.2 mM KCl, 0.1 mM $MgCl_2$, 0.2 mM $CaCl_2$, 0.5 mM HEPES, pH 7.5. Prepare a 10X stock solution, adjust to pH 7.5 with NaOH. Sterilize the 10X solution by autoclaving.
7. 1X MMR.

Materials for in vitro fertilization in X. tropicalis

1. Cryoprotectant premix: 0.4 M sucrose, 10 mM $NaHCO_3$, 2 mM pentoxifylline (SIGMA P1784).
2. Egg yolk (chicken).
3. L-15 medium (L4386).
4. Fetal bovine serum (SIGMA F9665).
5. Dissection tools and forceps.
6. 0.1% MS-222/0.1% sodium bicarbonate.
7. 90 mm petri dish.
8. Pellet pestles (SIGMA Z359947) and 1.5 mL Eppendorf tube.
9. 0.1X Marc's modified ringer (MMR).

Materials for Manipulation of Embryos

1. 2% L-cysteine (SIGMA, W326305) in 0.1X MMR, adjusted to pH 7.9–8.0 with 10N NaOH. The solution should be prepared freshly before the experiment.
   2 g of L-cysteine is dissolved in 100 mL of 0.1X MMR and add 0.4 mL 10N NaOH. pH can be checked using pH test paper strips.
2. 90 mm plastic petri dish.
3. 0.1X MMR.
4. 0.01X MMR for *X. tropicalis*.
5. 90 mm petri dish coated with 1% agarose/0.1X MMR for *X. tropicalis*.

Materials for Microinjection

1. Flaming/Brown micropipette puller (e.g., model P-87, Sutter Instruments Co.).
2. Borosilicate glass capillaries (e.g., GC100TF-10, 1.0 mm OD × 0.78 mm ID, Harvard Apparatus), for pulling needles using these settings, $p = 50$, $v = 100$, and $t = 5$.
3. Microinjector (e.g., Picospritzer II, General Valve Corporation).
4. Dissection microscope (e.g., MZ6, Leica) and eyepiece with a scale bar.
5. Manipulator.
6. 60 mm petri dish with mesh. Ideally the mesh should be such that each square is slightly smaller than an egg/embryo, so that the embyos are held during injection. The exact size is not important, as the mesh often holds the embryos simply through its stickness.
7. Microloader pipette tip (Eppendorf, 930001007).
8. Injection buffer: 2% Ficoll PM400 (SIGMA, F4375) in 0.4% MMR for *X. laevis*, or in 0.1X MMR for *X. tropicalis*. Ficoll is optional. If one uses a fine needle, one may eliminate Ficoll, and use 0.4X or 0.1X MMR.
9. RNA, morpholino oligonucleotide (MO), or CRISPR/Cas9 mix.

Induction of Ovulation and Mating in X. *laevis*

1. Preprime females by injection with 50 units of PMSG about 3–5 days prior to hCG injection into the posterior lymph sac in the back (Fig. 14.8B). Covering the head with a clean paper towel helps to immobilize the frogs. Prepriming with PMSG is optional. Females can ovulate after hCG injection alone, but the responsiveness (period after injection to ovulation) varies more if the females are not preprimed with PMSG. Thus, PMSG injection increases the responsiveness and consistency of the timing of ovulation following hCG injection.
2. Inject females with 250 units of hCG, 12–15 h before planned egg collection.
3. On the day of experiment, females are placed in 1X MMR and allowed to lay eggs in the high salt solution. Unfertilized eggs are stable in 1X MMR for a few hours at 16°C.
4. Proceed to in vitro fertilization as below.

For natural matings, males are injected with 25 units of PMSG 3–5 days prior to hCG injection and with 125 units of hCG, 12–15 h before planned matings. One female and one male are placed in a tank together, where the male will amplex and mate (as in Fig. 14.7C).

Induction of Ovulation and Mating in X. *tropicalis*

1. *X. tropicalis* frogs are more active outside of water than *X. laevis*, therefore, covering them with a clean paper towel is essential. Females are injected with 15 units of PMSG and males are injected with 10 units of PMSG one day prior to hCG injection. For *X. tropicalis* PMSG injections are critical for a successful mating. It is also important to select males that have visible black pads in their forearm to improve the chance of successful matings. As *X. tropicalis* frogs are smaller, the needle should remain inserted in the frogs following injection for 10–20 s, otherwise the hormone may flow out when the needle is withdrawn from the body.
2. Inject females with 75 units of hCG and males with 50 units of hCG, 3–4 h before planned egg collection. Put one or two pairs of frogs in the same tank.
3. 3–4 h later, females should start laying eggs. If males amplex with females, one should find fertilized eggs in the container. Transfer the amplexed pair carefully to a new tank and collect fertilized eggs (see embryo manipulation section).

Isolation of Testes for in vitro Fertilizations

While natural matings routinely produces more embryos, in vitro fertilizations provide well-synchronized developing embryos, and it works very well in *X. laevis*. For this reason, in vitro fertilizations are commonly done in

X. laevis. However, in vitro fertilization requires that a male must be sacrificed. Therefore, in vitro fertilizations are less suitable if the males are precious and/ or genetically modified. In *X. tropicalis*, natural matings are more common to obtain fertilized embryos, as in vitro fertilization using fresh testes does not work as efficiently in *X. tropicalis*. However, in vitro fertilization using frozen sperm works better in *X. tropicalis* than using a fresh testis, therefore a method for cryopreservation of sperm will also be described. In vitro fertilization using frozen sperm is also useful for transferring transgenic lines between laboratories. Freezing sperm can also help reduce cost of maintenance of males. While frozen sperm can be prepared from *X. laevis* males, the efficiency of fertilization is much worse using frozen sperm than using a small piece of fresh testis, so frozen sperm is used less often in *X. laevis*, except when transferring genetically modified lines between laboratories.

1. Sacrifice male by overdose in anasthesia by immersing male in 0.1% MS-222 (pH 7.0).
2. After confirmation of death, place the frog on a paper towel ventral side up.
3. Sterilize dissection tools in 70% ethanol.
4. Wipe the ventral region of the frog with 70% ethanol. Cut skin and body wall as shown in Fig 14.8C using dissection scissors.
5. Pull white, yellowish fat body with forceps, and find a testis at the attachment region of the fat body. Testis is a bean-shaped organ, which has a harder consistency than the fat tissue. Dissect the testis from the nearby connective tissue and fat body. There are two testes on each side of the body, so to dissect the other testis, begin by pulling out the contralateral fat body, and proceed like before.
6. For *X. laevis*, put each testis into a vial with 1X MMR and store at 4°C. Each testis can be placed separately in each vial to avoid contamination. If there is no bacterial contamination, an intact testis will remain fresh for at least one week at 4°C.
7. For *X. tropicalis*, proceed to the protocol for freezing sperm section below.

Method for in vitro Fertilization in *X. laevis*

1. Transfer about 1/8 piece of a testis into a 1.5 mL Eppendorf tube containing ~50 µL 1X MMR.
2. Macerate the piece of testis using a microfuge pellet pestle.
3. Collect unfertilized eggs laid in 1X MMR into a dry 90 mm petri dish using a wide pore pipette.
4. Remove as much liquid from the petri dish as possible.
5. Apply testis solution onto the eggs and mix well using the pellet pestle or tip one used to apply testis solution.
6. Leave for 5 min and pour 0.1X MMR over eggs.

7. Leave at least 15 min for eggs to be fertilized. If fertilization occurs efficiently, most or all the eggs should rotate, such that the animal pole faces upward after 15 min.
8. Dejelly embryos as in the embryo manipulation section.

Method for in vitro Fertilization in *X. tropicalis*

Preparation of Cryoprotectant

1. Egg yolk is diluted with the equal volume of water.
2. Mix 10 mL of 50% egg yolk with 40 mL of cryoprotectant premix.
3. Centrifuge for 20 min at 13,000 rpm, 4°C.
4. 1 mL aliquots can be stored at −80°C (up to a few years).

Freezing Sperm

1. Preprime males a day before (PMSG. Do NOT inject with hCG) as described in the section for hormone injection.
2. The day after PMSG injection, cull the males and isolate testes, taking care not to puncture the testes during dissection and isolation.
3. Put one testis into a 1.5 mL tube with chilled 250 μL L15/10% FBS (stored in −20°C), keep on ice.
4. Macerate well using a microfuge pellet pestle.
5. Add 250 μL chilled cryoprotectant and mix well by gentle pipetting.
6. Divide into 4 tubes (0.2 mL tube), 125 μL each, keep on ice (2 testes give 8 tubes).
7. When all tubes are ready, put them in a polystyrene box in −80°C to allow slow freezing.

In vitro Fertilization Using Frozen Sperm

1. Thaw sperm 30 s by hand.
2. Add 125 μL water and mix briefly by pipetting.
3. Add sperm onto the squeezed eggs in a petri dish and mix using the tip.
4. After 5 min flood with 0.1X MMR.
5. After at least 15 min dejelly eggs.

Method for Dejellying Embryos

1. Discard all solution after fertilization is completed. If the fertilization occurs, contraction of pigment in the animal pole should be observed.
2. Add 2% L-cysteine/0.1X MMR pH 7.8–8.0 and incubate for 5–7 min. with an occasional gentle agitation. One may transfer embryos with the cysteine solution into a beaker for more efficient gentle mixing.
3. Pour off the cysteine solution when jelly coat is dissolved and embryos become compact by contacting each other (Fig. 14.8D).
4. Wash embryos at least three times with 0.1X MMR and transfer embryos into a dish.

Method for Culturing Embryos

X. laevis embryos are cultured in plastic petri dishes with 0.1X MMR at 14–23°C. *X. tropicalis* embryos are quite sticky, therefore they are best cultured in petri dishes coated with 1% agarose filled with 0.01X MMR until they hatch at stages 26–30 at 22–28°C. Dead embryos should be removed from the dish as soon as they appear. If the medium is cloudy, embryos should be transferred to a new dish with fresh medium. Cultured embryos should be meticulously checked daily and maintained clean, to ensure continued healthy embryos and undisturbed development.

Method for Microinjection

Fig. 14.9A shows a typical injection setup. The microinjector is connected to a nitrogen cylinder through a pressure regulator (Fig. 14.9A right panel). The microinjector controls the pressure and duration of the gas to be sent to a glass needle held by a manipulator. The injection volume can be calibrated using a scale bar in the microscope eyepiece.

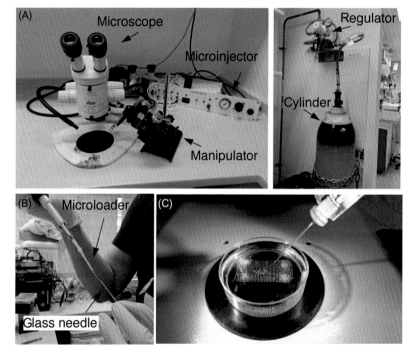

FIGURE 14.9 Microinjection in Xenopus. (A) Microinjection setup. (B) Backfilling an injection needle. (C) Injecting *X. laevis* embryos on the mesh of petri dish.

RNA can be generated in vitro using mMessage mMachine transcription Kit (Thermo Fisher Scientific). For details of in vitro transcription, please refer to the manual from the manufacturer available on line. pCS2+ plasmid and its derivatives are a popular backbone for RNA synthesis in Xenopus [16]. These plasmids contain an SP6 promoter before a multicloning site (MCS) followed by an SV40 polyA addition signal and another MCS. For CRISPR/Cas9 system, Cas9 can be introduced as either RNA or protein (NEB, M0646) [11,12].

1. Put embryos onto a 60 mm petri dish with mesh.
2. Back fill a needle with the RNA, morpholino antisense oligonucleotide or guide RNA and Cas9 mixture using a pipetman with a microloader tip (Fig. 14.9B) and set the needle into a needle holder of manipulator (Fig. 14.9C).
3. Clip the tip of the needle using a fine forceps and calibrate the volume of injection to 10 nL (*X. laevis*) or 2 nL (*X. tropicalis*) per embryo, by adjusting the time and pressure of the injector.

APPLICATIONS

Developmental Biology

Xenopus has been a powerhouse system in developmental biology for more than 50 years, and during this time, it has continually reinvented itself. Initially it became a convenient model for embryology, due to the ease by which embryos could be obtained anytime of the year, and the ease by which experimental embryological techniques could be employed, including tissue explants and transplants. It also rendered itself convenient for microinjection experiments, ranging from nuclear transplantations and eventually to the injection of nucleic acids and oligonucleotides, such as antisense MOs. Thus the ability to inject embryos has provided a convenient method for assessing gene function during development in Xenopus. More recently, the adoption of advanced genomic, transgenic, and gene-editing tools in Xenopus have propelled this system into the 21st century, and it is these techniques, when combined with the more traditional embryological strengths of the system, which allow one to model human disease conditions in this organism.

For example, it is very easy using this system to overexpress a gene product, via injection of in vitro transcribed mRNA, or to knock down the function of a gene product, via injection of antisense MOs [17,18]. In addition, one can use gene-editing tools, such as TALENs or the CRISPR/Cas9 system, to mutate or alter any gene of interest [10–12]. These techniques are not necessarily unique to this system. However, what is unique is to be able to combine these approaches with careful embryological surgical techniques, such as tissue transplantations, in order to generate very specific genetic chimaeras. Such approaches can be particularly valuable when addressing tissue interaction/induction mechanisms or when gene knockouts are embryonic lethal, or when one wishes to address how a gene product expressed in a tissue affects the fate

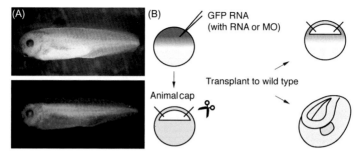

FIGURE 14.10 Examples of transplantation experiments. (A) Transgenic embryo expressing GFP ubiquitously are useful to mark cells during tissue transplantation studies. (B) Animal cap explants injected with GFP RNA with or without other RNA/MO can be transplanted to uninjected host embryos.

of other tissues, in vivo. Furthermore, one can use transplantation of tissues from transgenic lines expressing green fluorescent protein (GFP) ubiquitously, in order to easily distinguish donor from host cells in such transplantation studies (Fig. 14.10A). Alternatively, one can transplants cells isolated from embryos injected with GFP RNA along with RNA or MO of the gene of interest into unlabeled uninjected host embryos (Fig. 14.10B). Using this approach, we were able to show that animal cap tissues expressing *cebpa* and GFP RNA are able to differentiate into myeloid lineages in the host embryo, even though animal cap cells normally are only fated to become ventral epidermis. Thus, we used this approach to show that C/EBPα is a key transcriptional regulator capable of initiating the entire myelopoiesis pathway in naïve pluripotent animal cap cells in *Xenopus* [19]. Another example of this sort of approach can be seen in the work of Zygar and colleagues (1998), on lens ectoderm competence [20]. Such experiments rely on the fact that Xenopus embryos have remarkable capacities to heal following transplantation experiments, as evident by the ability to seemlessly fuse two halfs of separate embryos; for example, a pigmented anterior half of one embryo to the posterior half of an albino embryo, or vice versa (Fig. 14.11A), resulting in the fused halves healing completely and developing into adulthood (Fig. 14.11B).

Genetics and Disease Modeling

Although *X. tropicalis* is a quite new genetic model system, genome sequencing revealed that its organization is very similar to that of humans [7]. Many disease genes have been found in Xenopus, thus employing gene-editing technique in this system will facilitate the establishment of disease models in Xenopus [7,8]. The ex-utero development and the large number of embryos that can be produced from a single mating are powerful advantages to the use *X. tropicalis* for disease modeling. The system is also advantageous for forward genetics screens,

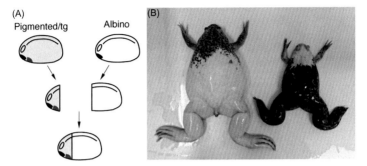

FIGURE 14.11 Another example of transplantation experiment. (A) An anterior part of a pigmented embryo including the anterior ventral blood islands (shown in *red*) is joined to a posterior part of an albino embryo. (B) Chimeric animals generated by the transplantation after reaching adulthood.

such as ethyl-N-nitrosourea (ENU) mutagenesis screens, which led to the identification of many genes involved in embryonic development and organogenesis [21]. A reverse genetic approach, TILLING (Targeting Induced Local Lesion In Genomes) has also been applied in *X. tropicalis*, where a mutation in a gene of interest is screened in sperm libraries containing mutations induced by ENU using extensive polymerase chain reaction (PCR) and sequencing. One mutation identified by TILLING in *X. tropicalis* was found in the retinal anterior homeobox (*rax*) gene, whose mutation in humans causes congenital anophthalmia and microphthalmia [22]. Analyses using this *rax* mutant in combination with transplantation techniques revealed the molecular mechanism responsible for how the eye field is specified [23].

Wound Healing and Regeneration

As described previously, Xenopus embryos have remarkable capacities to heal wounds completely and rapidly. It is for this reason that tissue transplantation experiments have been successfully performed in this model for many decades. More recently, Xenopus has also begun to be used to investigate the molecular and cellular mechanisms of scar-free wound healing and tissue regeneration [24].

Small Compound Screening

Amphibians are highly sensitive to enviromental changes. This characteristic combined with the large number of embryos one can obtained in this system, as well as their external development, make Xenopus an attractive system for in vivo chemical screening [25]. Particularly, Xenopus is ideal to screen small compounds that modulate signaling pathways, as phenotypes and molecular markers for various signaling pathways are well established. Large chemical screening using Xenopus carried by Kälin and colleagues (2009) identified 32 active compounds that affected vascular and/or lymphatic development [26].

Scenario—Assessing the Role of FGF Signaling During Development and Organogenesis

Fibroblast growth factor (FGF) signaling has been shown to be involved in many biological processes, such as cell proliferation, cell fate determination, cell migration, and patterning during embryonic development and organogenesis. There are 18 canonical FGFs and four FGF receptors. Gene knockouts of several *Fgfs* die during development and the others have subtle phenotypes, suggesting functional redundancy in these *Fgfs* [27]. In Xenopus FGFs and receptors have been showed to have dynamic and diverse expression patterns with some overlap during development and organogenesis [28]. Since their expression is diverse in wide range of cell types and tissues during embryonic development and organogenesis, simple knockout of their genes may be insufficient to study their roles in particular cell types or tissue. Therefore to explore the role of FGF signaling in various periods of embryonic development, inducible knockdown strategies are necessary.

Dominant Negative Construct

The structure of FGF receptor (FGFR) is shown in Fig. 14.12A. FGFR is a transmembrane protein consisting of an extracellular ligand-binding domain

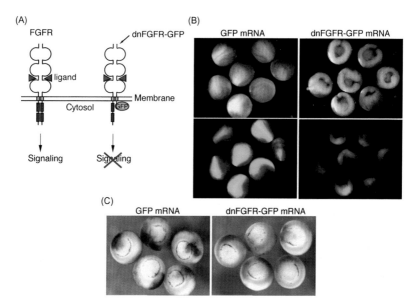

FIGURE 14.12 **Mesoderm induction by FGF signaling.** (A) dnFGFR-GFP disrupts the signaling pathway by dimerizing with endogenous receptor. (B) Embryos injected with dnFGFR-GFP shows gastrulation defects. (C) dnFGFR-GFP expression (*magenta*) inhibits brachyury (*blue*) expression, a marker of mesoderm.

and a cytoplasmic tyrosine kinase domain. FGFRs form dimers upon activation and consequential autophosphorylation in the cytoplasmic domain forming docking sites for signal transduction molecules, which are then involved in transducing downstream signaling. Therefore, a mutant receptor that lacks the cytoplasmic domain acts in a dominant negative manner, failing to activate downstream signaling [29]. To dissect the temporal roles for FGF signaling, we generated a transgenic *X. tropicalis* line expressing a heat shock-inducible dominant negative FGFR (dnFGFR). The dnFGFR1 construct was generated by replacing the cytoplasmic kinase domain with GFP, to help visualize the dnFGFR (Fig. 14.12A). To test whether the dnFGFR-GFP construct works as a dominant negative, dnFGFR-GFP RNA was synthesized in vitro and injected into the dorsal marginal zone of 4-cell embryos. As shown in Fig. 14.12B, dnFGFR-GFP expressing embryos failed to gastrulate forming an open blastopore. Double in situ hybridization using a Digoxigenin (DIG)-labeled Xenopus brachyury probe (*blue* in Fig. 14.12C), which is a marker of mesoderm, and a fluorescein-labeled GFP probe (*magenta* in Fig. 14.12C) confirmed that the gastrulation defect was caused by a disruption in mesoderm formation in the injected embryos, confirming that this construct works as had been previously shown with a dnFGFR construct, lacking the GFP fusion [29].

Transgenesis in Xenopus

There are two major methods to generate transgenic frogs. One is a method that utilizes restriction enzyme-mediated integration (REMI), mainly used in *X. laevis* [30]. This transgenic method in Xenopus involves the integration of transgenes into sperm nuclei, which are then transplanted into unfertilized eggs. Therefore, this technique gives rise to many transgenic F0 animals. It is suitable not only for generating transgenic founders but also for quick promoter analyses. However, the transgenes form concatemer, which may cause problems in some genetic analyses. The other transgenic method often used in Xenopus, is the meganuclease method, which relies on the use of I-SceI [31]. The disadvantage of the meganuclease method is that integration occurs after fertilization, and since the early embryos divide quickly, the resulting embryos are mosaic in terms of the integration events, thus the F0 founder animals must be raised to the F1 generation, before the transgene integration is nonmosaic. The advantage, however, is that this method leads to simpler and cleaner integration events. Also this method is the preferable transgenesis method for *X. tropicalis*.

We generated a transgenic *X. tropicalis* line using the I-SceI method, containing the dnFGFR-GFP construct showed in Fig. 14.13A. For details on the I-SceI meganuclease method, see Ishibashi et al. [32]. The construct is comprised of the heat shock protein 70 (HSP70) promoter driving dnFGFR-GFP and αCE2X6 (the enhancer element to drive lens-specific expression) driving red fluorescent protein (RFP). A founder frog was crossed to wild type and some offspring

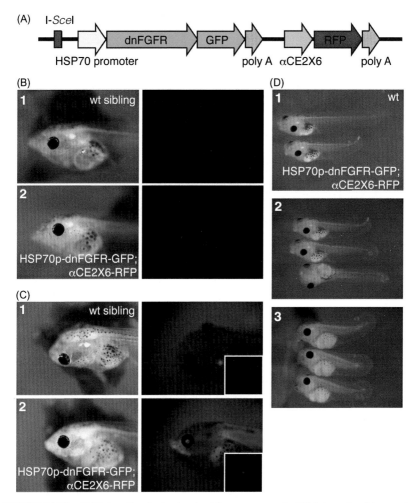

FIGURE 14.13   FGF signaling is also involved in organogenesis. (A) Schematic of transgenic construct. (B) F1 *X. tropicalis* tadpoles, wt sibling (1) and transgenic tadpole expressing RFP in the eye (2). (C) Heat shock treatment induced the expression of dnFGFR-GFP in the tadpole with *red* eyes (2), but not in wt (1). (D) F2 embryos were treated with heat shock. Wt embryos developed normally (1, top), transgenic embryos treated at st. 18–20 developed to tadpoles with bent at the tip of tail (2) and embryos treated at st. 14 showed shortened tails with hemorrhage (3).

showed the RFP expresion in the eye showing that they were transgenic for the dnFGF-GFP construct (Fig. 14.13B). These tadpoles were incubated at 37°C for 1 h to induce dnFGFR-GFP expression, and only the tadpoles expressing RFP in the eye showed GFP expression 22 h later (Fig. 14.13C).

To explore the roles of FGF signaling, heat shock was applied on F2 embryos at stage 14 or stage 18–20 and transgenic offspring were confirmed by RFP signal in the eye. The wild-type sibling embryos developed normal after the heat shock. However, the transgenic embryos treated at stage 18–20 developed into normal tadpoles except the tip of tail (Fig. 14.13D1); 16 out of 41 RFP positive tadpoles showed this bent-tail phenotype (Fig. 14.13D2). The low penetrance of the phenotype may be caused by the difference of the stages of embryos treated. When the transgenic embryos were treated with heat shock earlier at stage 14, 11 out of 13 tadpoles showed abnormalities on their blood vessels as well as tail development (Fig. 14.13D3). These data suggest that FGF signaling may play roles during later stages, during tail formation.

KEY LIMITATIONS

Imaging

Xenopus has been used for in situ hybridization, fluorescent immunohisto-chemistry, and live imaging on embryos or tadpoles. However, because of yolk content, signals inside the embryos are difficult to detect. Although the fixed samples can be cleared using Murray's clear (2:1 of benzyl benzoate and benzyl alcohol), live imaging deep inside early embryos is not possible using conventional imaging methods. However, at later stages of the development, the embryo clears, such that by the tadpole stages, the larvae are virtually transparent, see Movie 5 of stage 43 and stage 47 tadpoles; https://youtu.be/h-uwyJGvT2w.

Generation Time and Reproduction

Although the generation time of X. tropicalis is 4–6 months, which is much shorter than the 1–2 generation time of X. laevis, it is still longer than most other model organisms. One spawn gives many hundreds of embryos in X. laevis and thousands in X. tropicalis, but one needs to wait for 3 months before a female can be ovulated again. This may affect the frequency by which experiments can be performed.

TROUBLESHOOTING

Problem	Potential Solutions
Failure of female to lay eggs	They may not be sexually matured yet. Feed more and allow them to rest another 3 months.
Stringy eggs	Produced oocytes have been accumulated in ovary. Use the female on regular basis (every 3–6 months).
Soft eggs	Soft eggs can be fertilized, but they often become flat during development, resulting in abnormal develoment. It may be caused by not getting enough nutrition. Feed more.

Problem	Potential Solutions
Death after ovulation	This is sometimes seen in *X. tropicalis*. The possible problem here is egg retention, following ovulation, which can lead to bloating and then death. This can occur if the female has not been ovulated often enough, or was allowed to become sexually mature for a long time before first ovulation. The jelly produced in the oviduct becomes too thick, thus clogging the oviduct, such that eggs are unable to proceed through it and out of the body. Females should be ovulated within a year after they become sexually mature, and every 3–6 months thereafter to diminish the chance of getting egg retention issues.
No fertilization after in vitro fertilization (*X. laevis*)	High salt prevents fertilization. Make sure that 1X MMR is completely aspirated from the petri dish before adding sperm solution.
	If the testes are in cloudy solution, this might suggest that they have become contaminated with bacteria. Prepare fresh testes.
No fertilization after in vitro fertilization (*X. tropicalis*)	Most likely problem is improper sperm preparation. Since testes in *X. tropicalis* are small, one must be especially careful how they are isolated, to ensure that they are not damaged during the dissection protocol.
	Freeze-thawing of PMSG may affect hormone activity. Prepare small aliquots of PMSG and use them only once for *X. tropicalis*.
	Do not inject with hCG before sperm preparation.
No fertilization after natural mating (*X. tropicalis*)	In the case where the male does not grab the female, the male may not be sexually mature or it may not have been properly injected with hormone. Check whether the male has black arms. Freeze-thawing of PMSG may also affect hormone activity. Prepare small aliquots of PMSG and use them once for *X. tropicalis*.
	In the case where no fertilization is obtained although the male grabs the female, there may be two possible reasons. One would be that the egg quality is not good enough. The other would be that the male is sterile. In both cases, another mating with another female may solve the problem.

CONCLUSION

Over the past half century, there have been many important findings in the biomedical sciences that have grown out of studies using Xenopus as a model organism. As we move further into the 21st century, the advancement and applications of novel technologies in Xenopus means that this well established and powerful model organism will continue to prove an attractive model for human disease mechanisms for many decades in the future.

ACKNOWLEDGMENT

Nick Love generated the embryo fusions, shown in Fig. 14.11B.

REFERENCES

[1] Amaya E, Offield MF, Grainger RM. Frog genetics: Xenopus tropicalis jumps into the future. Trends Genet 1998;14:253–5.

[2] Gurdon JB, Hopwood N. The introduction of Xenopus laevis into developmental biology: of empire, pregnancy testing and ribosomal genes. Int J Dev Biol 2000;44:43–50.

[3] Harland RM, Grainger RM. Xenopus research: metamorphosed by genetics and genomics. Trends Genet 2011;27:507–15.

[4] Gurdon JB. The developmental capacity of nuclei taken from intestinal epithelium cells of feeding tadpoles. J Embryol Exp Morphol 1962;10:622–40.

[5] Gurdon JB. Adult frogs derived from the nuclei of single somatic cells. Dev Biol 1962;4: 256–73.

[6] Colman A. Profile of John Gurdon and Shinya Yamanaka, 2012 Nobel laureates in medicine or physiology. 2013.

[7] Hellsten U, Harland RM, Gilchrist MJ, et al. The genome of the Western clawed frog Xenopus tropicalis. Science 2010;328:633–6.

[8] Duncan AR, Khokha MK. Xenopus as a model organism for birth defects—Congenital heart disease and heterotaxy. Semin Cell Dev Biol 2016;51:73–9.

[9] Session AM, Uno Y, Kwon T, et al. Genome evolution in the allotetraploid frog Xenopus laevis. Nature 2016;538:336–43.

[10] Ishibashi S, Cliffe R, Amaya E. Highly efficient bi-allelic mutation rates using TALENs in Xenopus tropicalis. Biol Open 2012;1:1273–6.

[11] Blitz IL, Biesinger J, Xie X, Cho KWY. Biallelic genome modification in F(0) Xenopus tropicalis embryos using the CRISPR/Cas system. Genesis 2013;51:827–34.

[12] Nakayama T, Fish MB, Fisher M, Oomen-Hajagos J, Thomsen GH, Grainger RM. Simple and efficient CRISPR/Cas9-mediated targeted mutagenesis in Xenopus tropicalis. Genesis 2013;51:835–43.

[13] Nieuwkoop PD, Faber J. Normal table of Xenopus laevis (Daudin). A systematical and chronological survey of the development from the fertilized egg till the end of metamorphosis. Normal table of Xenopus laevis (Daudin) … 1956.

[14] Kimelman D. Mesoderm induction: from caps to chips. Nat Rev Genet 2006;7:360–72.

[15] Beck CW, Christen B, Slack JM. Molecular pathways needed for regeneration of spinal cord and muscle in a vertebrate. Dev Cell 2003;5:429–39.

[16] Gilchrist MJ, Zorn AM, Voigt J, Smith JC, Papalopulu N, Amaya E. Defining a large set of full-length clones from a Xenopus tropicalis EST project. Dev Biol 2004;271:498–516.

[17] Amaya E. Xenomics. Genome Res 2005;15:1683–91.

[18] Zhao Y, Ishibashi S, Amaya E. Reverse genetic studies using antisense morpholino oligonucleotides. Methods Mol Biol 2012;917:143–54.

[19] Chen Y, Costa RMB, Love NR, et al. C/EBPalpha initiates primitive myelopoiesis in pluripotent embryonic cells. Blood 2009;114:40–8.

[20] Zygar CA, Cook Jr TL, Grainger RM. Gene activation during early stages of lens induction in Xenopus. Development 1998;125:3509–19.

[21] Goda T, Abu-Daya A, Carruthers S, Clark MD, Stemple DL, Zimmerman LB. Genetic screens for mutations affecting development of Xenopus tropicalis. PLoS Genet 2006;2:e91.

[22] Voronina VA, Kozhemyakina EA, O'Kernick CM, Kahn ND, Wenger SL, Linberg JV, et al. Mutations in the human RAX homeobox gene in a patient with anophthalmia and sclerocornea. Hum Mol Genet 2004;13:315–22.

[23] Fish MB, Nakayama T, Fisher M, Hirsch N, Cox A, Reeder R, et al. Xenopus mutant reveals necessity of rax for specifying the eye field which otherwise forms tissue with telencephalic and diencephalic character. Dev Biol 2014;395:317–30.

[24] Li J, Zhang S, Amaya E. The cellular and molecular mechanisms of tissue repair and regeneration as revealed by studies in Xenopus. Regeneration (Oxf) 2016;3:198–208.

[25] Wheeler GN, Brändli AW. Simple vertebrate models for chemical genetics and drug discovery screens: lessons from zebrafish and Xenopus. Dev Dyn 2009;238:1287–308.

[26] Kälin RE, Bänziger-Tobler NE, Detmar M, Brändli AW. An in vivo chemical library screen in Xenopus tadpoles reveals novel pathways involved in angiogenesis and lymphangiogenesis. Blood 2009;114:1110–22.

[27] Ornitz DM, Itoh N. The Fibroblast Growth Factor signaling pathway. WIREs Dev Biol 2015;4:215–66.

[28] Lea R, Papalopulu N, Amaya E, Dorey K. Temporal and spatial expression of FGF ligands and receptors during Xenopus development. Dev Dyn 2009;238:1467–79.

[29] Amaya E, Musci TJ, Kirschner MW. Expression of a dominant negative mutant of the FGF receptor disrupts mesoderm formation in Xenopus embryos. Cell 1991;66:257–70.

[30] Kroll KL, Amaya E. Transgenic Xenopus embryos from sperm nuclear transplantations reveal FGF signaling requirements during gastrulation. Development 1996;122:3173–83.

[31] Ogino H, McConnell WB, Grainger RM. Highly efficient transgenesis in Xenopus tropicalis using I-SceI meganuclease. Mech Dev 2006;123:103–13.

[32] Ishibashi S, Love NR, Amaya E. A simple method of transgenesis using I-SceI meganuclease in Xenopus. Methods Mol Biol 2012;917:205–18.

[33] James-Zorn C, Ponferrada VG, Jarabek CJ, et al. Xenbase: expansion and updates of the Xenopus model organism database. Nucleic Acids Res 2013;41:D865–70.

[34] Karpinka JB, Fortriede JD, Burns KA, et al. Xenbase, the Xenopus model organism database; new virtualized system, data types and genomes. Nucleic Acids Res 2015;43:D756–63.

SUGGESTED FURTHER READING

Xenopus Protocols, Cell Biology and Signal Transduction (Methods in Molecular Biology 322), 2006 by X. J. Liu, Humana Press.

Xenopus Protocols, Post-Genomic Approaches, Second Edition (Methods in Molecular Biology 917), 2012 by S. Hoppler and P. D. Vize, Humana Press.

A Xenopus community database has been established, called Xenbase, where one can find much information about the model, including EST databases, RNAseq data throughout development, genome information, and many other resources, including transgenic animals and plasmids (http://www.xenbase.org/) [33,34].

GLOSSARY

Ploidy The number of chromosomes present in a single cell of an organism

Diploid An organism that contains two copies of each chromosome in most of its cells

Tetraploid An organism that contains four copies of each chromosome in most of its cells

Polyploid An organism that has more than two copies of each chromosome in most of its cells

Allotetraploid   An organism that contains four copies of each chromosome in most of its cells, following hybridization of two related species

Homologous genes   Genes that share a common ancestral gene

Paralogous genes   Duplicated genes within a single genome

Orthologous genes   Same gene in different organisms

Morphogenesis   Developmental process by which new form is generated

Gastrulation   Morphogenetic process by which the endoderm, mesoderm, and ectoderm each reach their final position in the embryo

Neurulation   Morphogenetic process by which the central nervous system begins to form, especially the formation of the neural tube and brain

Organogenesis   The developmental process by which organs form

Blastomere   A cell in the early embryo

Ectoderm   Tissue (germ layer) on the outside of the embryo after gastrulation; gives rise to the nervous system, neural crest, and epidermis

Mesoderm   Tissue (germ layer) between the ectoderm and endoderm after gastrulation; gives rise to the notochord, most of the muscles, kidney, blood, skeleton, dermis

Endoderm   Tissue (germ layer) on the inside of the embryo after gastrulation; gives rise to the internal organs such as the digestive tract, liver, lungs, pancreas, gall bladder

LIST OF ACRONYMS AND ABBREVIATIONS

Cas9   CRISPR associated protein 9
CRISPR   Clustered regularly interspaced short palindromic repeats
dnFGFR   Dominant negative fibroblast growth factor receptor
FGF   Fibroblast growth factor
FGFR   Fibroblast growth factor receptor
G1 phase Gap 1 phase
G2 phase Gap 2 phase
GFP   Green fluorescent protein
hCG   Human chorionic gonadotropin
MBT   Mid-blastula transition
MO   Morpholino oligonucleotide
PMSG   Pregnant mare serum gonadotropin
RFP   Red fluorescent protein
tg   Transgenic
wt   Wild type

Chapter 15

Basic Mouse Methods for Clinician Researchers: Harnessing the Mouse for Biomedical Research

Laurens J. Lambert[1], Mandar D. Muzumdar[1,2,3], William M. Rideout III[1] and Tyler Jacks[1,4]

[1]David H. Koch Institute for Integrative Cancer Research, Massachusetts Institute of Technology, Cambridge, MA, United States; [2]Dana-Farber Cancer Institute, Boston, MA, United States; [3]Harvard Medical School, Boston, MA, United States; [4]Howard Hughes Medical Institute, Massachusetts Institute of Technology, Cambridge, MA, United States

Chapter Outline

Objectives
- Learn about the history of the mouse as a model organism in biomedical research.
- Examine the key biological and anatomical features of the mouse.
- Explain proper mouse handling, breeding, injection, and sampling techniques.
- Review advanced mouse modeling approaches, applications, and limitations.

INTRODUCTION

The common house mouse (plural: mice), *Mus musculus*, belongs to Rodentia, the largest mammalian order [1]. The Rodentia are defined by continuous growth of teeth during the lifespan of the organism. Mice are excellent model organisms, as they are small, inexpensive to house, relatively short-lived, and capable of reproducing large numbers of offspring [1]. Crucially, mice are amenable to genetic manipulation and recapitulate human biology. Therefore, the mouse has become the model organism of choice for researchers interested in physiology and disease.

The common house mouse has a long history of domestication, dating back to European and Asian hobbyists who bred "fancy" mice for different coat styles and colors [2]. As scientists began to apply Mendel's genetic inheritance rules to mammals at the start of the 20th century, they turned to the mouse. At Harvard University, Clarence Cook Little and William Castle began an ambitious mouse breeding program that laid the foundation for modern-day mouse genetics [2,3]. Specifically, they sought to minimize genetic variability between *outbred strains* to study the transmission of a transplantable tumor [2,3]. Little and Castle succeeded in genetically fixing a mouse strain (DBA), the first *inbred strain*. Later in his career, Little derived the C57BL/6 strain, a precursor to the most used C57BL/6J inbred strain. Today, researchers have generated hundreds of inbred strains to model wide-ranging aspects of human physiology and disease.

A number of important biomedical advances have been made with the laboratory mouse (for an overview, see [4]). In 1928, Alexander Fleming discovered the antibiotic penicillin while working on bacterial cultures of *Staphylococcus*. However, it was not until Howard Florey and Ernst Chain infected large cohorts of mice with bacteria and cured them with penicillin that Fleming's discovery was fully appreciated. The three researchers shared a Nobel Prize in 1945 [4]. At Jackson Laboratory in Maine, USA, George Snell pioneered the concept of *congenic* mouse strains; crosses between inbred strains to fix small genetic loci [2]. Through these studies, Snell identified the major histocompatibility locus, forming much of the basis of our understanding of the human immune system and paved the way for organ transplantation. Snell shared a Nobel Prize in 1980. When molecular technology became available to perform gene targeting with recombinant DNA in the 1980s, Mario Capecchi, Martin Evans, and Oliver Smithies derived the first mouse *embryonic stem cells* (ES cells) from the 129/Sv strain [5]. Their efforts led to the first knockout mouse, affording genetic manipulation of the mouse genome. The three researchers shared a Nobel Prize in 2007 [4]. With our current understanding of the mouse genome, transgenic and knockout mice are now routinely generated by laboratories around the world.

IN PRINCIPLE

Key Biological Features of the Laboratory Mouse

Lifecycle of the Laboratory Mouse

The mouse has a relatively short life cycle and can reproduce large numbers of progeny. Typical litter size varies from 6 to 8 mice [2]. A mouse embryo has a developmental period (*gestation*) of about 18–21 days depending on the mouse strain [2,5]. Implantation in the uterus occurs approximately 4.5 days after fertilization [5]. The formation of the three primary germ layers (*gastrulation*) and development of early organs occurs around embryonic day 10 (E10). Between E14 and E19, internal organs are formed (Figure 15.1), and shortly before birth, skin, and lungs mature. At birth, the newborn mouse (*neonate*) is completely hairless and blind [5]. The development of the neonate continues for several weeks. For the first 21 days, neonates suckle milk from a lactating mother. Mice have 10 mammary glands and increase milk production based on litter

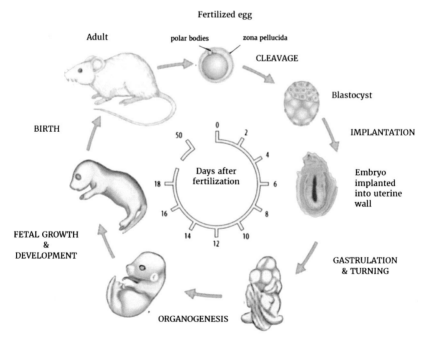

FIGURE 15.1 Mouse development.

size. The neonate develops hair about five days after birth (P5). Around P16, the neonate's eyes fully open and they begin to eat solid food [6]. At P21, the young mouse can be separated from the lactating mother and introduced to an adult diet (*weaning*). Weaned mice are typically separated by sex and allowed to gain adult size. Males reach sexual maturity by six weeks, while females reach maturity between four and six weeks [2]. Fecundity in female mice decreases between 6 and 12 months of age, while male mice typically retain reproductive capacity longer [2]. However, males of many inbred strains can become sedentary and reduce their sexual activity. The median lifespan of a mouse is highly dependent on genetic background, varying from 10 months (AKR/J) to 27–28 months (C57BL/6J) and 42 months (interstrain breeds) [2].

Social Behavior

The laboratory mouse is a social animal and lives in hierarchical groups. A mouse is generally not aggressive, and will try to evade rather than attack [7]. However, improper handling or stressful environmental conditions can cause a mouse to bite. Adult males can be aggressive to other mice to establish hierarchy and territory (*barbering*).

Mice are nocturnal, and are most active during the night [7]. Eating, drinking, and mating commonly takes place during the night. Laboratory mice display burrowing behavior and build nests, which helps maintain proper body temperature [7]. Disturbance of mothers with young litters should be avoided, as stress during this period can lead to neglect of neonates or even cannibalism.

Anatomy of the Laboratory Mouse

External Anatomy

Adult mice are covered in a hairy coat, except on their extremities. Coat color varies by inbred strain; common colors are black (C57BL/6J), beige (agouti) (129/Sv), and white (BALB/c). The upper and lower jaw incisors of the mouse are continuously growing, a feature shared among all members of the order Rodentia [1,7]. Laboratory mice are provided with solid, pellet food ("chow") formulated to wear down incisors. The mouse's 15 whiskers (vibrissae) grow from the mystacial pad, the facial muscular tissue [8]. The mystacial pad can generate rapid movement of the whiskers to explore and discriminate objects in the environment. The mouse uses both fore limbs and hindlimbs during locomotion. When balancing or standing upright on hindlimbs, the tail will often be used to provide additional support.

The external urogenital system allows sex determination of mice at weaning or during adulthood. The *anogenital distance* is shorter in females than in males [7] (Figure 15.2). Additionally, males also develop a protruding penis and a scrotal sac containing the testes. Females develop three pairs of nipples over the ventral thorax (pectoral nipples) and two pairs over the abdomen (inguinal

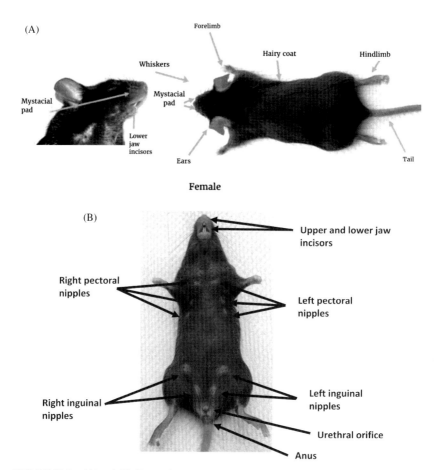

FIGURE 15.2 (A) and (B): External mouse anatomy.

nipples). Swelling of the nipples can allow determination of lactating mothers in a cage with neonates (Figure 15.3).

Internal Anatomy

The internal anatomy of the mouse is grossly similar to human internal anatomy and most organs are shared among species. This makes the mouse an excellent system to model human physiology. For an exhaustive discussion of anatomical and histological differences, refer to [9]. Highlighted below are key differences between human and mouse internal anatomy relevant to practical manipulations of the mouse.

The mouse skeletal system includes a skull adapted for a smaller brain, larger olfactory structures, and mandibles supporting growing incisors [9]. Additionally, the mouse tail is an osseous structure and the pelvis and shoulders

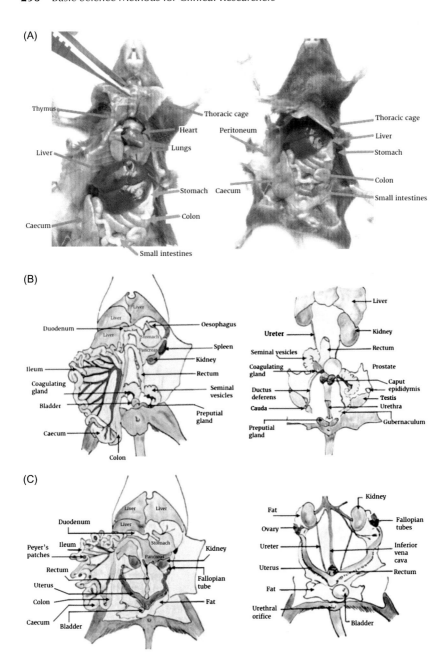

FIGURE 15.3 (A), (B), and (C): Internal mouse figure.

are adjusted to allow quadruped locomotion. The mouse thoracic cage consists of 13 pairs of ribs, the 6 most cranial "true" rib pairs bridging the ventral sternum and dorsal vertebral column [10]. The most caudal true rib pair is connected to three "false' rib pairs. Three most caudal "floating" ribs are only connected to the dorsal vertebral column.

The respiratory system of the mouse is organized similarly to the human respiratory system. However, mice have a higher basal metabolic and respiratory rate [9]. Therefore, the mouse's airway lumens are relatively large to decrease airway resistance. The lungs consist of five lobes, one lobe on the left side and four lobes (cranial, middle, caudal, and accessory) on the right side of the thoracic cage [9]. The mouse and human respiratory tissue share many of the specialized cell types involved in conducting and oxygen uptake.

Many features of cardiovascular system are shared between mice and humans. Both species have a four-chambered heart, consisting of two atria, a thicker left ventricle, and thinner-walled right ventricle [9]. The heart to body weight is similar in both species. However, the mouse heart is able to move more freely in the pericardial sac, and tends to be more oval shaped than the human heart [9]. Additionally, the mouse pulmonary veins fuse before connecting to the left atrium and are wrapped by cardiomyocytes, whereas human pulmonary veins connect separately and do not contain cardiomyocytes. Furthermore, the mouse artery walls are relatively thinner than human arteries.

The mouse has a gastrointestinal system very similar to human gastrointestinal system, and consists of the esophagus, the stomach, the small intestine (duodenum, jejunum, and ileum), and the large intestine (cecum, colon, and rectum) [9]. The mouse lacks an appendix. Although sphincters at either end of the esophagus are shared among species, mice are incapable of vomiting and retrograde movement of food is prevented. Both humans and mice possess abundant microbiota in their colon; however, the mouse relies more heavily on fermentation and vitamin production by commensal bacteria in its gut [9]. Mice will often try to reabsorb nutrients by consumption of their feces. Finally, compared to the human gastrointestinal system, the mouse's transition from the colon to the rectum is abrupt and the short rectum can easily prolapse in several inbred strains [2,9].

The mouse and human liver share central hepatic functions, such as nutrient metabolism, toxin removal, and compensatory regeneration upon tissue injury [9]. The mouse liver covers the entire abdominal space under the diaphragm and is not connected to surface ligaments as in the human liver. The basic hepatic lobular structure is conserved between mouse and human.

In contrast, the mouse pancreas is more dispersed than the human pancreas [9]. Additionally, the mouse pancreas contains relatively smaller lobules and is surrounded by larger fat and lymphoid tissues. The exocrine-secreting islets are also found interlobular, whereas human islets are intralobular. Despite these differences, the basic functions and overall histology of the pancreas are conserved between species.

Both the mouse and the human urinary systems function in osmoregulation, blood pressure regulation, and excretion [9]. However, the anatomy of urinary system is reversed in the species. The mouse retroperitoneal kidneys are caudal (left) and cranial (right; close to the liver), whereas the human kidneys are found in the opposite orientation [9].

The mouse and human lymphoid organs share the basic functions in immune response; however, a few anatomical differences should be noted [9]. Mouse lymph nodes are generally small and found in internal locations, whereas human lymph nodes are larger (palpable) and can be found in externally visible locations. Human lymph nodes are also surrounded by thicker connective tissue than in the mouse, but the overall organization of cortex, paracortex, and medulla is conserved. The thymus has similar functions in mice and humans, and decreases in size with age. However, the decrease is greater in humans [9]. In contrast to the human thymus, the mouse thymus does not contain sublobules. The anatomy of the mouse and human spleen is broadly shared; however, the spleen also functions in hematopoiesis in adult mice, whereas the human spleen is solely involved in immune responses [9].

Finally, anatomical and functional differences exist in the organization of the reproductive system between mice and humans. The human prostate exists as a single organ and lacks lobular structure. In contrast, the mouse contains four histologically distinct lobes and multiple accessory sex glands, including ampullary and preputial glands absent in the human male reproductive system [9]. Furthermore, the mouse has an ossified bone (baculum) in the penis. The female reproductive system also varies anatomically between mouse and human, while histology and function is shared between species. The two mouse ovaries are located caudal to the kidneys and are almost completely closed off from the abdominal cavity [9]. Female mice have an estrus of four to six days during which multiple follicles in the ovaries mature that will ovulate. Rather than fallopian tubes, mice have oviducts. The mouse uterus is divided into two horns that join above the cervix and the vagina.

IN PRACTICE: BASIC MOUSE METHODS

Mouse Handling

Proper handling is essential to the success of all experimental procedures performed on mice. To move mice, adult animals can be grasped by the base of the tail and supported with a cupped hand [7,11]. Lifting the mouse by the tip of the tail can result in injury to the skin and should be avoided. Handling neonates should be done carefully, by grasping the loose skin around the shoulders or lifting neonates in a cupped, gloved hand [7,11]. Mice can also be moved using devices such as forceps. This can be advantageous to avoid exposure to pathogens when handling immunocompromised or specific pathogen-free mice.

Experimental procedures usually require prolonged animal restraint. To begin restraint, lift the mouse by the base of the tail and let the mouse grip the metal cage top [7,11]. Move the tail backward to straighten the body of the mouse, and firmly grasp the loose skin behind the ears with the thumb and index finger of the other hand. This will fix the mouse head and prevents biting injuries. The tail can be further restrained with the ring and little finger. For prolonged or repeated experimental procedures, it can be helpful to use restraining devices [11]. Device usage should be justified in animal protocols.

Breeding and Colony Management

Efficient breeding schemes are essential to the maintenance and management of the mouse colony. Culling aged animals will keep the colony healthy, as many inbred strains can develop diseases, suffer complications, or experience reduced fecundity at older age [2]. Breeding pairs should be set up as soon as animals reach sexual maturity. A sexually mature female mouse has a gestation period of 18–21 days [2,5] and can produce progeny every three weeks in continuous breeding schemes.

There are several schemes that can be followed for breeding mice. In continuous breeding schemes, the male is kept in the cage with the female(s) at all times. Males can either be set up in *pair mating* or *harem mating* [2]. Continuous breeding schemes take advantage of female postpartum estrus, a fertile period immediately following delivery of neonates. In rotating breeding schemes, experienced male breeders will move between cages with cohoused female(s) mice. Rotating breeding schemes take advantage of the *Lee-Boot effect* and the *Whitten effect* [7]. In this rotating scheme, one male can impregnate several breeding cages and limited number of males are necessary to generate large numbers of progeny. The investigator can check breeding females for *vaginal plugs* to determine mating timing [2].

Injection and Sampling Techniques

Substances, such as chemical or biological materials, can be administrated to the animal by various routes. The common routes are discussed in this section. However, other routes may be necessary for experimental procedures. It is important to minimize the frequency of administration, administer no more than the maximal dose per route (Table 15.1), and avoid rapid injections, because this can cause cardiovascular failure and death [11].

Compounds, such as drugs or antibodies, should be prepared in water, phosphate-buffered saline (PBS), or corn-oil/vegetable oil (nonpolar compounds) [11]. Solutions should be within the pH range of 4.5–8.0, although solutions with pH 3.0 can be given with oral administration. Alkaline substances (pH > 8.0) are poorly tolerated by the mouse [11].

TABLE 15.1 Maximal Dose Per Administration Route

Administration Route	Maximal Dosage (mL)	Recommended Needle Size
Oral	0.2	22 G
Subcutaneous	2–3 (scruff/ flank)	25 G
	0.2 (inguinal)	
Intraperitoneal	2–3	23 G
Intravenous	0.2	25 G
Intratracheal	0.05–0.075	22–24 G
Intramuscular	0.05	25–27 G
Intradermal	0.05	26 G
Intracerebral	<0.03	27 G

Source: [11].

Oral Administration

The easiest oral route is including substances with food or drinking water. However, this offers little control over dosage. Most investigators administer substances by oral gavage [7,11]. The gavage consists of a ball-tipped 22 G injection needle, which minimizes damage to the esophagus, attached to 1–2 mL syringe. Oral gavage can be performed on a conscious animal. It requires immobilization of the head and straightening the neck to pass the gavage needle through the mouth and esophagus in a single line for injection [7,11]. If the mouse starts to choke, or exhibits nasal passage of fluids, this may indicate that gavage has entered the trachea instead, and euthanasia is recommended.

Subcutaneous Administration

Compounds can be administered subcutaneously by lifting the loose skin during injection [7,11]. Common injection sites include the interscapular skin (upper back and shoulders), the flank, or the inguinal skin (ventral abdomen). To administer under the interscapular skin, lift the scruff of the neck skin (around the shoulders) with thumb and index finger (forming a "tent") and inject upto 2–3 mL. To administer in the subcutaneous space in the dorsal flanks, lift the skin with thumb and index finger and inject upto 2–3 mL. To administer under the inguinal skin, restrain the mouse and insert the needle firmly through the skin left or right in the lower abdomen, while avoiding insertion in the abdominal midline. Administer upto a maximum of 0.2 mL [11]. If resistance is felt during the subcutaneous injection, the needle likely has not fully penetrated the skin.

Intraperitoneal Administration

Intraperitoneal administration allows quick reabsorption of large volumes of substances and is the preferred injection route for nonirritant, isotonic solutions [7,11]. For intraperitoneal administration, restrain the mouse with one hand and tilt the head backward to expose the abdomen. Insert the needle at a shallow angle through the musculature in the lower left abdomen. Push the needle 2–3 mm further to avoid leakage at the injection site. Retract the syringe plunger slightly prior to injection and observe whether any fluids are drawn into the needle, as would indicate penetration of peritoneal organs or blood vessels. Reposition the needle as necessary. Inject slowly but steadily upto the maximum volume of 2–3 mL [11]. Intraperitoneal injections into the lower right abdomen can potentially damage the cecum and should be avoided.

Intravenous Administration

Intravenous administration provides an excellent route for the injection of highly concentrated, acidic, and irritant solutions [7,11]. The blood will buffer and rapidly circulate substances. The two common injection sites are the lateral tail vein or into retro-orbital vein. It is recommended to anesthetize the animal during intravenous administration.

For lateral tail vein injection, heat the tail with a lamp or warm water to dilate the blood vessels and disinfect the tail [7,11]. Insert a 27 G needle almost parallel to the tail vein with the bevel facing up, and push the needle 2–4 mm further before slowly injecting upto a maximum of 0.2 mL. Resistance during the intravenous injection likely indicates injection in the surrounding tissue. Firmly press on the injection site to prevent retrograde flow of substances and to stop potential bleeding. For retro-orbital vein injection, gently protrude the eyeball by applying pressure around the eye [7,11]. Insert the needle at a shallow angle with the bevel down (facing away from the eye) into the retro-orbital vein behind the eye and slowly inject upto a maximum of 0.2 mL. There should be no bleeding behind the eye after smooth retraction of the needle.

Blood Collection

Body fluids can be collected through various routes, from either live or euthanized animals [7,11]. Collection of mouse blood, urine, feces, or vaginal material might be necessary based on tissue of interest. This section will focus on collection of mouse blood. Owing to the small size of the mouse, collection of excess blood may endanger the animal's health, particularly with repeated collections [7,11]. The most common routes include puncture of the retro-orbital sinus, the tail vein, or the heart. Small-bore pipettes are recommended for a retro-orbital sinus punctures. Restrain the animal, shallowly insert the pipette and twist the pipette to draw the animal's blood by capillary force. Applying mild pressure on the collection site should prevent bleeding. For tail blood collection, hold the animal in a restraining device and dilate the tail vein. Pierce the lateral tail vein with an 21 G needle and collect upto 0.2–0.3 mL of blood. Collection of blood by cardiac puncture allows the collection of large volumes of blood [11]. It is a terminal procedure, and should be performed on anesthetized animals.

Anesthesia and Euthanasia

During many experimental procedures anesthesia is necessary for success of the procedure and minimization of the animal's discomfort. There are many types of anesthetics with different pharmacological properties, and dosage must be adjusted based on the mouse strain, body weight of the animal, and desired duration of anesthesia [11]. Anesthetics can be either injectable or inhalable. Injectable anesthetics can be administered via routes discussed above. The anesthetic will be absorbed more slowly with subcutaneous injections than intravenous or intraperitoneal injections and overdosing the animal must be avoided [11]. Inhalable anesthetics can be administered in an incubation chamber or through a face mask, which can provide extended anesthesia during a surgical procedure [11]. It is important to continually monitor the animal health during anesthesia, as mice may suffer hypothermia due to their high surface-to-volume ratio.

According to guidelines of the American Veterinary Medical Association, laboratory animal euthanasia methods must fulfill certain criteria: (1) painless, (2) rapid unconsciousness and death, requires minimal restraint, (3) appropriate for the age of the animal, (4) removed from the social setting, and (5) safe to administer by the investigator [12]. Approved euthanasia methods will be specified in animal protocols. For adult animals, euthanasia with barbiturates and carbon dioxide are recommended methods [11,12]. Euthanasia with carbon dioxide is not recommended for neonates, while cervical dislocation is an acceptable method [11,12].

IN PRACTICE: ADVANCED IN VIVO MODELING IN THE MOUSE

Genetically engineered mouse models (GEMMs) can be generated to study physiological and pathological processes. Several approaches can be used for in vivo modeling. In the *transgenic* approach, exogenous genetic material is introduced into the mouse genome to allow expression of a novel gene or enhance expression of an endogenous gene [13,14]. In the *knock-out* or *knock-in* approach, the genomic location of a specific gene is altered to create a functionally null allele (knock-out) or add extra genetic material (knock-in) [15]. Addition of regulatory genetic components, such as tissue-specific promoters, doxycycline responsive elements, or Cre-sensitive LoxP-Stop-LoxP cassettes, allow both temporal and spatial control of *conditional alleles* [16]. With recently developed *CRISPR-Cas9* technology, these main approaches can be easily executed in ES cells [17]. Furthermore, researchers have now started to model precise single nucleotide changes or large genomic deletions that were previously challenging. This section will provide a broad outline for generating GEMMs. The generation of a GEMM requires several steps, including the design of a transgene or targeting vector, microinjection into zygotes or transfection into ES cells, selection of positive ES cell clones, creation of chimeric animals, and breeding of germline animals. These steps are outlined below, and suggestions are given for further in-depth reading where appropriate (Figure 15.4).

Construction of the Targeting Vector

The first step in the generation of a GEMM is constructing the vector that can be microinjected into the mouse embryo or transfected into ES cells. In transgenic approaches, the DNA plasmid vector contains all necessary regulatory elements such as promoters and enhancers for either ubiquitous, tissue-specific, or inducible gene expression in the desired tissue. The transcribed sequences of the vector can be *cDNAs* or full genomic sequences (including introns), and polyadenylation sites [13,14]. Addition of a unique restriction enzyme site allows linearization of the transgene before microinjection into zygotic pronuclei. Knock-out/knock-in approaches utilize vectors that contain DNA regions homologous to the insertion site in the mouse genome (homology arms), flanking a gene modification of interest or a positive resistance marker [15]. These *targeting vectors* take advantage of the process of homology directed repair (HDR) to integrate into a specific locus of the genome and swap out the endogenous sequences. In the case of a knock-out or knock-in mouse model, special attention needs to be paid to the exonic and intronic structure of the targeted genomic gene to achieve the desired loss-of-function disruption [15]. In addition to the elements described above, traditional targeting vectors for HDR should also include a suicide gene for negative selection and a unique restriction enzyme site to linearize the plasmid before electroporation into ES cells [15].

Microinjection into Zygotic Pronuclei or ES Cell Injection into Blastocysts

Once a vector has been constructed and verified by sequencing, the plasmid is linearized using a unique restriction enzyme site and introduced into the mouse germline via two principal injection strategies into early mouse embryos. Transgenic models are usually generated by direct injection of the transgene into the pronuclei of fertilized 1-cell embryos [13,14]. Knock-out or knock-in models are usually generated through in vitro manipulation of ES cells that are then injected into 3.5 days post fertilization blastocysts [5,15].

During pronuclei injection, fertilized zygotes (1-cell stage embryos) are harvested from donor mothers impregnated by fertile males [5,13]. The timing of zygote harvest is crucial, and requires specialized dissection tools. Purified and concentrated transgenic DNA is injected into the pronucleus using an inverted microscope equipped with micromanipulators, embryo holding needle, and a

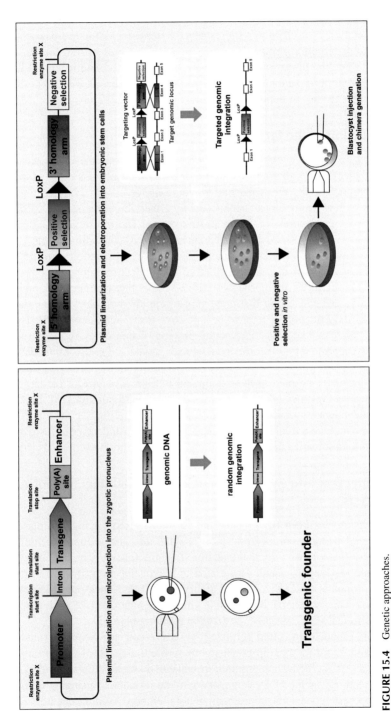

FIGURE 15.4 Genetic approaches.

pressurized injection needle [5]. After DNA plasmid injection, it can be helpful to leave successfully manipulated 1-cell stage embryos overnight at 37°C and 2.5% CO_2. During this time, viable embryos will undergo their first division and the next day 2-cell embryos can be selected for oviduct transfers into pseudo pregnant surrogate mothers (0.5 days post coitus with vasectomized males) [5].

In contrast to transgenic approaches, the knock-out/knock-in approach involves manipulation and selection of targeted ES cells prior to embryo injection. Mouse ES cells, originally derived from the inner cell mass of the developing blastocyst stage embryo, are capable of giving rise to all tissues in the organism, but lack the ability to form extra-embryonic tissues. They can be cultured over extended periods, if provided with the correct media and supporting feeder cells [5,13]. Additionally, ES cells can be genetically manipulated using standard molecular biology techniques. DNA uptake is most efficient when a linearized targeting vector is electroporated into the ES cells [15]. The vector will then be integrated into the genome through homologous recombination. The efficiency of homologous recombination varies and may be affected by the specific locus targeted [15]. Meganuclease-based approaches, including CRISPR-Cas9, which cleave the mouse genome at the target locus, have recently been used to improve recombination efficiency. After targeted vector electroporation, culture of manipulated ES cells in selective antibiotic-containing medium allows outgrowth of clonal cells containing integrated vector (positive selection) and lacking the suicide gene (negative selection) [5,13]. Negative selection eliminates ES cell clones that have randomly integrated the vector. Genomic DNA should be isolated from clonal cell populations and screened for proper integration using PCR or Southern blotting. Once a pure population of positive ES cells has been obtained, blastocysts (3.5 days post coitus) can be harvested from previously impregnated donor mothers [13]. The modified ES cells (typically between 12 and 15 cells) are then directly injected into the blastocoel (hollow cavity of the developing blastocyst). In order to estimate the contribution of the injected ES cells to overall organism, ES cells derived from a black mouse are typically injected into a blastocoel isolated from a white strain of mouse.

Chimeric Animals and Germline Transmission

Successfully re-implanted 2-cell or blastocyst-stage embryos will be carried by the surrogate mother to birth [13]. If a low number of littermates is expected, it is useful to provide foster parent care for the neonates. Pronuclei injected embryos are usually expected to give rise to fully transgenic P1 (parental) animals which, if fertile, will always transmit the transgene to their progeny. For blastocysts injected with ES cells, the percentage contribution of the ES cells to the organism will vary from animal to animal and possibly vary per tissue in an individual; these mice are called *chimeras*. The determination of the percentage of chimerism in these animals is generally estimated by coat color. Usually, the greater the percentage of coat color from the ES cells the more likely the mouse is to transmit the transgene through the germline to their progeny. Germline transmission of the allele can be difficult to achieve in some ES cells, and performing several implantations in pseudo pregnant surrogate mothers increases the probability of successfully obtaining germline chimeras [5,13].

Maintenance of Genetically Engineered Mouse Lines

Once the transgenic or chimeric founder has successfully transmitted the transgene or knock-out/knock-in allele to their progeny, a mouse line can be created. In transgenic approaches it will be necessary to screen multiple transgenic lines, as locus-specific effects can influence the penetrance and fidelity of transgene expression [13,14]. In targeted knock-out/knock-in approaches, variability in allele penetrance is not an issue. Once a suitable mouse line has been started, genetic background differences that were acquired with the ES cells can be further eliminated by breeding the GEMMs to mice from the desired inbred genetic background (*backcross*) [13]. It may also be necessary to obtain two transgene or knock-out/knock-in alleles, which can be achieved with sibling mating. It is helpful to genotype potentially transgenic neonates as early as possible (between day 10 and day14), as only a fraction of the progeny will harbor the allele of interest [13]. Therefore, culling nontransgenic littermates can significantly reduce transgenic mouse line housing costs. Finally, once the GEMM has been fully established, transgene expression or knock-out should be validated in tissues of choice by histological analysis and biochemical assays of choice.

APPLICATIONS OF MOUSE MODELS

The mouse has been applied to study numerous human physiological and pathological processes, such as neurological diseases, immunology, metabolic diseases, and cancer. A comprehensive discussion of applications of these diverse models is beyond the scope of this chapter. Instead, this section will highlight the applications, strength, and limitations of mouse models in the field of cancer biology.

Mouse Models of Cancer

Mouse models of cancer aim to delineate the environmental and genetic contributions to the development of human cancers and can serve as an in vivo platform for preclinical testing of therapeutics [16,18].

In its simplest form, these models can be inbred strains that suffer from certain spontaneously arising cancers or *xenograft* models. Recently, there has been a rising number of GEMMs of cancer that more accurately recapitulate human cancer progression. These models offer precise control over the timing of oncogenic events to reduce latency and experimental variability in cancer models.

In Cancer Genetics

Mouse models of cancers have been invaluable in elucidating contributions of genetic mutations to virtually all aspects of cancer biology. Researchers have used mouse models to investigate cancer initiation, the role of oncogenes and tumor suppressor genes in cancer progression, factors driving metastasis, and therapeutic response during cancer treatment [16,18]. In recent years, these modeling approaches have been bolstered by the availability of large cancer genomic data sets. DNA sequencing consortia, such as The Cancer Genome Atlas (TCGA), have extensively profiled genetic mutations in more than 30 human cancers [19]. These genetic mutations can then be precisely modeled in the mouse genome to further dissect how these affect protein function and ultimately impact cancer progression [18]. Table 15.2 lists several mouse models that recapitulate human cancer.

In Preclinical Testing, Drug, and Biomarker Discovery

Mouse models can also be extremely valuable in preclinical drug testing and biomarker discovery [16]. Preclinical drug testing has traditionally been performed on xenograft mouse models in which human cancer cells are transplanted into immunocompromised mice. This provides a flexible "platform" for high-throughput screening of candidate compounds and establishing dose-response relationships. As the grafts are surrounded by normal tissue, the pharmacodynamics of compounds are more accurately modeled than in cell culture systems [16]. Furthermore, efficacy and timing of combinations of compounds can also be tested to accelerate translation in clinical trials. Finally, as xenograft models are more homogenous than complex human cancer tissues, they can also be applied for biomarker discovery [16]. Serum biomarkers can be extremely valuable tools for earlier diagnosis of cancer and monitoring response to treatment in the clinic. Several efforts are underway to translate research findings from mouse models to clinical settings.

Scenario: Modeling *p53* Restoration in Cancer

In the following case study, we describe an application of mouse modeling to cancer genetics.

Ventura and colleagues were interested in the effect of the tumor suppressor gene *p53* on tumor maintenance [20] and took a conditional knock-in modeling approach. They constructed a targeting vector for the *p53* locus, containing a transcription stop cassette before exons 2–11 and diphtheria toxin as a negative selection marker. The targeted allele was functionally null and resulted in tumor-prone mice. These mice were crossed to mice harboring tamoxifen-inducible Cre recombinase (Cre-ER^{T2}), which would remove the stop cassette before exon 2 upon tamoxifen administration and reactivate *p53* expression at will. By examining the effects of *p53* restoration, the authors concluded that *p53* has cancer type-specific functions, inducing apoptosis in lymphomas while

TABLE 15.2 Cancer Genetics of Selected Genetically Engineered Mouse Models

Organ	Histopathology	Genetics
Lung	Adenocarcinoma	Kras; Trp53
	Small cell carcinoma	Rb1; Trp53
Colon	Polypoid adenocarcinoma	Kras; Apc
	Hereditary nonpolyposis carcinoma	Msh6
Breast	Ductal carcinoma	Brca2; Trp53;
	Lobular carcinoma	Cdh1; Trp53
Pancreas	Ductal adenocarcinoma	Kras; Cdkn2a
		Kras; Trp53
	Mucinous cystic neoplasm	Kras; Dpc4
Prostate	Prostate carcinoma	Pten
		Pten; Nkx.1
		Rb1; Trp53
Liver	Hepatocellular carcinoma	Apc
		Myc; Trp53
		Myc; TGFA
Ovary	Endometrioid carcinoma	Kras; Pten
		Apc; Pten
Oesophagus	Squamous cell carcinoma	Pten; Dpc4
		Ccnd1; Trp53
Bladder	Transitional cell carcinoma	Hras
Kidney	Renal cell carcinoma	Apc; Trp53
Brain	Astrocytoma	Pten; Rb1
	Glioblastoma	Nf1; Trp53
Stomach	Gastric carcinoma	Wnt; Ptgs2; Ptges
Skin	Melanoma	Hras; Ink4a
	Squamous cell carcinoma	Xpd

Source: [18].

suppressing growth in sarcomas. Importantly, this study highlighted the concept of p53 pathway reactivation as a therapeutic strategy.

Key Limitations of Mouse Models of Cancer

Mouse models closely recapitulate human cancer progression, particularly GEMMs. These models are powerful in dissecting complex cancer genetics and can be applied in preclinical testing. However, several key technical and biological limitations exist [16].

Mouse models require significant resources, time, and labor to generate and maintain. Despite their small size and relatively low housing requirements, mouse models necessitate large cohorts to produce conclusive data. Historically, mouse models have also been limited by the genetic complexity that can be engineered. Human cancer is a complex, heterogeneous disease, and cannot always be accurately captured in a mouse model. The advent of rapid in vivo modeling with CRISPR-Cas9 technology promises to overcome some technical barriers [17,21]. This technique allows very precise engineering of mammalian genomes in vivo and can rapidly generate new cancer models [21]. The technology has been successfully applied to model mutations in liver and lung cancer, and is poised to greatly impact the study of human cancer genetics in mouse models.

Inherent species-specific differences between mice and humans exist [16]. The mouse has a higher metabolic rate, which can result in more rapid breakdown or clearance of compounds affecting preclinical testing. Most mouse cells also actively maintain telomeres, in contrast to human cells, which may predispose them to become precancerous. Furthermore, the immune system differences between mouse and human may shape the host response to a growing cancer. Efforts have been made to "humanize" mice, by reconstituting the human immune system in the mouse. Further development will focus on overcoming these technical and biological limitations by engineering increasingly advanced models.

Troubleshooting

Common Problems	Optimization Strategies
Breeding pair does not produce litters or generates small numbers of pups	• Set up harem breeding for cross-fostering of pups by mothers. • Switch to younger males and females of desired genotype, if possible. • Perform timed matings and check vaginal plugs to determine coitus.

Common Problems	Optimization Strategies
Targeting vector does not integrate into ES cells	• Perform quality control steps of the vector generation/purification process. Prepared vector should be run on an agarose gel to ensure sufficient quality and concentration for targeting in ES cells. • Ensure good-quality ES cells by maintaining them at subconfluency and providing appropriate medium to prevent differentiation. • Switch to meganuclease-based (TALEN, Cas9) methods to generate more efficiently targeted DNA double-stranded breaks for vector integration.
ES cells clones die during positive selection	• Double check quality and quantity of prepared vector prior to transfection. • Perform kill curve to properly titrate the concentration of the selection antibiotic.
ES cells survive positive selection but lack transgene expression	• Verify that homology arms are properly designed for integration into the desired locus. Avoid excessive repetitive elements in the homology arms as this can increase random genomic integration. • Target transgene to well-characterized genomic sites, such as the Rosa26 locus to ensure consistent transgene expression.
Blastocyst injected ES cells do not give rise to chimeras or do not transmit to the germline	• Start experiments with a high quality, germline validated parental ES cell line. • Inject other positive ES cell clones generated during the antibiotic resistance selection. • Seek advice of a local expert to further optimize blastocyst injection setup.
The transgene/targeted allele is embryonically lethal	• Target transgene to well-characterized locus (see above) to avoid position effects. • Consider generating tissue-specific or conditional knock-out/knock-in alleles of the gene of interest.

CONCLUSIONS

In summary, the mouse has a long history of successful contributions to scientific research. It's particularly well-suited for biomedical research, as the mouse recapitulates human physiology and is amenable to sophisticated genetic manipulation. However, species-specific biological and anatomic differences do exist and must be kept in mind when modeling disease in the mouse. The mouse can be experimentally manipulated to study many aspects of pathophysiology. The convergence of mouse developmental biology, embryonic stem cell culture and manipulation, and precision editing of complex genomes through rapidly improving molecular biological techniques allows the unprecedented generation of mouse models that recapitulate disease etiology, and can be applied to dissecting human genetics, preclinical research, and other areas of biomedical research. As the development of powerful in vivo modeling technologies accelerates, we anticipate that mouse models will continue to be critically important in translating biomedical research.

REFERENCES

[1] Database JSE. Essentials of Biology 2: Mouse, Zebrafish, and Chick. An Introduction to the Laboratory Mouse: Mus musculus. Journal of Visualized Experiments 2015.

[2] Flurkey K, Currer JM, Leiter EH, Witham B. The Jackson Laboratory handbook on genetically standardized mice: Jackson Laboratory; 2009.

[3] Paigen K. One hundred years of mouse genetics: an intellectual history. I. The classical period (1902–1980). Genetics 2003;163(1):1–7.

[4] AnimalResearch.Info: Nobel Prizes. http://www.animalresearch.info/en/medical-advances/nobel-prizes/ (accessed January 2, 2016).

[5] Behringer R. Manipulating the mouse embryo: a laboratory manual, Fourth Edition: Cold Spring Harbor Laboratory Press; 2014.

[6] Silver LM. Mouse genetics: concepts and applications: Oxford University Press; 1995.

[7] Suckow MA, Danneman P, Brayton C. The laboratory mouse: CRC Press Inc.; 2001.

[8] Bosman LW, Houweling AR, Owens CB, et al. Anatomical pathways involved in generating and sensing rhythmic whisker movements. Frontiers in integrative neuroscience 2011;5:53.

[9] Treuting PM, Dintzis SM. Comparative Anatomy and Histology: A Mouse and Human Atlas (Expert Consult): Academic Press; 2011.

[10] Cook MJ. The anatomy of the mouse. *The anatomy of the laboratory mouse* 1965.

[11] Hedrich H. The laboratory mouse: Academic Press; 2004.

[12] Leary S, Underwood W, Anthony R, et al. AVMA guidelines for the euthanasia of animals: 2013 edition. 2013.

[13] Hofker MH, van Deursen J. Transgenic mouse: methods and protocols: Springer; 2011.

[14] Haruyama N, Cho A, Kulkarni AB. Overview: Engineering Transgenic Constructs and Mice Current Protocols in Cell Biology: John Wiley & Sons, Inc.; 2009.

[15] Hall B, Limaye A, Kulkarni AB. Overview: generation of gene knockout mice. Current protocols in cell biology 2009 19.2. 1-.2. 7.

[16] Cheon D-J, Orsulic S. Mouse models of cancer. 2011.

[17] Singh P, Schimenti JC, Bolcun-Filas E. A mouse geneticist's practical guide to CRISPR applications. Genetics 2015;199(1):1–15.

[18] Frese KK, Tuveson DA. Maximizing mouse cancer models. Nat Rev Cancer 2007;7(9):654–8.

[19] The Cancer Genome Atlas Data Portal. https://tcga-data.nci.nih.gov/tcga/tcgaHome2.jsp (accessed December 30, 2015).

[20] Ventura A, Kirsch DG, McLaughlin ME, et al. Restoration of p53 function leads to tumour regression in vivo. Nature 2007;445(7128):661–5.

[21] Sánchez-Rivera FJ, Jacks T. Applications of the CRISPR-Cas9 system in cancer biology. Nature Reviews Cancer 2015.

SUGGESTED FURTHER READING

[1] Behringer R. Manipulating the mouse embryo: a laboratory manual, Fourth Edition : Cold Spring Harbor Laboratory Press; 2014. **In-depth reading on generating traditional genetically-engineered mouse models**.

[2] Hedrich H. The laboratory mouse: Academic Press; 2004. **In-depth description of common animal techniques, including administration and fluid collections**.

[3] Singh P, Schimenti JC, Bolcun-Filas E. A mouse geneticist's practical guide to CRISPR applications. Genetics 2015;199(1):1–15. **Practical tips for using CRISPR-Cas9 in mouse genome engineering**.

[4] Treuting PM, Dintzis SM. Comparative Anatomy and Histology: A Mouse and Human Atlas (Expert Consult): Academic Press; 2011. **Exhaustive discussion of mouse anatomy and histology**.

[5] Yang H, Wang H, Jaenisch R. Generating genetically modified mice using CRISPR/Cas-mediated genome engineering. Nat Protoc 2014;9(8):1956–68. **Detailed protocol on embryo manipulation with CRISPR-Cas9 technology**.

GLOSSARY

Anogenital distance   Distance between the anus and the external genitalia.

Backcross   Genetic cross between hybrid progeny and homozygous parent (or genetically identical individual).

Barbering   Focal loss of a subordinate's hair or whiskers by dominant male.

cDNA   Complementary DNA synthesized from messenger RNA.

Chimera   Individual consisting of heterogeneous genetic background.

Conditional allele   Genetically manipulated allele that can be activated in inducible fashion.

Congenic   Genetically identical individual except for a single linked locus.

CRISPR-Cas9   Clustered Regularly Interspaced Short Palindromic Repeats, a bacterial defense system adapted for precise genomic engineering.

Embryonic stem cells   Early blastocyst-derived stem cells giving rise to all organismal tissues.

Gastrulation   Reorganization of the embryos cells into three distinct tissue layers.

GEMM   Genetically engineered mouse model.

Gestation   Developmental period of an embryo.

Harem mating   Cross between one male and two females.

Inbred   Genetic fixation of a strain after 20 generation of sibling mating.

Knock-in   Addition of extra genetic material into a targeted genomic locus.

Knock-out   Disruption of targeted genomic locus to create a functionally null allele.

Lee-Boot effect   Suppression of the female estrus cycle.

Neonate   Young, newborn individual.

Oncogene   Mutated or overexpressed cancer-causing gene.

Outbred   A maximal genetically diverse strain.

Pair mating   Cross between one male and one female.

Targeting vector   DNA plasmid containing homology arms for introduction of genetic material into a genomic locus.

Transgene   Introduction of extra genetic material into an individual.

Tumor suppressor gene   Gene with protective properties against cancer.

Xenograft   Transplantation (grafting) of human cancer cells into immunocompromised mice.

Vaginal plugs   White, coagulated excretions of the male reproductive system.

Weaning   Introduction of an individual to the adult diet.

Whitten effect   Female estrus induction by exposure to male pheromones.

Appendix A

Legal Framework on the Scientific Use of Animals in Research

David Pettitt[1], Adam Pettitt[2], MacKenna Roberts[1], James Smith[1] and David Brindley[1,3,4,5]

[1]University of Oxford, Oxford, United Kingdom; [2]Royal Veterinary College, London, United Kingdom; [3]University College London, London, United Kingdom; [4]Harvard Stem Cell Institute, Cambridge, MA, United States; [5]USCF-Stanford Center of Excellence in Regulatory Science and Innovation (CERSI), San Fransisco, CA, United States

AN INTRODUCTION TO ANIMAL RESEARCH

Animal models serve a fundamental role in scientific research and the translation of basic science research into clinical applications—from "bench to bedside." A variety of animal models are now available to researchers including both small and large animals, spanning a diverse variety of species.

Animals have long been used in medical research and over the last century, and every Nobel Prize for medical research has been dependent on animal research [1]. They have successfully facilitated the discovery of numerous drugs that are still utilized in modern-day medical practice (see Table A.1).

An excess of 100 million animals are used for research purposes around the world on an annual basis, with the majority being used in the European Union (EU) and United States (US) [2]. Presently, rats, mice, and other types of rodents constitute approximately 90% of all research animals—dogs and cats constitute approximately 1–1.5% of all research animals and nonhuman primates account for under 1% of all research animals [2]. In the EU, the total number of animals used by member states for experimentation and other scientific purposes has been decreasing year-on-year (see Fig. A.1), with the most recent report (2011) citing that just under 11.5 million animals were used in 2011—a reduction of approximately half a million animals compared to the number reported in 2008 [3]. Rodents and rabbits accounted for approximately 80% of this total [3].

TABLE A.1 Key Drug Discoveries Using Animal Models [1]

Key Drug Discoveries Using Animal Models

Year	Drug	Animal Model
1901	Diptheria/tetanus antitoxin	Horse
1922	Insulin	Dog
1939	Prontosil	Mouse
1941	Penicillin	Rat
1952	Streptomycin	Chicken

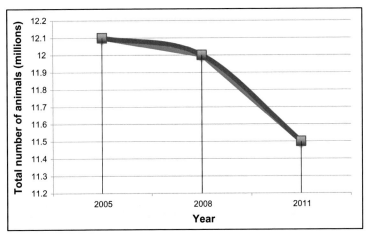

FIGURE A.1 Number of animals used by EU member states for experimentation. Trends in the statistical data taken from the 2005, 2008, and 2011 Reports from the Commission to the Council and the European Parliament on the Statistics on the number of animals used for experimental and other scientific purposes in the member states of the European Union [3].

Despite animals being conventionally employed in preclinical research, a number of challenges remain with translating animal data to in-human clinical trials. Several factors account for translational difficulties, including disparities between species, the internal validity of the research in question and variability in experimental design [4]. The extent of such challenges is convincingly described by O'Collins et al. (2006) [5], who in their research discovered that despite over 500 postischemic stroke neuroprotective interventions being demonstrably effective in animal models, not a single intervention was subsequently found to be clinically effective in humans.

Concern has also been expressed in the published literature that both the methodological quality of systematic reviews reporting animal data and reproducibility of results is poor [6,7]—and data is also subject to publication bias [7]. It is reported that fewer than 2% of investigators who use animal models in their research justify their sample size calculations—and furthermore, had they actually done so, the majority of calculations would have demonstrated a need for larger quantities of animal subjects [8]. Additional concerns include the internal validity of animal studies—often cited as "inadequate" [6], insufficiencies in study design and questionable experimental conduct [9]—all of which contribute to the difficulties seen when translating animal research into human subjects [6]. These issues are however widely acknowledged and substantial improvements are likely to be seen as regulation is refined and harmonized [9].

Animal Models

Animal models allow for a variety of pathobiological and therapeutic processes to be studied in vivo, and play an important role in the development of novel diagnostic and therapeutic applications [10]. Uncertainty surrounds the etiology, pathophysiology, and natural history of many diseases and valid, sensitive animal models enable investigators to study animal specimens exhibiting similar pathological phenomenon as humans [11]. Therapeutic drug applications require preclinical studies on animal models to validate proofs and concept and help determine underlying mechanisms of action, toxicology, safety, and efficacy [11]. They also provide a practical means of studying rare diseases [12], where patient recruitment may be difficult, and provide a basis for orphan drug applications [13].

It is essential that appropriate animal models are selected to optimally replicate human disease processes (see Box A.1). Certain animal models may be particularly suited to a specific disease type or researchers may be able to select from a variety of animal models to investigate a particular condition—Parkinson's disease, for example, can be studied in species ranging from fruit flies (*Drosophila melanogaster*) to Marmoset monkeys (*Callithrix jacchus*) [11].

Box A.1 Selection of Appropriate Animal Model (Davidson 1987)

1. Appropriateness as an analog
2. Transferability of information
3. Genetic uniformity of organisms, where applicable
4. Background knowledge of biological properties
5. Cost and availability
6. Generalizability of the results
7. Ease of and adaptability to experimental manipulation
8. Ecological consequences
9. Ethical implications

It can however be particularly challenging to develop representative animal models; even if being reproduced genetically, animals may display different disease phenotypes to humans [14]. Such differences can become considerably pronounced and limit the applicability and transferability of findings. Additional factors to consider include ease of experimental manipulation, availability, cost, and ethical implications.

Common Animal Models Are Outlined Below

Murine

Murine (mouse and rat) models have historically proven indispensable to scientific research and they continue to be used to effectively investigate a myriad of biological processes. They possess a short life span, can reproduce efficiently, require lower financial investments compared to larger animal models, and are easier to house and handle [10]. Technological advancements also mean that transgenic and knockout mice are readily available, making murine models scientifically, logistically and economically favorable [15].

Porcine

Porcine (pig) models serve as valuable preclinical animal models and have traditionally been used in laboratory research owing to their anatomical and physiological similarities with humans. They have similar immune systems to humans and certain inbred pigs have genetically defined and fixed major histocompatibility complexes, facilitating reproducible studies of immunologic mechanisms [16]. More recently, miniature pigs have gained popularity due to their ease of handling and suitability to medical research (however are more expensive than their normal-sized counterparts) [16].

Nonhuman Primates

Nonhuman primates are widely considered to most closely resemble human subjects genetically, physiologically, and immunologically [15]. They are often comparably susceptible to infectious diseases and offer distinct advantages over human subjects whereby researchers can more closely control variables such as dosing regimes and inoculation routes. They also demonstrate analogous clinical indicators to human subjects, including serum C-reactive protein and blood cell counts [15]. Their use in biomedical animal research is tightly restricted and permitted only in very specific circumstances [2]. In the EU for example, there has been significant decreases in the use of nonhuman primates and no great apes have been used in EU since 1999 [3].

REGULATION

The first law intended to protect animal welfare was enacted by the British Parliament in 1822 and later followed by the *Cruelty to Animals Act* in 1876 [1].

Box A.2 The 3Rs

W.M.S. Russell and R.L. Burch formulated the 3Rs in their 1959 publication, *The Principles of Humane Experimental Technique* [17]. Their innovative concept is based on the notion that the humanitarian use of animals serves as the underlying basis for excellence in scientific research and covers three key domains (the 3Rs), outlined below:

1. *Replacement*
   Refers to the methods, strategies or approaches that do not involve the use of live animals.
2. *Reduction*
   Refers to approaches that will result in fewer animals being used.
3. *Refinement*
   Refers to the modification of any procedures or husbandry and care practices from the time the experimental animal is born until its death, which minimize pain, suffering, and distress experienced by animals and enhance well-being.

Since then, various legislation governing animal research and welfare has been instituted globally and in 1985 the Council of International Organizations for Medical Sciences (CIOMS) translated Russell and Burch's "3Rs" (see Box A.2) [17] into 11 fundamental international principles that govern animal experimentation. However, despite such efforts, considerable variation across international jurisdictions remains.

European Union (EU)

The EU currently consists of 28 member states and pan-European legislation applies to the EU member states (however, it is not necessarily applicable to all countries of the European continent) [18].

Throughout the 20th century there was considerable variation between European member states with regards to laboratory animal welfare [18]. The Council of Europe formally instituted legislation specifically governing laboratory animal welfare and experimentation in 1986, with the intention of eliminating disparities between national laws and the local regulations of member states. This legislation (EC Directive 86/609) served to [19]:

- Improve controls on the use of laboratory animals.
- Define minimum standards for animal care and housing.
- Address training requirements for personnel handling animals and supervising experiments.
- Utilize nonanimal research methods where possible.
- Support the development and validation of animal model alternatives.

In addition to providing guidance on the protection of animals used for experimental and other scientific purposes, Directive 86/609/EEC also provided direction on data collection. Article 13 recommended that member state should collect and make publicly available any statistical data derived from animal experimentation (it did not however specify a common format or the depth of data expected). Although this legislation delivered a number of key recommendations and detailed minimum laboratory animal welfare standards, EU Directive 2010/63/EU superseded it in 2010, with further modifications to address animal protection in experimentation, breeding, transport, and slaughter [19]. This was largely driven by advancements in biomedical technology, heightened public awareness and sensitivity, and continued widespread variation in animal welfare legislation across the EU [18].

EU Directive 2010/63/EU was formally adopted on September 22, 2010 and took full effect on January 1, 2013. It contained the following key objectives [20]:

- Firmly advocate the principle of the 3Rs.
- Regulate the use of animals through a systematic project evaluation requiring inter alia assessment of pain, suffering distress, and lasting harm caused to the animals.
- Enforce risk-based European Commission-supported inspections of procedures and facilities on a regular basis.
- Improve research transparency through measures including:
  - Publication of nontechnical project summaries
  - Retrospective assessments.
- Further promote the development, validation and implementation of alternatives to animal research methods.
- Utilize the Union reference laboratory for the validation of alternative methods supported by laboratories within member states.

The legislation seeks to both harmonize regulation among EU member states and extend its legislative scope to include more animal species. It now adopts the following definition of "animal" [20]:

a. *Live nonhuman vertebrate animals, including:*
  - *Independently feeding larval forms and*
  - *Fetal forms of mammals as from the last third of their normal development.*
b. *Live cephalopods.*

This ensures that the legislation is now applicable to almost any laboratory animal [2] and includes the class *Cephalopoda*, which gives them the same legal status as vertebrates [21]. Furthermore, it specifically tightens regulation for certain species (such as nonhuman primates) and aims to increase the transparency of animal experimentation to the general public.

To comply with EU legislation, institutions (including universities, hospitals, public and private entities) using animals for research purposes must have research proposals examined by a competent ethics committee—which is usually locally or university-based [2]. They will subsequently supervise the research and provide implementable advice as required. The legislation, although not an exhaustive list of best practice, details acceptable, and discouraged practices, and directs readers to relevant published literature if further details are required [2].

United Kingdom (UK)

In the UK specifically, the 1986 Animals (Scientific Procedures) Act [22] was introduced to protect and regulate animals that were subject to experimental procedures for scientific advancement. The UK also has a universal Animal Welfare Act (2006) [23]; however, research animals are governed by the Animals (Scientific Procedures) Act 1986 (ASPA).

The European Directive 2010/63/EU (that replaced the 1986 Directive) was transposed into the ASPA via the ASPA 1986 Amendment Regulations 2012 [24]. The ASPA protects:

a. All living vertebrates, other than man, and any living cephalopod. In addition:
   • Embryonic and fetal forms of mammals, birds, and reptiles are protected during the last third of their gestation or incubation period;
   • Fish and amphibia once they can feed independently; and
   • Cephalopods at the point they hatch.

It stipulates that every breeder, supplier, and scientific procedure establishment must have an animal welfare and ethical review body (AWERB) and the place where regulated procedures are carried out, or animals are supplied or bred for use in these procedures, must have an established (Section 2 C) license. The issuance of a project license is dependent on a number of factors, including adherence to the 3Rs and cost–benefit justification, and can be issued to either those supervising research programs or to individuals performing scientific procedures.

United States (US)

The US employs multiple interrelated and dynamic systems to regulate animal research (see Table A.2). In 1966, Congress enacted the federal Animal Welfare Act (AWA) (7 U.S.C. § 2131), which is enforced by the Animal and Plant Health Inspection Service (APHIS) of the US Department of Agriculture (USDA). It serves to provide a foundation for the protection of nonhuman animals in a variety of environments, including laboratory-based research [25]. This was essentially not only focused on cats and dogs but also endeavored to regulate hamsters, guinea pigs, rabbits, and nonhuman primates. This initial legislation was somewhat problematic—entities were only protected once registered or licensed and could only obtain this if they conducted government-funded research or participated in

TABLE A.2   Key Components of US Animal Research Regulation [18]

Mandatory	Voluntary
Animal Welfare Act (US Department of Agriculture (USDA))	Association for the Assessment and Accreditation of Laboratory Animal Care International (AAALAC)
Public Health Service (PHS) Policy on Humane Care and Use of Laboratory Animals (Office of Laboratory Animal Welfare (OLAW) of the National Institutes of Health (NIH))	Standards set and maintained by individual users
Good Laboratory Practices (GLPs) (US Food and Drug Administration (FDA))	
Private funding agency requirements	

business activity that crossed state lines [25]. Consequently, it has been subject to various amendments over the last 50 years—1970, 1976, 1985, and 1981. In 1970, congress amended the definition of "animals" from a succinct list to include all warm-blooded animals used in research (with the exception of farm animals). The AWA now defines animals as:

Any live or dead dog, cat, monkey (nonhuman primate mammal), guinea pig, hamster, rabbit, or such other warm-blooded animal, as the [USDA] Secretary [of Agriculture] may determine is being used, or is intended for use, for research, testing, experimentation, or exhibition purposes, or as a pet; but such term excludes:

(1) birds, rats of the genus Rattus, and mice of the genus Mus, bred for use in research,

(2) horses not used for research purposes, and

Notably, the AWA does not regulate the most commonly used laboratory animals—rats, mice, and birds (nor invertebrates such as insects and fish), which remains a considerable topic of concern for animal protection communities. A degree of protection is offered under the Public Health Service (PHS) Policy, but only to those animals in government-funded research facilities [26].

(3) other farm animals, such as, but not limited to livestock or poultry, used or intended for use as food or fiber, or livestock or poultry used or intended for use for improving animal nutrition, breeding, management, or production efficiency, or for improving the quality of food or fiber. With respect to a dog, the term means all dogs including those used for hunting, security, or breeding purposes.

The 1985 amendment is considered to be the most extensive. This amendment established a database for researchers that provides suitable alternatives

TABLE A.3   Key Requisites for AWA Compliance [29]

1. Establish a defined protocol for inspections.

2. Identify an inspection walkthrough team familiar with operational areas (and who can accompany an inspector).

3. Notify key persons (such as animal care and IACUC support staff) who may be involved in the inspection.

4. Determine what the inspector plans to review during the process and arrange for the availability of pertinent records.

5. Ensure inspection team members have a detailed working knowledge of the institution's animal care and use program and Animal Care Resources Guide and the information contained in Section 7 of the Research Facility Inspection Guide [30,31].

to "painful animal experiments" (known as the Animal Welfare Information Center (www.nal.usda.gov/awic)), and also requires US-based research facilities to register with the USDA and establish an Institutional Animal Care and Use Committee (IACUC) to review all experimental protocols involving live, warm-blooded animals [27]. The IACUC oversees an institution's animal care and use program and will review animal research facilities, inspect animal laboratories (usually on a biannual basis), evaluate research protocols (and approve, disapprove, or modify them), monitor any public complaints or alerts concerning noncompliance and report any insufficiencies in animal care [28]. Table A.3 highlights requisite areas for AWA inspection compliance.

The Animal Welfare Act is the only US federal law covering animal research. The Public Health Service (PHS) Policy on Humane Care and Use of Laboratory Animals provides policy recommendations for National Institutes of Health (NIH)-funded research; however, these are not legal requirements. In addition to US federal law, individual states can regulate animal research within their area of jurisdiction.

Japan

In Japan, laboratory research using animals is regulated by law and employs a regulatory system that resembles that of the US (see Table A.4). Basic considerations for laboratory animal handling originate from the Law for the Humane Treatment and Management of Animals (Law No. 105, 1973) and Standards Relating to the Care and Management of Experimental Animals (Notice No. 6 of the Prime Minister's Office 1980).

The uppermost regulation is the Law for the Humane Treatment and Management of Animals, which was enacted in 1973 and most recently revised

TABLE A.4 Key Components of Japanese Laboratory Animal Research Regulation [32]

Law for the Humane Treatment and Management of Animals

Standards Relating to the Care and Management, and Alleviation of Pain and Distress of Experimental Animals

Guidelines

- Ministry of Education
- Ministry of Health
- Ministry of Environment
- Ministry of Agriculture and Fisheries
- The Science Council of Japan

Institutional rules in animal experimentation

in 2012. Article 41 regulates "Methods When Animals are Used for Scientific Purposes and Subsequent Disposal of Such Animals." Section 1 of Article 41 (added to the 2005 revision) encourages the application of suitable alternative research methods that do not require animals, while Section 2 of Article 41 advocates the use of methods that cause minimal pain and distress to animals. Additionally, and in parallel to EU and US regulation, the Law requires implementation of the 3Rs for animal experimentation [32].

Within the Japanese Law, there are four standards and one guide concerning animal welfare. The four standards pertain to a type of animal category (household, exhibition, farm, or experimental animals) and the guide governs methods of euthanasia [32]. The 1980 Standards Relating to the Care and Management of Experimental Animals was revised in 2006 to include the alleviation of pain and distress (Standards Relating to the Care and Management, and Alleviation of Pain and Distress of Experimental Animals) [33].

Further to the Law and Standard, the Science Council of Japan issued the Guideline for Proper Conduct of Animal Experiments in 2000, and later revised it in 2006. This serves to guide research institutions on how to self-regulate animal experimentation and designate an Institutional Animal Care and Use Committee (IACUC) who are responsible for general oversight and ethical review of the research protocol [33]. Most medical schools, pharmaceutical organizations, and approximately one-third of breeders have established IACUCs [34]. The overall responsibility of a research institution resides with its dedicated president, who oversees all animal experiments conducted at the institution.

Regulatory Snapshot: Cosmetics

As of 2013, the EU prohibits animal testing for cosmetic products, which includes the finished product, individual ingredients and product marketing under the Cosmetics Directive 76/768/EEC 1976 [35]. The EU, with its 28 member states, is paving the way for a global ban, with support from international nations including India, Israel, and Brazil [36]. China, who still require in vivo animal testing of all foreign imported cosmetics, has assented somewhat to international pressures to change current practice and the Association of South East Asian Nations (ASEAN) is encouraging the remainder of the region to do so [36]. Equally, the US is progressing in a similar direction following the introduction of the Humane Cosmetics Act in 2015 [25].

ETHICAL CONSIDERATIONS

Despite implementation of the 3Rs, which are often legally enforced, ethical concerns remain a substantial international issue and tension exists between animal activists and research communities. Contemporary laboratory research practice has seen a shift away from animal research in certain areas, for example toxicology, and decreased utilization of more "morally sensitive" animals such as chimpanzees or dogs [8].

The ethical dilemma surrounding the trade-off of demonstrable or expected benefit versus potential animal pain, suffering, and death has long-existed and more recent laws and policies have attempted to address both animal welfare and public disconcert [37]. The EU Directive, for example, demands score-based estimations for potential levels of pain, suffering, and distress inflicted to animals and recommends a cut-off limit that prohibits any proposals or procedures that score above it [2]. In certain regions, such as Japan, religion plays a more influential role in governing animal welfare with strong beliefs in good *Karma*, which effects both those engaged in animal research and even experimental outcomes [34]. However, current scientific practice and legislation reflect a perceived necessity for animal models in translational research with the overriding assumption that the benefits to human health outweigh any associated animal costs [2].

It has also been reported that few research ethics committees stipulate increasing the quantity of animals required in experimental research programs—even when proposals are markedly underpowered—which could adversely affect research findings [8]. Although there is no simple solution to this irresolvable ethical dilemma, considerable progress is being made in establishing viable alternatives to animal models.

ALTERNATIVES TO ANIMAL MODELS

In recognition of the ethicolegal implications concerning the use of animal models in research and the associated financial costs (particularly for larger animals), research communities are examining newer avenues of scientific study. These include areas such as in silico testing and Quantitative structural activity relationship (QSAR) research, which utilize computational modeling to simulate or predict functions and reactive processes.

> **Key Definitions**
>
> *In vivo*: Experimentation using a whole living organism
> *In vitro*: Experimentation in a controlled environment outside a living organism
> *In silico*: Performed by computer or computer simulation
>
> *Source: [38]*

In silico computational models offer novel experimental tools to investigate virtual cells, tissues, and organs. In silico technologies have been successfully employed by the toxicology community to understand and predict in vivo toxicity [39]. Further developments in micro- and nanofluidic technologies have enabled the so-called organ-on-a-chip systems to emerge and these systems, along with engineered human 3D organotypic culture models (OCMs), can accurately replicate biochemical microenvironments and provide excellent platforms for drug screening, tissue engineering, and delivery testing [40]. Some of the most advanced platforms can accurately mimic complex organs such as the liver, and effectively emulate metabolic processes [39]. QSAR is able to map and compare the unique structural designs of chemicals to predict how they may react biologically. This enables scientists to predict potential adverse effects such as hypersensitivity reactions or malignant transformations [25].

Although these applications are encouraging, challenges remain with regard to their wider applicability and the degree of engineering expertise required producing them. Consequently, they are unlikely to replace animal models in the immediate future. Further research will focus on overcoming these challenges and strategically optimizing production to facilitate efficient and scalable production—which will translate into efficacious and accessible applications for the wider research community.

CONCLUSION

In the interim period while nonanimal-based models continue to be developed and validated, regulatory authorities must continue with their endeavors to both protect animal welfare and promote rigorous animal research. Efforts should be made to reduce publication bias, increase the reliability of preclinical animal models, and utilize globally available instruments such as the ARRIVE guidelines to enhance the internal and external validity of animal studies [41]. The international research community must also make a concerted effort to collaborate transnationally and attempt to harmonize current regulatory frameworks.

REFERENCES

[1] Badyal DK, Desai C. Animal use in pharmacology education and research: The changing scenario. Indian journal of pharmacology 2014;46(3):257.

[2]   Ruiz-Meana M, Martinson EA, Garcia-Dorado D, Piper HM. Animal ethics in Cardiovascular Research. Cardiovasc Res 2012;93(1):1–3.

[3]   European Commission. Seventh Report from the Commission to the Council and the European Parliament on the Statistics on the number of animals used for experimental and other scientific purposes in the member states of the European Union COM(2013)859/final. 2011; Available at: http://ec.europa.eu/environment/chemicals/lab_animals/reports_en.htm. Accessed March, 2016.

[4]   Howells DW, Sena ES, Macleod MR. Bringing rigour to translational medicine. Nature Reviews Neurology 2014;10(1):37–43.

[5]   O'Collins VE, Macleod MR, Donnan GA, Horky LL, van der Worp Bart H, Howells DW. 1,026 experimental treatments in acute stroke. Ann Neurol 2006;59(3):467–77.

[6]   van Luijk J, Bakker B, Rovers MM, Ritskes-Hoitinga M, de Vries RB, Leenaars M. Systematic reviews of animal studies; missing link in translational research? PloS One 2014;9(3):e89981.

[7]   Hooijmans C, Ritskes-Hoitinga M. Progress in using systematic reviews of animal studies to improve translational research. PLoS Med 2013;10(7):e1001482.

[8]   Salman RA, Beller E, Kagan J, Hemminki E, Phillips RS, Savulescu J, et al. Increasing value and reducing waste in biomedical research regulation and management. Lancet 2014;383(9912):176–85.

[9]   Macleod M. Why animal research needs to improve. Nature 2011;477(7366) 511–511.

[10]   Milani-Nejad N, Janssen PM. Small and large animal models in cardiac contraction research: advantages and disadvantages. Pharmacol Ther 2014;141(3):235–49.

[11]   Vitale A, Chiarotti F, Alleva E. The use of animal models in disease research. Rare Dis Orphan Drugs 2015;2(1):1–4.

[12]   Committee for Medicinal Products for Human Use Guideline on clinical trials in small populations. London: EMEA; 2006.

[13]   Vaquer G, Dannerstedt FR, Mavris M, Bignami F, Llinares-Garcia J, Westermark K, et al. Animal models for metabolic, neuromuscular and ophthalmological rare diseases. Nat Rev Drug Discov 2013;12(4):287–305.

[14]   Martić-Kehl MI, Schibli R, Schubiger PA. Can animal data predict human outcome? Problems and pitfalls of translational animal research. Eur J Nucl Med Mol Imaging 2012:1–5.

[15]   Kaushal D, Mehra S, Didier P, Lackner A. The non-human primate model of tuberculosis. J Med Primatol 2012;41(3):191–201.

[16]   Kobayashi E, Hishikawa S, Teratani T, Lefor AT. The pig as a model for translational research: overview of porcine animal models at Jichi Medical University. Transplant Res 2012;1(1):8.

[17]   Russell WMS, Burch RL, Hume CW. The principles of humane experimental technique. 1959.

[18]   Pankevich DE, Wizemann TM, Mazza A, Altevogt BM. International Animal Research Regulations: Impact on Neuroscience Research: Workshop Summary. National Academies Press; 2012.

[19]   Passantino A. Application of the 3Rs principles for animals used for experiments at the beginning of the 21st century. Annu Rev Biomed Sci 2008;10:T27–32.

[20]   Directive E. DIRECTIVE 2010/63. EU of the European Parliament and of the Council of 2010;22.

[21]   Smith JA, Andrews PL, Hawkins P, Louhimies S, Ponte G, Dickel L. Cephalopod research and EU Directive 2010/63/EU: Requirements, impacts and ethical review. J Exp Mar Biol Ecol 2013;447:31–45.

[22]   Hollands C. The Animals (scientific procedures) Act 1986. Lancet 1986;328(8497):32–3.

[23]   Act AW. Animal Welfare Act, 2006. The Stationery Office Ltd.Printed under the authority and superintendence of Carol Tullo, controller of Her Majesty's Stationary Office and Queen's Printer of Acts of Parliament.UK 2006.

[24] ASPA 1986 Amendment Regulations. 2012.

[25] Lee CG. The Animal Welfare Act at Fifty: Problems and Possibilities in Animal Testing Regulation. Nebraska Law Rev 2016 Forthcoming.

[26] Metzger MM. Knowledge of the animal welfare act and animal welfare regulations influences attitudes toward animal research. J Am Assoc Lab Anim Sci 2015;54(1):70–5.

[27] National Research Council (US) Committee to Update Science, Medicine, and Animals. 2004.

[28] Kuwahara SS. Functions of the Institutional Animal Care and Use Committee (IACUC) and GLP regulated studies. J GXP Compliance 2010;14(4):34.

[29] Cardon AD, Bailey MR, Bennett BT. The Animal Welfare Act: from enactment to enforcement. J Am Assoc Lab Anim Sci 2012;51(3):301–5.

[30] United States Department of Agriculture. Animal care resources guide policies; 2011.

[31] United States Department of Agriculture. Research facility inspection guide; 2001.

[32] Japanese regulation of laboratory animal care with 3Rs. Proceedings of 6th World Congress on Alternatives & Animal Use in the Life Sciences; 2007.

[33] Guillen J. Laboratory Animals: Regulations and Recommendations for Global Collaborative Research. Academic Press; 2013.

[34] Japanese Regulations on Animal Experiments: Current Status and Perspectives. The Development of Science-based Guidelines for Laboratory Animal Care: Proceedings of the November 2003 International Workshop; 2004.

[35] Buzek J, Ask B. Regulation (ec) no 1223/2009 of the European Parliament and of the Council of 30 november 2009 on cosmetic products. Off J Eur Union L 2009:342.

[36] Laquieze L, Lorencini M, Granjeiro JM. Alternative methods to animal testing and cosmetic safety: an update on regulations and ethical considerations in Brazil. Appl In Vitro Toxicol 2015;1(4):243–53.

[37] Goodman J, Chandna A, Roe K. Trends in animal use at US research facilities. J Med Ethics 2015;41(7):567–9.

[38] Tunev SS. Differences between in vitro, in vivo, and in silico studies.

[39] Knudsen TB, Keller DA, Sander M, Carney EW, Doerrer NG, Eaton DL, et al. FutureTox II: in vitro data and in silico models for predictive toxicology. Toxicol Sci 2015;143(2):256–67.

[40] Wang Z, Samanipour R, Kim K. Organ-on-a-chip platforms for drug screening and tissue engineering. biomedical engineering: frontier research and converging technologies: Springer; 2016.209.33

[41] Kilkenny C, Browne WJ, Cuthill IC, Emerson M, Altman DG. Improving bioscience research reporting: the ARRIVE guidelines for reporting animal research. Animals 2014;4(1):35–44.

Appendix B

Regulatory Frameworks for Stem Cell Research

Eleanor Jane Budge and Morteza Jalali
University of Oxford, Oxford, United Kingdom

INTRODUCTION

Stem cell research is one of the most fascinating and promising areas of 21st century science. The opportunity and hope afforded by the use of human embryonic stem cells (hESCs) to produce disease-relevant cell models in which to study early development and disease progression, to develop regenerative therapeutics and to screen potential therapies, is truly unique [1,2].

While rapid advances in stem cell research have generated much excitement, this field of scientific investigation has also posed many challenging scientific, political, and ethical questions, attracting much attention from the research community and beyond [3,4]. These issues largely concern hESCs due to the method by which they are obtained and as such are the main focus here.

WHO MAKES THE RULES?

Internationally many jurisdictions have passed legislation, ethical guidelines, and established oversight bodies to govern hESC research. There exists to be much diversity between these guidelines, owing to differing social and cultural beliefs internationally [5,6]. The global variety in legislative stances is depicted in Table B.1.

Countries listed as having permissive legislation permit derivation of hESCs from donated embryos generated via in vitro fertilization (IVF) or somatic cell nuclear transfer (SCNT). SCNT describes the process by which the nucleus of a somatic cell from a donor is implanted into a recipient enucleated oocyte. This technique can be used to generate embryonic stem cells [8]. Those classed

TABLE B.1   Nations Categorized by Level of Legislative Restriction on Human Embryonic Stem Cell Research [7]

Type of Stem Cell Research Legislation

Permissive	Intermediate	Prohibitive	No Legislation
United Kingdom	Canada	Denmark	Luxemburg
Sweden	United States	Estonia	Latvia
Belgium	Ireland	Germany	Slovakia
Spain	Norway	Czech Republic	Hungary
Israel	Finland	Austria	Slovenia
Singapore	Netherland	Italy	Turkey
South Korea	Switzerland	Poland	Malta
Japan	France	Lithuania	Tunisia
Australia	Portugal	Cyprus	Iceland
	Greece	Bulgaria	Panama
	Romania		Colombia
	Georgia		Peru
	China		Argentina
	India		Thailand
	Brazil		Vietnam
	South Africa		
	New Zealand		
	Taiwan		
	Philippines		

as having intermediate legislation have some restrictions with regard to hESC derivation and stem cell research. Those countries with prohibitive laws restrict embryo research and prohibit derivation of new hESCs, limiting research to the use of imported cell lines. Countries lacking legislation may have published guidelines but have no formal laws on hESC research regulation.

ADVANCES AND CORRESPONDING LEGISLATION

Pioneering contributions to the field of hESC research have resulted from an international effort, with key studies coming from laboratories all over the world. In 1981, Cambridge scientists Evans and Kaufman first derived mouse embryonic stem cells in the United Kingdom [9]. Martin, a scientist at the University of California, San Francisco (USCF) in the United States, also achieved this great feat in the same year [10]. In 1998, Thomson et al., from the University of Wisconsin in the United States, isolated cells from the inner cell mass of early human blastocysts and demonstrated their differentiation into multiple lineages [11]. In 2006, Japanese scientists Takahashi and Yamanaka demonstrated the induction of pluripotent stem cells from mouse embryonic and adult fibroblasts by introducing defined reprogramming factors to cell culture conditions. These induced pluripotent stem cells (iPSCs) functioned like embryonic stem cells, creating a valuable resource for stem cell research and future therapeutic applications [12]. Following this discovery and further studies demonstrating the generation of iPSCs from adult human dermal fibroblasts with the same factors, Yamanaka was awarded the Nobel Prize in Physiology or Medicine in 2012 [13].

Countries have often drafted legislation in response to their discoveries and therefore a lag time has been inevitable. As a result, contributing nations have created policy at different times relating to various aspects of stem cell research. Some countries have passed formal legislation, while others use restrictive national funding to control stem cell research. Owing to the early role the United Kingdom took in stem cell research, it has among the most stringent and long-held laws guiding hESC research, and as such is of particular relevance.

UNITED KINGDOM

The United Kingdom has comprehensive and well-established regulations on hESC research governed by Human Fertilization and Embryology Authority (HFEA) [14]. The authority acts to enforce the Human Fertilization and Embryology Act 1990, which outlines legislation surrounding hESC research [15].

The Act permits destruction of human embryos to derive embryonic stem cells for research purposes and allows for SCNT [16]. However, the Human Reproductive Cloning Act 2001 explicitly prohibits reproductive cloning via the process of nuclear transfer [17].

Research Licenses

All research using human embryos in the United Kingdom is subject to obtaining a research license from the HFEA, which can be valid for upto three years [18]. This will only be approved if it can be shown that the research objectives may only be achieved by research on hESCs and that the aim of the research adheres to criteria approved by the HFEA [19] (Table B.2).

The Scientific and Clinical Advances Advisory Committee monitors research developments worldwide and acts to inform the HFEA Licence Committee on whether or not other research techniques have developed sufficiently, rendering the use of hESCs unnecessary [14].

TABLE B.2 Research Objectives Approved by the HFEA for a Research Grant [20]

Research Goals Approved for HFEA Licensing

- to promote advances in the treatment of infertility
- to increase knowledge about the causes of congenital disease
- to increase knowledge about the causes of miscarriages
- to develop more effective techniques of contraception
- to develop methods for detecting the presence of gene or chromosome abnormalities
- to increase knowledge about the development of embryos
- to increase knowledge about serious disease
- to enable any such knowledge to be applied in developing treatments for serious disease

Details of licenses that have been awarded and those that are currently active can be viewed on the HFEA website [21]. Licenses are specific to both a research supervisor (Person Responsible) and to a location, which will be inspected before a license can be granted by the license committee [22,23]. The Authority has the liberty to inspect, receive six-monthly reports and ask for records at any point and retains the right to suspend or revoke a license for a three-month period during which a review will take place [24]. If research centers use therapeutic cloning to establish new hESC lines, a sample of the stem cell line derived must be deposited in the UK Stem Cell Bank [16].

The Act also states policy regarding the practicalities of stem cell research and sets out the code of practice in relation to payment for donors. The Act states that no financial or other benefit must be given or received in respect to any supply of gametes, embryos, or human admixed embryos unless authorized [25]. Furthermore, the Human Tissue Authority regulates the procurement, processing, testing, storage, distribution and import/export of tissues, and cells for human application and research under the Human Tissue Act 2004 [26].

14-Day Limit on Blastocyst Research

If granted, licensed research on embryos older than 14 days is prohibited and stem cells must be isolated from blastocysts at 5–6 days [15]. At 14 days, the primitive streak appears in the embryo. This creates patterning axes and marks the beginning of gastrulation, an early phase in embryonic development from a single-layered blastula to a three-layered (tri-laminar) structure known as a gastrula. This change marks the formation of the embryo's central nervous system [27,28].

This limit was defined by the Committee of Inquiry into Human Fertilization and Embryology in 1982 addressing experimentation on naturally produced embryos [29]. Following the advent of therapeutic cloning, a second committee was established to re-examine ethical issues [30]. This committee sought inputs from researchers, religious groups, and the general public. After this review, the limit of 14 days was upheld and applied to both naturally produced embryos and those produced by therapeutic cloning.

Chimera Research

Following a number of applications from research teams, in 2008 the HFEA amended its powers to grant research-only licenses to teams creating human admixed embryos, also known as chimeras [31]. Admixed embryos are created by transferring human nuclei into animal eggs that have had almost all of their genetic information removed. This amendment offers another option if the number of donor eggs are scarce. Other countries have not legalized this process due to the large ethical debates raised by research of this kind [32].

Europe

European Union (EU) law holds a prohibitive position banning therapeutic cloning as stated in Article 18 of the Council of Europe Convention on Human Rights and Biomedicine [33]. In addition to EU laws, many individual countries within Europe have passed their own national legislation on stem cell research.

The EU Tissues and Cells Directives (EUTCD) acts as the Human Tissue Authority across the EU and was passed into UK Law in 2007 [34,35]. These directives were introduced to ensure standards of quality, safety, and a record of tissue and cells for human application. The EUTCD also works to engender a collaborative support and exchange system internationally, encouraging global development within the field.

United States

Legislation surrounding stem cell research in the United States has a complex history and remains to be polarized. There are no federal laws regarding hESC research; however, several states have their own laws.

In 2001, President George W. Bush prohibited federal funding for any research using hESC lines derived after August 9, 2001, claiming that this still permitted the use of 60 stem cell lines [36]. This limited research within the field as little information regarding the population origin of these existing stem cell lines was available, preventing researchers from selecting specific lines most representative of the population affected by their diseases of interest [37]. This policy did not affect privately or state-funded research but resulted in a mass exodus of researchers in the field to countries where this research was encouraged and facilitated.

In 2005, congress passed the Stem Cell Research Enhancement Act (H.R 810) to expand federal funding to include research on hESCs created from embryo donation. Unfortunately, Bush vetoed this bill and although the House voted with a majority in favor of the Act, it did not reach a two-thirds majority required to overrule a presidential veto. These events occurred a second time with the same result in 2007 [38].

Finally in 2009, President Barack Obama lifted these restrictions at the National Institutes of Health (NIH) and extended federal NIH funding for research on embryos donated by couples using IVF treatment and the use of old stem cell lines created for research purposes [39].

However, research on new hESC lines created for research purposes only and research introducing hESCs or iPSCs into nonhuman blastocysts, remains ineligible for federal funding [40].

Japan

Japan was one of the first countries to explicitly prohibit reproductive cloning via enacting the Human Cloning Regulation Act [41]. Originally the Ministry of Education, Culture, Sports, Science and Technology (MEXT) published guidelines for research on hESCs in 2001 [42]. These guidelines set out standards for the derivation of hESCs, putting a moratorium on SCNT. The original guidelines stated regulations on the donation and use of human embryos, the domestic distribution (but not export) of hESCs, and approval requirements for institutions and research projects [43]. These guidelines declared that in order to acquire a license researchers must undergo two stages of review, at institutional and ministry levels. Furthermore, researchers using hESCs were obliged to use separate facilities from other stem cell research. This made hESC research approval a protracted process and even minor changes to protocol had to be approved [43].

Some Japanese researchers expressed concern over the substantial approval process required for research on pre-existing hESC lines [44,45]. In response, the ban on SCNT was lifted in 2009 [46] and the guidelines were drawn up into two parts: one part for the derivation and distribution of hESCs [47] and another for their use [48]. Guidelines on utilization simplified the review system, requiring only an institutional review [49].

Most recently, in 2014 these guidelines were amended [50,51]. These amendments revised: (1) the method of protection of personal information; (2) procedures of re-consent, where needs arise to use hESCs for a purpose not initially consented for; and (3) guidance surrounding disclosure of incidental findings, in situations where researchers discover concerning information regarding the donor's health [52].

International Guidelines

In order to create international parity with regard to stem cell research regulation, and to ease collaboration and aid the progression of the research field on a global scale, the International Society for Stem Cell Research (ISSCR) was founded in 2002. This nonprofit organization has created ethical guidelines to advise researchers who are using hESCs [53]. Although these guidelines have not been passed as legislation, international funding bodies and journals help to put pressure on researchers to adhere.

These guidelines state that all research teams using hESCs will be subject to review, approval, and ongoing monitoring by a stem cell research oversight (SCRO) process, in order to evaluate the distinctive aspects of their research [3]. On an international platform, an SCRO process acts as an informal licensing committee.

An SCRO process considers the justification of using hESCs, weighing up the importance of the potential discovery and the lack of a suitable alternative method to answering the research question [3]. This is necessary before all research that aims to create new hESC lines and research generating human–animal chimeras. As in the United Kingdom, no experiments are permitted to extend beyond a 14-day limit at the initiation of primitive streak formation.

However, some research is exempt from an SCRO. This includes research using pre-existing hESC lines or involving routine research practice and research using induced pluripotent stem cells (iPSCs).

IPSCs are adult stem cells, which have been genetically reprogrammed back to an embryonic stem cell-like state, regaining pluripotency. This technology can, in several cases, bypass the need for hESCs in research, without posing such difficult ethical questions, as they are autologous to the adult donor. These cells also provide exciting prospects for autologous treatment, where recipients receive their own differentiated cells as treatment avoiding risks of immune rejection and of course are unlimited in their supply. However, concerns regarding the tumorigenicity of iPSCs have been a major limiting factor in therapeutic progression [54].

The ISSCR has made sample documents to guide researchers to gain valid informed consent from donors, sensitive to language barriers and educational levels [55].

RESOURCES FOR RESEARCHERS

The following websites contain additional information on stem cell policy around the world.

International Society for Stem Cell Research (ISSCR)

Established in 2002, the ISSCR is an independent, nonprofit organization whose goal is to foster the corroborative exchange of information and ideas in the field of stem cell research. In February 2007, The ISSCR published Guidelines for the Conduct of Human Embryonic Stem Cell Research, which are available on the ISSCR website. (http://www.isscr.org)

International Stem Cell Registry (ISCR)

This is a comprehensive database of hESC and iPSC lines for prospective researchers, and presents published and validated unpublished information on over 1,200 cell lines. The registry also has a literature database, collating all relevant research topics. (http://www.umassmed.edu/iscr/)

StemGen—This is a source of legal and socioethical issues in stem cell research. The website also has a world map resource representing policies and laws on stem cell research worldwide. (http://www.stemgen.org/)

UK Stem Cell Tool Kit

A compilation of research regulations for those conducting human stem cell research in the United Kingdom. (http://www.sc-toolkit.ac.uk/)

UK Stem Cell Bank (UKSCB)

An international collection of all stem cell lines available for research use and clinical application. (http://www.nibsc.org/ukstemcellbank)

National Institutes of Health

Research regulations for stem cell research in the United States. (http://www.nih.gov/).

The following resources are global online networks providing internationally accessible platforms for collaboration and coopetition facilitating interdisciplinary communication in order to expedite discovery within the field [56].

Hinxton Group

A website facilitating collaborative communication among scientists, policymakers, journal editors, and the public about international stem cell research. (http://www.hinxtongroup.org)

International Stem Cell Forum (ISCF)

This is an international forum consisting of global funders to encourage international collaboration and to provide financial support for researchers. (http://www.stem-cell-forum.net)

CONCLUSION

Despite the variety in stem cell regulations worldwide and the highly political and ethical debates they create, their importance is paramount. Internationally, the ISSCR remains to be a very reliable and culturally sensitive source for code of practice guidance, which should be referenced by all those undertaking stem cell research.

Owing to the promising nature of stem cell research and the potential therapeutic discoveries it proffers, it is of great importance to ensure the application of these existing ethical frameworks. Furthermore, frequent reviews of new ethical implications must be instigated to keep up with this broad and diverse field as it advances further.

REFERENCES

[1] Teo AKK, Vallier L. Emerging use of stem cells in regenerative medicine. Biochem J 2010;428(1):11–23.

[2] Jalali M, Kirkpatrick WNA, Cameron MG, Pauklin S, Vallier L. Human stem cells for craniomaxillofacial reconstruction. Stem Cells Dev 2014;23(13):1437–51.

[3] Daley GQ, Ahrlund Richter L, Auerbach JM, Benvenisty N, Charo RA, Chen G, et al. Ethics. The ISSCR guidelines for human embryonic stem cell research. Science 2007;315(5812):603–4.

[4] Vogel G, Holden C. Stem cells. Ethics questions add to concerns about NIH lines. Science 2008;321(5890):756–7.

[5] Jain KK. Ethical and regulatory aspects of embryonic stem cell research. Expert Opin Biol Ther 2002;2(8):819–26.

[6] Jain KK. Ethical and regulatory aspects of embryonic stem cell research. Expert Opin Biol Ther 2005;5(2):153–62.

[7] STEMGEN. International Database on the Legal and Socio-Ethical Issues in Stem Cell Research. STEM CELL WORLD MAP [Internet]. [cited 2016 Jan 16]. Available from: http://www.stemgen.org/stem-cell-world-map.

[8] Tachibana M, Amato P, Sparman M, Gutierrez NM, Tippner-Hedges R, Ma H, et al. Human Embryonic Stem Cells Derived by Somatic Cell Nuclear Transfer. Cell 2013;153(6):1228–38.

[9] Evans MJ, Kaufman MH. Establishment in culture of pluripotential cells from mouse embryos. Nature 1981;292(5819):154–6.

[10] Martin GR. Isolation of a pluripotent cell line from early mouse embryos cultured in medium conditioned by teratocarcinoma stem cells. Proc Natl Acad Sci USA 1981;78(12):7634–8.

[11] Thomson JA, Itskovitz-Eldor J, Shapiro SS, Waknitz MA, Swiergiel JJ, Marshall VS, et al. Embryonic stem cell lines derived from human blastocysts. Science 1998;282(5391):1145–7.

[12] Takahashi K, Yamanaka S. Induction of Pluripotent Stem Cells from Mouse Embryonic and Adult Fibroblast Cultures by Defined Factors. Cell 2006;126(4):663–76.

[13] Takahashi K, Tanabe K, Ohnuki M, Narita M, Ichisaka T, Tomoda K, et al. Induction of pluripotent stem cells from adult human fibroblasts by defined factors. Cell 2007;131(5):861–72.

[14] Human Fertilisation and Embryology Authority S and ID. HFEA - How we regulate human embryo research [Internet]. [cited 2016 Jan 14]. Available from: http://www.hfea.gov.uk/161.html.

[15] Human Fertilisation and Embryology Authority S and ID. HFE Act 1990 [Internet]. [cited 2016 Jan 14]. Available from: http://www.hfea.gov.uk/2070.html.

[16] Human Fertilisation and Embryology Authority S and ID. HFEA Code of Practice: Guidance Note 22. Research and training (version 4.0) [Internet]. [cited 2016 Jan 14]. Available from: http://www.hfea.gov.uk/3468.html#guidanceSection5017.

[17] Human Fertilisation and Embryology Authority S and ID. Cloning issues in reproductive science - HFEA policy reviews [Internet]. [cited 2016 Jan 16]. Available from: http://www.hfea.gov.uk/518.html.

[18] Human Fertilisation and Embryology Authority S and ID. After you are licensed [Internet]. [cited 2016 Jan 16]. Available from: http://www.hfea.gov.uk/3391.html.

[19] Participation E. Human Fertilisation and Embryology Act 2008 [Internet]. [cited 2016 Jan 14]. Available from: http://www.legislation.gov.uk/ukpga/2008/22/contents.

[20] Human Fertilisation and Embryology Authority S and ID. Applying for a research licence [Internet]. [cited 2016 Jan 16]. Available from: http://www.hfea.gov.uk/3388.html.

[21] Human Fertilisation and Embryology Authority S and ID. Human embryo research we have approved - HFEA [Internet]. [cited 2016 Jan 16]. Available from: http://www.hfea.gov.uk/166.html.

[22] Human Fertilisation and Embryology Authority S and ID. HFEA Code of Practice: Guidance Note 1. Person Responsible (version 2.0) [Internet]. [cited 2016 Jan 16]. Available from: http://www.hfea.gov.uk/383.html#mandatoryAct.

[23] Human Fertilisation and Embryology Authority S and ID. After you apply for a research licence – clinic inspections [Internet]. [cited 2016 Jan 16]. Available from: http://www.hfea.gov.uk/3389.html.

[24] Human Fertilisation and Embryology Authority S and ID. HFEA inspections – HFEA [Internet]. [cited 2016 Jan 16]. Available from: http://www.hfea.gov.uk/6672.html.

[25] Human Fertilisation and Embryology Authority S and ID. HFEA Code of Practice: Guidance Note 13. Payments for donors (version 2.0) [Internet]. [cited 2016 Jan 10]. Available from: http://www.hfea.gov.uk/500.html.

[26] Human Tissue Act 2004 [Internet]. [cited 2016 Jan 16]. Available from: http://webarchive.nationalarchives.gov.uk/20120405095111/http://www.legislation.gov.uk/ukpga/2004/30/contents.

[27] Tam PP, Behringer RR. Mouse gastrulation: the formation of a mammalian body plan. Mech Dev 1997;68(1–2):3–25.

[28] Roger A, Pedersen KW, Pedersen RA, Wu K, Balakier H. Origin of the inner cell mass in mouse embryos: cell lineage analysis by microinjection. Dev Biol 1986;117(2):581–95.

[29] Brahams D. Warnock Report on Human Fertilisation and Embryology. Lancet 1984;324(8396): 238–9.

[30] Stem Cell Research: Medical Progress with Responsibility: Executive Summary. Cloning. 2000;2(2):91–6.

[31] Human Fertilisation and Embryology Authority S and ID. Hybrids and chimeras - HFEA policy reviews [Internet]. [cited 2016 Jan 16]. Available from: http://www.hfea.gov.uk/519.html.

[32] Hug K. Research on human-animal entities: ethical and regulatory aspects in Europe. Stem Cell Rev 2009;5(3):181–94.

[33] Full list [Internet]. Treaty Office. [cited 2016 Jan 10]. Available from: http://www.coe.int/web/conventions/full-list.

[34] Human Fertilisation and Embryology Authority S and ID. European Union Tissues and Cells Directives [Internet]. [cited 2016 Jan 17]. Available from: http://www.hfea.gov.uk/2072.html.

[35] Human Fertilisation and Embryology Authority S and ID. EU Standards – FAQs for Fertility Clinic Staff – HFEA [Internet]. [cited 2016 Jan 27]. Available from: http://www.hfea.gov.uk/fertility-clinic-questions-eu-standards.html.

[36] Human Embryonic Stem Cell Policy Under Former President Bush (Aug. 9, 2001–Mar. 9, 2009) [Internet]. [cited 2016 Jan 16]. Available from: http://stemcells.nih.gov/policy/Pages/2001policy.aspx.

[37] Mosher JT, Pemberton TJ, Harter K, Wang C, Buzbas EO, Dvorak P, et al. Lack of Population Diversity in Commonly Used Human Embryonic Stem-Cell Lines. N Engl J Med 2010;362(2):183–5.

[38] Timeline of major events in stem cell research policy [Internet]. Research!America. [cited 2016 Jan 16]. Available from: http://www.researchamerica.org/advocacy-action/ issues-researchamerica-advocates/stem-cell-research/timeline-major-events-stem-cell.

[39] Executive Order 13505 – Removing Barriers to Responsible Scientific Research Involving Human Stem Cells [Internet]. whitehouse.gov. [cited 2016 Jan 16]. Available from: https://www.whitehouse.gov/the-press-office/removing-barriers-responsible-scientific-research-involving-human-stem-cells.

[40] National Institues of Health, National Institutes of Health, U.S. Department of Health and Human Services, 2015. National Institutes of Health Guidelines on Human Stem Cell Research [Internet]. [cited 2016 Jan 17]. Available from: http://stemcells.nih.gov/policy/ pages/2009guidelines.aspx.

[41] National Diet, Japan. Law concerning regulation of human cloning techniques and other similar techniques. [Internet]. 2001 [cited 2016 Jan 27]. Available from: http://www.cas.go.jp/jp/ seisaku/hourei/data/htc.pdf.

[42] Minister of Education, Culture, Sports, Science and Technology (MEXT) Japan. The Guidelines for Derivation and Utilization of Human Embryonic Stem Cells [Internet]. 2001 [cited 2016 Jan 27]. Available from: http://www.lifescience.mext.go.jp/files/pdf/32_90.pdf.

[43] Kawakami M, Sipp D, Kato K. Regulatory impacts on stem cell research in Japan. Cell Stem Cell 2010;6(5):415–8.

[44] Nakatsuji N. Irrational Japanese regulations hinder human embryonic stem cell research. Nat Rep Stem Cells [Internet]. 2007 Aug 9 [cited 2016 Jan 27]; Available from: http://www. nature.com/stemcells/2007/0708/070809/full/stemcells.2007.66.html.

[45] Cyranoski D. Japan's embryo experts beg for faster ethical reviews. Nature 2005;438(7066) 263–263.

[46] Cyranoski D. Japan relaxes human stem-cell rules. Nat News 2009;460(7259) 1068–1068.

[47] Japan's Ministry of Education, Culture, Sports, Science and Technology. Guidelines for derivation and distribution of human ES cells [Internet]. 2009 [cited 2016 Jan 27]. Available from: http://www.lifescience.mext.go.jp/files/pdf/56_229.pdf.

[48] Japan's Ministry of Education, Culture, Sports, Science and Technology. Guidelines for utilization of human ES cells [Internet]. 2009 [cited 2016 Jan 27]. Available from: http://www. lifescience.mext.go.jp/files/pdf/57_232.pdf.

[49] Caulfield T, Scott C, Hyun I, Lovell-Badge R, Kato K, Zarzeczny A. Stem cell research policy and iPS cells. Nat Methods 2010;7(1):28–33.

[50] Ministry of Education, Culture, Sports, Science and Technology (MEXT). Guidelines on the Distribution and Utilization of Human Embryonic Stem Cells [Internet]. 2014 [cited 2016 Jan 27]. Available from: http://www.lifescience.mext.go.jp/files/pdf/n1553_02.pdf.

[51] Ministry of Education, Culture, Sports, Science and Technology (MEXT) and the Ministry of Health, Labour and Welfare (MHLW). Guidelines on the Derivation of Human Embryonic Stem Cells [Internet]. 2014 [cited 2016 Jan 27]. Available from: http://www.lifescience.mext. go.jp/files/pdf/n1553_01.pdf.

[52] Mizuno H. Ethical Issues for Clinical Studies That use Human Embryonic Stem Cells: The 2014 Revisions to the Japanese Guidelines. Stem Cell Rev 2015;11(5):676–80.

[53] Daley GQ, Richter LA, Auerbach JM, Benvenisty N, Charo RA, Chen G, et al. The ISSCR Guidelines for Human Embryonic Stem Cell Research. Science 2007;315(5812):603–4.

[54] Ben-David U, Benvenisty N. The tumorigenicity of human embryonic and induced pluripotent stem cells. Nat Rev Cancer 2011;11(4):268–77.

[55] International Society for Stem Cell Research. ISSCR Sample Informed Consent Documents [Internet]. [cited 2016 Jan 17]. Available from: http://www.isscr.org/home/publications/guide-clintrans/sample-consent-documents.

[56] Budge EJ, Tsoti SM, Howgate DJ, Sivakumar S, Jalali M. Collective intelligence for translational medicine: Crowdsourcing insights and innovation from an interdisciplinary biomedical research community. Ann Med 2015;47(7):570–5.

Appendix C

Using Multiple Experimental Methods to Address Basic Science Research Questions

Justyna Zaborowska and Morteza Jalali
University of Oxford, Oxford, United Kingdom

INTRODUCTION

The Human Genome Project was a global collaborative project and remarkable scientific feat, which determined the nucleotide sequences that make up our genome. It has permitted the identification and mapping of human genes and is now indispensible as a resource for human biomedical research. Investigators around the world conduct biomedical research to determine how genes are regulated and how the proteins they encode function, in order to advance our knowledge of human biology in health, disease, and crucially, to illuminate new therapeutics.

Conducting biomedical research involves key initial steps, which begin with the broad assimilation of the literature surrounding the topic under consideration, in order to decide which scientific questions are most relevant and can be addressed experimentally. A hypothesis is constructed and experiments are designed to test the hypothesis. In the following steps, experiments are conducted and resulting data analyzed prior to the biomedical scientist drawing conclusions based on the evidence before them.

WHY IS DETERMINING PROTEIN FUNCTION IMPORTANT?

One of the big challenges in biology is to assign function to the thousands of gene products discovered by genome sequencing. Learning when, where, and under what conditions genes are expressed can be informative of protein function. One can compare protein and RNA transcript levels of a specific gene of interest in different cell types, different development states and in response to different treatments. Experimental data can be used to develop a theory of how a protein functions and correspondingly, how it malfunctions in a pathological state. Consequently, this could provide targets for treatment of the disease.

ANALYZING PROTEIN X'S FUNCTION

In order to provide an overview of the expression patterns of a given protein (in this scenario Protein X) between different cell types and physiological states of cells, a researcher can use the workflow presented in Fig. C.1. The cells are cultured as described in Chapter 9, Cell Culture: Growing Cells as Model Systems In Vitro. Most cell lines can be grown using Dulbecco's Modified Eagle Medium (DMEM) or RPMI culture media with 10% fetal bovine serum, 2 mM glutamine, and antibiotics. Before starting "downstream" experiments, the researcher verifies on a microscope that the cells are healthy and growing as expected.

To start with, the investigator can examine the expression of messenger RNA (mRNA) X. For that, the researcher will use the *reverse transcription*-polymerase chain reaction (RT-PCR) method described in Chapter 1, The Polymerase Chain Reaction: PCR, qPCR, and RT-PCR. The isolated RNA is converted to cDNA with reverse transcriptase. cDNA can then be used as a template in PCR reactions, using primers for the gene X. A more abundant transcript will yield more product than a weakly transcribed gene (Fig. C.1A).

The next step is to investigate whether a particular protein is present in a certain cell line by looking for a band of the correct size by performing Western blot using a specific antibody. Protein lysates are prepared and subjected to Western blot analysis as described in Chapter 6, Western Blot. Following immunodetection, a band corresponding to the size of the target protein should become visible in the positive control, in comparison to a known null cellular sample. As seen in the example (Fig. C.1B), Protein X is expressed in cell types 1 and 2. Furthermore, cell type 2 has higher levels of Protein X in comparison to cell type 1 (Fig. C.1B).

INVESTIGATING THE SUBCELLULAR LOCALIZATION OF PROTEIN X

Investigating the subcellular localization of proteins using imaging techniques such as immunofluorescence (IF) is of great importance, as it leads to a better understanding of protein function. As described in Chapter 8, Immunofluorescence, proteins can be visualized by the use of specific antibodies. However, if a specific antibody is not available for the protein of interest, an N- or C- terminal tag can be fused by cloning for IF detection (Fig. C.2A). It is important to sequence the PCR product to determine if the cloning was performed successfully. To check whether Protein X clones are expressed, cells are transfected (as described in chapter: Transfection) with plasmids encoding the Flag-tagged Protein X. Nontransfected cells are used as a negative control.

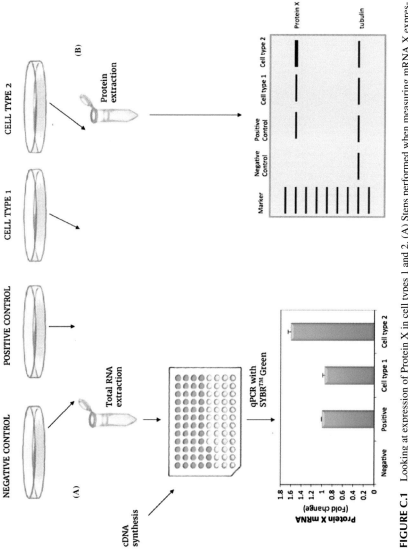

FIGURE C.1 Looking at expression of Protein X in cell types 1 and 2. (A) Steps performed when measuring mRNA X expression. RNA is first isolated; cDNA is then synthesized and used as a qPCR template. (B) Western blot analysis of Protein X. Tubulin was used as a loading control. A molecular weight marker (kDa) is indicated on the left.

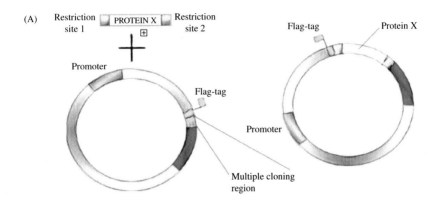

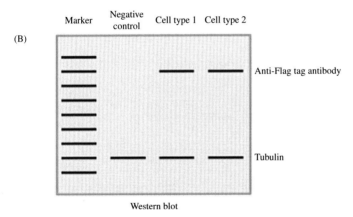

FIGURE C.2 Fusing a Flag tag to Protein X. (A) Vector Flag-tagging. A schematic diagram of the cloning technique used for the production of a Flag-tagged Protein X. (B) Western blot analysis with anti-Flag antibody in cells after transfection of Flag-tagged Protein X. Non-transfected cells are used as a negative control. Tubulin is used as a loading control. The molecular weight marker (kDa) is indicated on the left.

Western blot analysis detects the expression of the protein product of transiently transfected Flag-tagged Protein X (Fig. C.2B). Following that, an anti-Flag antibody can be used to perform an IF assay to assess the localization of Protein X (chapter: Immunofluorescence).

GENOME-WIDE MICROARRAY

The Protein X function can be further investigated on a genome-wide scale using a DNA microarray (chapter: Microarrays: An Introduction and Guide to Their Use). In particular, the generation of mutant libraries in which gene X has been deleted or disrupted provides an invaluable tool for exploring its role in different molecular pathways by analyzing its effect on the transcriptome. Such analysis may lead the researcher to new avenues, whereby Protein X's function impacts on developmental regulatory pathways. Following, the scientist can hypothesize how Protein X achieves this by looking at chromatin binding and protein complex formation.

Index

Printed in the United States
By Bookmasters